Alternative Logics.
Do Sciences Need Them?

Springer
Berlin
Heidelberg
New York
Hong Kong
London
Milan
Paris
Tokyo

Paul Weingartner (Ed.)

Alternative Logics. Do Sciences Need Them?

With 29 Figures

Springer

Professor Dr. Paul Weingartner
Internationales Forschungszentrum
für Grundfragen der Wissenschaften
Institut für Wissenschaftstheorie
Mönchsberg 2a
5020 Salzburg, Austria

Publications of the International Research Center Salzburg, Austria, No. 82

Library of Congress Cataloging-in-Publication Data.
Alternative logics: do sciences need them?/Paul Weingartner (ed.).
p.cm. – Includes bibliographical references and index.
ISBN 3-540-40744-8 (acid-free paper)
1. Logic. 2. Science–Philosophy. I. Weingartner, Paul.
Q175.32R45A48 2003 501–dc22 2003060771

ISBN 3-540-40744-8 Springer-Verlag Berlin Heidelberg New York

Springer-Verlag is a part of Springer Science+Business Media
springeronline.com

Printed in Germany

Typesetting: Data prepared by the authors/editor using a Springer TEX macro package
Final processing: LE-TeX, Leipzig
Cover design: *design & production* GmbH, Heidelberg

Printed on acid-free paper 55/3141/tr 5 4 3 2 1 0

Preface

The contributions in this book are based on the conference 'Alternative Logics: Do Sciences Need Them?' of the Académie Internationale de Philosophie des Sciences held at the Institut für Wissenschaftstheorie, Internationales Forschungszentrum Salzburg, May 1999. The papers given at the conference have been revised and partially extended, taking into account the discussion and scientific communication at the conference. Some who had originally been invited to participate in the conference could not attend at the last moment but have sent their contributions. A few other participants could not send their contributions because of other commitments.

Following a suggestion from Prof. Beiglböck of Springer-Verlag, which was agreed upon by the President of the Academy Prof. E. Agazzi, some additional scientists have been invited by the organizing committee to incorporate their contributions. The aim is to give an even more complete picture of recent research in *Alternative Logics* which have been developed for application in the sciences.

The editor would like to thank the President of the Academy, Prof. Agazzi, for cooperation and Prof. van Cangh concerning organisational matters of the conference, Springer-Verlag and Prof. Wolf Beiglböck for publishing this volume, Mag. Simon M. Huttegger for his help concerning proof-reading and the preparation of this volume, and Mr. Ferry Baumgartner for proof-reading and conversion of the texts into LaTeX 2_ε. Last, but not least, the editor would like to express his gratitude in the name of all participants of the conference to the three institutions who have sponsored this research conference: The Académie Internationale de Philosophie des Sciences, the University of Salzburg, and the Internationales Forschungszentrum Salzburg.

Salzburg, August 2003 *Paul Weingartner*

Contents

Part I General Topics

Why Is It Logical to Admit Several Logics?
Evandro Agazzi ... 3

1 The Issue of the Plurality of Logics ... 3
2 The Domain of Logic ... 4
3 The Notion of Logical Consequence ... 6
4 The Basic Status of Alethic Logic ... 8
5 Mathematical Logic and Its Double Meaning ... 9
6 The Variety of Logical Calculi ... 10
7 The Broadening of the Domain of Logic ... 12
8 Logic and the Object of Discourse ... 17
9 Meaning, Sense, and Reference ... 19
10 The Relevance of the Referents for Logic, or from Pure to Applied Logic ... 21
11 The Logic of Particular Sciences ... 22
12 Why Is It Logical to Admit Several Logics? ... 25
References ... 26

Does Metaphysics Need a Non-Classical Logic?
Francisco Miró Quesada ... 27

1 Trying to Clarify Concepts ... 27
2 Plotinus ... 28
2.1 Hypostases ... 29
3 Logic ... 30
3.1 Provisos ... 30
4 Hegel ... 30
4.1 Metaphysics and Logic ... 30
4.2 Holism ... 31
5 Empiricism ... 32
5.1 Empiricist Metaphysics ... 32
6 Noneism ... 33
6.1 Noneism Metaphysics ... 33
6.2 Noneist Universe ... 35

7 Noneist Logic ... 37
7.1 Quantification ... 37
References ... 37

Logic and the Philosophical Interpretation of Science
Dudley Shapere ... 41

1 The Promise of the Vienna Circle ... 41
2 Two Clues to the Roots of the Logistic Failure ... 43
3 Nature of and Contrasts Between the Two Programs ... 47
4 Roots of the Difficulties, and Their Solution ... 48
5 What Does It Mean to Say That Science Is Content-Driven? ... 52
References ... 53

How Set Theory Impinges on Logic
Jesús Mosterín ... 55

1 The Set-Theoretical Universe ... 55
2 Second-Order Logic and Set Theory ... 57
3 The Axiom of Choice ... 57
4 The Continuum Hypothesis ... 58
5 Inaccessible Cardinals ... 59
6 First-Order Logic and Set Theory ... 60
References ... 62

Geometries and Arithmetics
Graham Priest ... 65

1 Introduction: Geometry and Arithmetic ... 65
2 Non-Euclidean Geometry ... 66
3 Nonstandard Arithmetics ... 67
4 Solving Equations ... 68
5 Models of the Inconsistent ... 70
6 Inconsistent Arithmetic ... 71
7 The General Structure of Models ... 72
8 Empirical Applications ... 73
9 Non-Standard Logics in Science ... 76
References ... 77

Remarks on Criteria of Truth and Models in Science
Giuseppe Del Re ... 79

1 Premise ... 79
2 Logic for Science ... 79
2.1 What We Mean by Logic Here ... 79
3 Truth in Science ... 80
3.1 Metaphors, Models, and the Direct-Access Level ... 80
3.2 Science and Arguments by Analogy ... 82
3.3 Rational or Reasonable? ... 83

Significant? Not Significant? The Dilemma of Statistical Induction in Scientific Research
Italo Scardovi 87

References 92

Part II Alternative Proposals

Outline of a Paraconsistent Category Theory
Newton C.A. da Costa, Otávio Bueno, Analice Volkov 95

1 The Theory OB_0 95
1.1 Subcategories 96
1.2 Union of Subcategories 96
1.3 The Empty Category 96
1.4 Objects of a Category 96
1.5 Morphisms of a Category 97
1.6 The Category of Two Morphisms 98
1.7 The Postulates of Identity 98
1.8 The Postulate P_{17} 98
1.9 Functorial Categories 99
1.10 Functors 99
1.11 The Category of Subcategories of a Given Category 100
1.12 Image by a Functor 100
1.13 Functors and Their Images 101
1.14 Sets 101
1.15 Complete Categories of Sets 102
1.16 ZF and OB_0 103
2 The Paraconsistent Theory OB_1 104
2.1 The Calculus $C_1^=$ 104
2.2 The Theory OB_1 106
3 Philosophical Remarks 107
References 113

Combinatory Logic, Language, and Cognitive Representations
Jean-Pierre Desclés 115

1 Concatenative and Applicative Structures 118
2 Combinatory Logic as an Applicative Language 121
3 Combinators Are Abstract Operators 125
4 β-Reductions (β-Expansions) of Combinators 128
5 Operators in Combinatory Logic Are "Objects of Thought" 130
6 Combinators Used to Express Complex Concepts 133
7 Meaningless and Types 136
8 Cognitive Representations 139

9 Conclusion 142
References 146

Extending the Realm of Logic: The Adaptive-Logic Programme
Diderik Batens 149

1 The Problem 149
2 Adaptive Logics: A Step Within Symbolic Logic Towards Actual Reasoning 151
3 Possible Objections 154
4 Inconsistency-Adaptive Logics 156
5 Further Corrective Adaptive Logics 159
6 Some Other Adaptive Logics 160
7 In Conclusion 162
References 163

Comments on Jaakko Hintikka's Post-Tarskian Truth
Gerhard Heinzmann 165

References 172

Part III Alternative Logics Motivated by Problems of Application to Science

Applied Logics for Computer Science
Paul Gochet, Pascal Gribomont 177

1 Hoare Logic for Sequential Programs 177
  1.1 A Sorting Algorithm and Its Specification 177
  1.2 Hoare's Axiomatic System 179
  1.3 Formal Verification of the Sorting Algorithm 181
2 Using Logic for Program and Invariant Construction 184
  2.1 Mutual Exclusion 184
  2.2 Peterson's Algorithm 185
3 Logic Programming 189
  3.1 Propositional Logic Programming 189
  3.2 Predicate Logic Programming 192
  3.3 Beyond First Order Logic Programming 198
4 Conclusion 201
References 202

Stochastic *versus* Deterministic Features in Learning Models
Ion-Olimpiu Stamatescu 205

1 Introduction 205
2 Approaches to Learning in Artificial Systems 207

3 Some Examples Concerning Stochastic Learning Procedures 208
4 Conclusion 211
5 Appendix 213
References 213

Praxic Logics
David Ritz Finkelstein, James Baugh 215

1 Empirical Logic 215
2 Variant Semantics 218
3 Variant Epistemologies 221
4 Variant Logics 222
5 Simplicity 224
6 Relativity of the State 224
7 Work to Be Done 226
8 Variant Second-Order Logics 227
9 Cliffordonic Second-Order Logic 228
10 Acknowledgements 230
References 230

Reasons from Science for Limiting Classical Logic
Paul Weingartner 233

1 Introduction 233
2 The Problem of Commensurability Is Not Specific To Physics 233
2.1 Logic and Mathematics 234
2.2 Animal Behaviour 234
2.3 Human Actions 234
2.4 Social Help Among Animals 234
2.5 Incommensurable Pairs of Magnitudes in Physics 235
2.6 The Problem with Distributivity 235
3 Commensurability and Distributivity in Quantum Logic 236
3.1 General Considerations and Notation 236
3.2 Basic Definitions and Theorems of Quantum Logic 237
3.3 Implication 238
4 An Alternative Logic for QM 239
4.1 Preliminaries for the Alternative Logic 240
4.2 Replacement Criterion (R) 241
4.3 Properties of R ($\overset{\vdash}{cr}$) 242
4.4 Reduction Criterion (RD) 243
4.5 RD Solves Paradoxes 244
4.6 Application of R and RD to Quantum Logic 245
References 247

The Language of Interpretation in Quantum Physics and Its Logic
Roland Omnès ... 249

1 About the Language of Interpretation ... 249
2 Von Neumann's Predicates ... 250
3 Classical Properties ... 251
4 Consistent Histories ... 252
5 Recovering Standard Logic ... 253
6 Verifying Logical Consistency ... 254
7 Complementarity ... 254
8 Wider Issues ... 255
References ... 256

Why Objectivist Programs in Quantum Theory Do Not Need an Alternative Logic
Alberto Cordero ... 257

1 Introduction ... 257
2 Logic and Empirical Science ... 258
3 The Search for an Empirical Logic ... 261
4 Lingering Classical Intuitions ... 262
5 On the Credibility of Property Determinateness ... 263
6 The Conception of Properties in Three Leading Objectivist Programs ... 266
7 Conclusion ... 267
References ... 268

Does Quantum Physics Require a New Logic?
Peter Mittelstaedt ... 269

1 Introduction ... 269
2 The Logic of Classical Physics ... 270
  2.1 The Language ... 270
  2.2 Semantics and Syntax ... 270
  2.3 Classical Logic ... 273
3 The Quantum Physical Reality ... 275
  3.1 Complementarity ... 275
  3.2 Nonobjectifiability ... 275
  3.3 The Probability Argument ... 276
  3.4 Universality ... 277
4 The Logic of Quantum Physics ... 277
  4.1 Language and Reality in Quantum Physics ... 277
  4.2 The Syntax of Quantum Language ... 278
  4.3 Quantum Logic ... 280
5 Conclusion ... 283
References ... 284

Experimental Approach to Quantum-Logical Connectives
Ernst-Walther Stachow ... 285

1 Introduction ... 285
2 Experimental Proofs of Elementary Propositions ... 285
3 Experimental Proofs of Connected Propositions ... 288
4 Conclusion ... 297
References ... 298

From Semantics to Syntax: Quantum Logic of Observables
Vladimir L. Vasyukov ... 299

1 Introduction: Quantum Logic of Observables as a Converse Problem of Semantics ... 299
2 QLO: A Syntax ... 303
3 QLO: Semantics ... 307
4 A Normed Quantum Logic of Observables ... 312
5 A Temporal Quantum Logic of Observables ... 314
6 Final Remarks ... 319
References ... 320

An Unsharp Quantum Logic from Quantum Computation
Gianpiero Cattaneo, Maria Luisa Dalla Chiara, Roberto Giuntini ... 323

1 Introduction ... 323
2 Quantum-Logical Gates ... 327
3 Quantum-Computational Semantics ... 333
References ... 337

Quantum Logic and Quantum Probability
Enrico G. Beltrametti ... 339

1 Introduction ... 339
2 The Ordered Structure of the Events ... 340
3 Probability Measures on the Events ... 341
4 The Operational Framework ... 343
References ... 346

Operator Algebras and Quantum Logic
Miklós Rédei ... 349

1 The Idea of Quantum Logic: Logicization of Non-Boolean Algebras ... 349
2 Von Neumann Algebras and Their Projection Lattices ... 351
3 Von Neumann Lattices as Logics ... 354
4 Philosophical and Historical Comments ... 356
References ... 359

Index ... 361

List of Contributors

Evandro Agazzi
Séminaire de Philosophie,
Université de Fribourg
Faculté des Lettres, Miséricorde
1700 Fribourg, Switzerland
agazzi@nous.unige.it

Diderik Batens
Wijsbegeerte Universiteit Gent
Blandijnberg 2
9000 Gent, Belgium
Diderik.Batens@rug.ac.be

James E. Baugh
236 Gordon Hwy
Milledgeville, GA 31061
USA
gt5217c@prism.gatech.edu

Enrico G. Beltrametti
Department of Physics
University of Genoa &
Istituto Nazionale
di Fisica Nucleare
Sezione di Genova
Via Dodecaneso 33
16146 Genova, Italy
beltrametti@genova.infn.it

Otávio Bueno
Department of Philosophy
University of South Carolina
Columbia, SC 29208, USA
obueno@sc.edu

Gianpiero Cattaneo
Dipartimento di Informatica,
Sistemistica e Communicazione
Università di Milano - Bicocca
Via Bicocca degli Arcimboldi 8
20126 Milano, Italy
cattang@disco.unimib.it

Maria L. Dalla Chiara
Dipartimento di Filosofia
Università di Firenze
Via Bolognese 52
50139 Firenze, Italy
dachiara@
mailserver.idg.fi.cnr.it

Alberto Cordero
City University of New York
Queens College &
The Graduate Center
65-30 Kissena Boulevard
Flushing, New York 11367-0904,
USA
corde@prodigy.net

Newton C. A. da Costa
Department of Philosophy
University of Sao Paulo
P.O. Box 8105
São Paulo-SP, 05508-900, Brazil
ncacosta@usp.br

Jean Pierre Desclés
Sciences Cognitives de Paris
École Doctorale Sc Langage
ISHA
54 bd Raspail
75006 Paris, France
`jean-pierre.Descles@`
`paris4.sorbonne.fr`

David R. Finkelstein
School of Physics
Georgia Institute of Technology
Atlanta, Georgia 30332-0430, USA
`df4@`
`physics.prism.gatech.edu`

Roberto Giuntini
Dipartimento di Scienze
Pedagogiche e Filosofiche
Università di Cagliari
Via Is Mirrionis 1
09123 Cagliari, Italy
`giuntini@risc.idg.fi.cnr.it`

Paul Gochet
Université de Liège
Institut d'Electricité Montefiore
Sart Tilman, B 28
4000 Liège, Belgium
`pgochet@ulg.ac.be`

Pascal Gribomont
Université de Liège
Institut d'Electricité Montefiore
Sart Tilman, B 28
4000 Liège, Belgium
`gribomont@montefiore.ulg.ac.be`

Giuseppe Del Re
Università di Napoli "Federico II"
Dipartimento di Chimica
Via Mezzocannone 4
80134 Napoli, Italy
`G.Delre@agora.stm.it`

Gerhard Heinzmann
Department of Philosophy
H. Poincaré-Archives
University of Nancy 2
3, Place Godefroy de Bouillon
B.P. 3397, 54015 Nancy, France
`Gerhard.Heinzmann@`
`clsh.univ-nancy2.fr`

Jesús Mosterín
Consejo Superior de
Investigaciónes Científicas
Instituto de Filosofía
Calle Pinar, 25
28006 Madrid, Spain
`jesus@mosterin.com`

Graham Priest
Department of Philosophy
University of Melbourne,
Victoria 3010, Australia
`g.priest@unimelb.edu.au`

Francisco Miró Quesada
Av. Primavera 606
Chacarilla del Estanque
33 Lima 100, Peru
`fmiroqca@comercio.com.pe`

Miklós Rédei
Department of History and
Philosophy of Science
Faculty of Sciences
Loránd Eötvös University
Hungary
`redei@hps.elte.hu`

Peter Mittelstaedt
Institut für Theoretische Physik
Universität Köln
Zülpicher Strasse 77
50937 Köln, Germany
`mitt@thp.uni-koeln.de`

Roland Omnés
Laboratoire de Physique Théorique
Université de Paris-Sud
91405 Orsay, France
Roland.Omnes@th.u-psud.fr

Italo Scardovi
Doyen de la Fac. des Sciences Statis.
Université de Bologne
Via Bella Arti, 41
40126 Bologna, Italy
scardovei@stat.unibo.it

Dudley Shapere
Department of Philosophy
and History of Science
Wake Forest University
P. O. Box 7229
Winston-Salem, NC 27109, USA
shapere@wfu.edu

Ernst-Walther Stachow
Vorgebirgstrasse 35
50677 Köln, Germany
BGL@pateam.de

Ion-Olimpiu Stamatescu
Institut f. Theoretische Physik
Universität Heidelberg
Philosophenweg 16
69120 Heidelberg, Germany
stamatescu@
thphys.uni-heidelberg.de

Vladimir L. Vasyukov
Institute of Philosophy RAS
Department of Logic
Volkhonka 14
119842 Moscow, Russia
vasyukov@logic.ru
vasyukov@usa.net

Paul Weingartner
International Research Center
Institut f. Wissenschaftstheorie
Mönchsberg 2a
5020 Salzburg, Austria
paul.weingartner@sbg.ac.at

Part I

General Topics

Why Is It Logical to Admit Several Logics?

Evandro Agazzi

1 The Issue of the Plurality of Logics

The question of the plurality of logics has been the subject of a lively discussion, especially in the first decades of the twentieth century, mainly as a consequence of the construction of several logical calculi. These were the expression of the sharp formalistic outlook that inspired mathematical logic and brought into consideration logic itself as a purely syntactical discipline. The famous statement of Rudolf Carnap: "in logic there is no moral."[1] (this is considered as the motto of logical conventionalism) wanted to stress that the same mature "modern" awareness – that had obliged people to overcome, for example, the old view that there is just one "genuine" geometry (that is, *the* geometry), after the construction and the full acceptance of the non-Euclidean or non-Archimedean geometries – had to be applied to logic too. Therefore, there is no "genuine" logic (or *the* logic), but many possible logics, whose admission is suggested (but not imposed) by practical considerations. The defenders and the opponents of this view were often involved in a rather sterile dispute, for lack of sufficient analysis, and due to an excess of polemic spirit.

The polemic attitude was produced by the claim of certain defenders of the new logic (or logics), according to whom it had "falsified" or overcome the traditional logic (for example, Whitehead and Russell maintained, in the *Principia Mathematica*, that the Aristotelian syllogistic revealed several mistakes, if submitted to the exact formal tools of mathematical logic). Or, in the case of more tolerant positions, it was maintained that the new logical systems were "at variance" with the old logic. The opponents of the pluralistic view sometimes tried to show that the traditional logic was by no means wrong (for example, Łukasiewicz was able to clarify why the Aristotelian syllogistic was correct from the point of view of mathematical logic, provided certain philosophical presuppositions of this syllogistic were duly recognized).[2] Sometimes

---

[1] Cf. Carnap (1934) where the famous "tolerance principle" is posed from the beginning as a basic view in the interpretation of the spirit of logic. The quoted statement occurs in §17 of this work, specifically devoted to "The tolerance principle in syntax".

[2] Cf. Łukasiewicz (1957).

they tried to "recapture" the features of the new logics within the framework of the traditional one, or maintained that certain recalcitrant features of the new logics should be ascribed to the "methodology" of certain paticular sciences, and not to logic in a proper sense.

The lack of sufficient analysis we have mentioned consists of the fact that no preliminary clarification was carefully proposed as to the very *meaning* of logic. If such a clarification had been provided, the debated issue could have been seen not as an "aut-aut" question (one logic, or many logics), but as a question admitting a double answer: there is a *sense* according to which logic is unique, and another *sense* according to which several logics are legitimate. In this paper we shall try to outline such a diversified answer.

To give a rough idea of the solution we are going to propose, we could say that we want to present a view of logic comparable with the view of geometry maintained in the celebrated "Erlangen Programme" of Felix Klein.[3] The best known part of this programme consists in the group-theoretic treatment of the different geometries (elementary, metric, projective, affine, etc.) that permits us to uncover a strict and almost genetic relationship among these theories, which puts them on an equal footing as to their mathematical legitimacy. Nevertheless, this does not imply that whatever consistent mathematical theory is a "geometry", in spite of its being expressible in group-theoretic terms. Klein explicitly distinguishes between "geometry" and simple "mathematics", and maintains that only those theories are geometries, whose group "preserves" the "fundamental group" (that is, the group characterizing elementary geometry). In this way it is possible to maintain the unicity (or specificity) of geometry, with reference to its most standard part, and at the same time to justify the plurality of geometries. A serious exploration of this Kleinian analogy, however, would be technically too complicated, and lead us too far afield. Therefore, we shall be content with this general hint.

2 The Domain of Logic

The method we will adopt for approaching our issue is that of the analysis of language, or of the different linguistic uses of the term "logic". In our language we find expressions such as: "the logic of the events brought about that...", "the logic of power does not permit...", "the logic of economy, or of profit, entails that...", "the logic of research", "the logic of thought, or of discourse", and so on. It is very easy to say that the "proper" use of our term is that indicated as "logic of thought or discourse", since thought (and more precisely thought explicitly expressed in a discourse) is the traditional subject matter of the specific *discipline* historically called *logic* (as the definitions of the subject matter, and often even the titles of the treatises devoted to this

[3] Cf. Klein (1872).

discipline often declare).[4] One should refrain, however, from rejecting the other uses as spurious, confused, or even meaningless: they share a particular meaning of "logic", which could be indicated as the intrinsic nature, or the specific objective structure, of a given domain, from which the functioning of this domain, or the actions that are characteristic of it, can be more or less explicitly expressed by means of certain laws or rules. Therefore, we can safely decide that, when speaking of logic, we consciously intend to speak of a discipline that studies thought expressed in a discourse, and this determines its "domain". It must be admitted, however, that this delimitation is still too broad, for thought and discourse can be studied from several points of view (such as those of linguistic, grammar, etc.) that we do not consider as belonging to logic.

A simple analysis shows that we normally think of logic as being concerned with that particular aspect or part of thought that consists of "reasoning" or of constructing "arguments" (such that, even if other aspects or parts of thought are taken into consideration in logic, they are investigated only to the extent they are relevant to the construction and evaluation of arguments). In conclusion, we can say that the domain of logic is the study of *correct reasoning*.[5]

But what is reasoning? From a descriptive point of view we can say that a reasoning is a particular connexion of propositions or sentences (we speak of propositions when we consider thought, and of sentences when we consider the linguistic expression of thought).[6] Now we speak, for example, of an "eloquent" reasoning, of a "persuasive" reasoning, of a "correct" reasoning. In the first case we mean a reasoning having an emotional impact, in the second a reasoning capable of producing other people's assent, especially when action is concerned. What do we mean by a "correct" reasoning? A very plausible answer is that a reasoning is correct when its conclusions are a "logical consequence" of its premisses. This answer is interesting, since it surfaces in it that the term "logical" supports our choice of defining logic as the study

[4] Let us only mention two titles, belonging to different historical times, and also to rather different conceptions of logic: *La logique ou l'art de penser* (1662) published by A. Arnaud and P. Nicole (the famous "logic of Port-Royal"): and *An Investigation of the Laws of Thought on Which are Founded the Mathematical Theories of Logic and Probability*, published in 1854 by G. Boole.

[5] In the literature the term "*valid* arguments" is often used, meaning that they are formally correct, and then it is added that they must also be *sound*, hinting at certain additional requirements they must satisfy (cf. for example Haak 1978). We use *correct* as an adjective denoting at the same time validity and soundness.

[6] We are aware of the non-univocal meaning attributed by different authors to the terms "proposition", "sentence", and "statement", and to the mutual relations existing among them, especially in the discussions regarding the "truth-bearers" (cf. Haak 1978, pp. 74–85). We do not need, however, to enter this controversy and, for the sake of this paper, the distinction we have proposed will be sufficient.

of the correct reasoning; still it is not very informative, for it is by no means obvious what "logical consequence" means.

3 The Notion of Logical Consequence

If we try to explain what it means that a proposition B is a logical consequence of another proposition A (to consider the simplest case), we could say that one feels obliged to admit B, once one has admitted A. This kind of explanation considers the link of logical consequence as a particular form of psychic constraint, like an irresistible mental inclination (and this seems well in keeping with the idea that logic studies the movements or articulations of thinking). Nevertheless, anyone can certainly remember some occasions on which he or she had believed a certain B to be a logical consequence of a certain A, and then had recognized that B was not really a "logical consequence" of A. This indicates that the link of logical consequence has a psychic "counterpart", but does not precisely consist of this; that it is something much more "objective", whose nature should be looked for in "thought", rather than in "thinking". In order to capture this objective feature one could say that B is a logical consequence of A if "it is not possible" to admit A and reject B. This answer, however, is far from clear, since it is rather mysterious what this impossibility actually amounts to.

A very convenient way out of this impasse consists of considering the primary nature of propositions (whose linguistic expression are the "declarative sentences"): a proposition is the content of a judgment (that is, of a particular form of *thought*) whose primary characteristic is that of being either *true* or *false*. It is not essential for the sake of our investigation to find an agreement about a "definition" of truth: we can safely admit the definition that is implicitly presupposed by *whatever* "theory of truth" (since all such theories rather propose different "criteria" for truth)[7], and that consists of saying that a proposition is true if and only if that aspect or item of reality to which thought intends to *refer* is actually such as the proposition says. Now there are cases in which this aspect or item of reality is *immediately* present to our thought, but humans have spontaneously and unconsciously uncovered that, thanks to their thinking, they are endowed with the marvellous capability of attaining truth even in those cases in which this is not immediately present, since they are able to capture this truth by means of their *reasoning*. This reasoning is able to establish certain *links* between propositions that are "truth-preserving", that is, that lead from true propositions to other propositions that are also true, in spite of not being immediately true.

[7] Here, again, one should note that the distinction between definition and criteria for truth has been the subject of a large literature, in which very different positions have been advocated, but it is not important for the sake of this paper to enter this discussion.

The practice of intuitive reasoning, however, in particular when the conclusion is derived from a set including several premisses, easily shows two things: sometimes the reasoning seems impeccable, but the conclusion is patently false; sometimes a "form" of reasoning that has led to true conclusions in certain cases, leads to a false conclusion in a different case. In the effort of understanding these facts, and once we can exclude that some premisses admitted as true were in fact false, we may find out that the conditions under which the premisses (or some of them) were recognized as true are not the same as those under which the conclusion can be true. This leads to a refinement of the notion of logical consequence, and we can stipulate that a proposition C is a logical consequence of a given set of premisses P=$\{P_1 \ldots P_n\}$ if and only if in *all the cases* in which all the premisses are true, also the consequence C is true exactly under the same conditions.

This definition may be satisfactory, but is unmanageable, it offers no "criterion" for ascertaining if and when C is a logical consequence of the set P, since it would imply something like an infinite control of all *possible* truth conditions of P. A solution of this difficulty seems available: if C has been reached by means of a reasoning, and reasoning has been understood as a "truth-preserving" linking among propositions, the truth of C would be granted if P is true. This solution is viable, however, only if, among the many sorts of reasoning humans spontaneously adopt, we are able to single out the *correct reasonings*, that is, those links that *necessarily* lead from true propositions to true propositions. Therefore, if we find that a certain "form" of reasoning had permitted us to attain true conclusions from true premisses in several cases, but failed to do so in other cases, we must say that the true conclusions were not obtained *in virtue* of the reasoning, but only accidentally or contingently, since the reasoning was not *correct* (following Aristotle, we could say that it was simply a "sophistical" argument).

We have come, finally, to an acceptable characterization of logic: it is a theory of the correct reasoning and, for this reason, it cannot be understood simply as a *descriptive* enterprise. Sure, in order to concretely construct logic, we must take into consideration human reasoning or "thinking", but in this thinking we find quite commonly used incorrect reasonings, and in order to discard them we need a meta-reflection in which the necessarily truth-preserving forms could be singled out and explicitly codified. This is the *normative* aspect of logic, that entitles its being also called the investigation of the "laws of thought". The use of "thought" instead of "thinking" underlines its objective nature, while the term "laws" hints at its features of universality and necessity, which correspond to the level of *idealization* that is inherent to logic as well as to *any* intellectual construction (as we will better see in the following sections).

4 The Basic Status of Alethic Logic

The analysis outlined thus far has shown logic arising from the reflections on truth, and on the notion of logical consequence conceived as strictly related to the investigation of truth. This is tantamount to saying that "alethic logic"[8] has emerged as the basic framework of logic as such. This framework has the features of a programme that must be patiently and carefully developed, and for this reason it is necessary to pass from thought to its explicit linguistic expressions. We must proceed to several idealizations, for example by first summarizing under a few "sentential connectives" or operators the many *immediate* links among sentences we use in everyday language, or by singling out certain "quantifiers", and so on. In brief, we must outline a "logical grammar" of the (idealized) language. We then proceed to explore more complex links among propositions, that correspond to arguments or reasonings in a proper sense, and we begin finding certain general patterns of them. The most elementary are those in which the "truth-preserving" characteristic of the links results from considering sentences simply as linguistic entities that are either true or false, and the explicitation of such patterns constitutes the domain of *sentential logic*. It is easily found, however, that we use a lot of correct arguments whose patterns cannot be captured by sentential logic: not even a very elementary syllogism can be recognized as correct on the ground of this logic, since it presupposes a consideration of the "terms" entering a sentence, and the correct logical link among the sentences depends on certain correct links among the terms.

We can call *syllogistic* (or "categorical syllogistic") such a deeper-level logic, and try to determine all the patterns of it that correspond to correct reasonings: something Aristotle had done, believing to have captured in this way the ideal forms to which *every* correct reasoning is reducible. Already medieval logicians, however, had noticed that even such an elementary and obviously correct reasoning as the following: "the circle is a figure, hence who draws a circle draws a figure" is not fully compressible in any of the standard syllogistic "figures", in spite of having a shape vaguely resembling that of a syllogism. We know that the reason for this fact is that traditional syllogistic only considers properties, but not relations (in modern terminology we would say that it corresponds to a logic of classes or of monadic predicates), and this also explains why it was never concretely used in the exposition of mathematics, where the consideration of relations is no less essential than the consideration of properties in the proof of even the simplest theorems.

[8] We explicitly stress that we shall use the term *alethic* in a very *general* sense, immediately related to its etymology (*aletheia* = truth), in acccordance with our proposal of strictly linking logic with the investigation of truth conditions. In the literature this adjective has a much more restricted meaning, since it is used to indicate the basic *modal* logic, and distinguish it from other "structurally similar" modalities, such as, for example, "deontic" modalities (cf. von Wright 1951).

The situation just mentioned is quite interesting: during a very long historical period it was believed that logic had been completely developed (this was, for example, the opinion of Kant), though it only contained syllogistic with a few complements of sentential logic and modal logic (that were mainly related to the treatment of hypothetical and modal syllogisms). A great deal of correct reasonings adopted in everyday discourse and several sciences (including mathematics), however, followed patterns that essentially overstepped those of syllogistic, but people seemed unaware of this, being convinced that any correct reasoning could "in principle" be reshaped as a sequence of syllogisms, though this re-elaboration would be too cumbersome in practice to deserve being performed.

5 Mathematical Logic and Its Double Meaning

Things changed deeply with the construction of *mathematical logic* in the 19th century. Two meanings have been attached to this denomination. According to the first (that we can refer to Boole for simplicity), it indicates the fact that logic has been reformulated in a "mathematical way" (that is, essentially, in the form of symbolic *calculi*). According to the second (that we can refer to Frege in a similar way), it denotes the "logic of mathematics". From the point of view of the "external appearence", the first meaning expresses the most significant "novelty" with respect to traditional logic. Yet this is not so: exposing logic in the form of symbolic calculi was simply the concrete application of a point of view already present in the *Posterior Analytics*, and explicitly advocated by Leibniz. It was the programme of submitting the common reasonings of mathematics to a rigorous *logical* study, instead, that led to uncover the complexity of such reasonings being uncovered, and eventually to the development of new logics (such as the logic of relations, the logic of classes, the predicate logic, and so on), which included authentic "novelties" with respect to traditional logic (novelties that could be suitably expressed in certain calculi, as Frege and other logicians were able to do).

This story lends itself to a few interesting reflections. In the first place, one can see here an historical example that "science may need new logics", for it was in the effort of making explicit the correct reasonings applied in a particular science (that is, mathematics) that genuine *new* logics (new with respect to traditional logic) were found. Yet one can legitimately say that this was simply a *de facto* circumstance, since such new logics are indeed needed not only for rigorously analyzing mathematical reasonings, but also a great deal of reasonings of common sense. A more subtle remark is the following: is it really correct to say that "new logics" were discovered, instead of saying that "new branches of logic" were developed? The second alternative indeed seems more appropriate and, as a matter of fact, we usually consider, for example, sentential logic, first-order predicate logic, second-order predicate logic, and so on, as distinct but not separate chapters of one logic, corresponding to different "levels" of the logical analysis of language. Moreover, the "older" parts of logic are "included" in the new branches (for example, categorical syllogistic is contained in first-order predicate logic, that "in addition" is also able to consider polyadic predicates, or relations, and the correct reasonings in which they occur). In other words, the different logics behave like the different geometries in the approach of the Erlangen Programme: they must all contain something like the "fundamental group" (that is, traditional logic), and are characterized by properties that from a certain point of view are broader, and from another point of view are more restricted than those of the "inferior" levels (precisely as it happens in the case of the group-theoretic hierarchy of geometries). For example, sentential logic corresponds to a linguistic level endowed with a low "expressive power", but is correct, complete, and decidable; first-order logic is endowed with a greater expressive power (individuals, properties and relations are distinguished, quantification over individuals is admitted), but, while correctness and completeness are preserved, decidability is lost in general (that part which corresponds to traditional logic, that is, the logic of monadic predicates, however, is decidable); second-order logic is even more expressive (properties of properties, properties of relations, relations between properties are considered, quantification over properties is admitted), correctness is preserved, but completeness and decidability are lost.

Without going into additional details we can conclude: in spite of the fact that, historically speaking, several new branches of logic were discovered and developed as a consequence of having concentrated the attention on the correct reasonings of mathematics, these new branches cannot be said to be intrinsically *dependent* on the needs of mathematics. They rather constitute the completion of lacking chapters of logic as such: we can call "logic" any single chapter (and characterize it by means of an adjective or a specification), this, however, does not really mean a plurality of logics, but rather an articulation in parts or subdomains of the unique logic that has been characterized (we must not forget) as an effort for making explicit the links corresponding to the relation of *logical consequence*. Therefore, we can say in brief that the domain of logic has appeared to coincide, up to now, with that of *alethic logic*.

6 The Variety of Logical Calculi

The thesis of the plurality of logics , as we have already seen, was especially advocated, in the first decades of the twentieth century, as a consequence of the existence of a great deal of *logical calculi*, and of the unlimited possibility of constructing new ones. Owing to the strict syntactic and formalistic point of view prevailing at that time, any such calculus was considered to be a logic in itself, and the plurality of logics was indeed consistent with this view. Yet a rather simple reflection shows the inadequacy of such a perspective. It is enough to consider that a *calculus* (as has been clear since Leibniz, and repeatedly stressed since) is nothing but a system of rules for operating with symbols. This does not entail, however, that whatever calculus is a *logical* calculus (for example, chess and bridge are games whose very explicit rules justify considering them as calculi, but they are never listed among the logical calculi). Already Leibniz had said that, among the many varieties of calculi, it was possible to single out a particular one, which he called *calculus ratiocinator*, whose specific nature was that of reproducing the features of correct reasoning. At first sight one would be inclined to say: very well, Leibniz was convinced that logic is mirrored in a single special calculus, we accept his point of view and say that, if we nowadays admit several logical calculi, we are obliged to admit that each of them correspond to a definite logic, and therefore we admit several logics.

Things, however, are not so simple. The fact is that, once we have "freely" constructed a calculus, we must prove that it satisfies certain *conditions*, before recognizing it as a "logical" calculus, and these conditions are the *same* for all calculi. They consist essentially of a "faithfullness" with respect to the notion of "logical consequence", in the following sense. First of all we must be able to show that the symbols and rules of the calculus can be "interpreted" in such a way that (globally taken) they can be seen as expressing sentences, and links among sentences. Then a metatheorem must be proved: it must be shown that, by using this calculus, *only* logical consequences can be derived from any set of premisses. This *minimal* requirement amounts to recognizing that the calculus is *correct*, and a calculus not satisfying this requirement is excluded at the outset from the domain of logical calculi, since it would permit to derive from a set of premisses certain conclusions that *are* not a "logical consequence" of the premisses. Besides this *indispensable* condition, another appears as a highly *desirable* requirement for a calculus: that this permits to derive from any set of premisses the set of *all* its logical consequences. This requirement is indicated as the semantic *completeness* of a calculus. We know that only for calculi corresponding to sentential logic and first-order logic completeness can be proved, while higher order calculi fail to satisfy this requirement. The fact that these two metatheorems are taken as a criterion for admitting a calculus as a "logical" one, and that this criterion applies identically to all the calculi "corresponding" to a given logic, already indicates that the plurality of calculi does not mean a plurality of logics.[9] The analogy with geometry is again instructive for understanding this point. Elementary geometry and projective geometry are certainly *two* distinct geometries: each one of them, however, can be axiomatized in *many* different ways, and we know several axiomatizations of *the* elementary geom-

[9] We can note, incidentally, that certain authors (among them, Quine and Kneale) were so severe in requiring the full respect of the two stated conditions as to exclude from the realm of logic second-order logic.

etry, or of *the* projective geometry (mathematicians of the end of the 19th century, and in particular Peano and his school, have shown a great skill in producing such axiomatizations). We can express this fact by saying that, for example, elementary geometry, though being *one*, can be presented or formulated under *many* different "forms", and we can explain the meaning of this assertion by saying that all these forms are "equivalent", in the sense that precisely the *same* geometrical truths can be proved as theorems in each of these axiomatic systems. In the very same way we can say, for example, that all logical calculi corresponding to a given "logic" (sentential, first order, etc.) that are correct and complete are "equivalent" in the sense that they permit, from any set of premisses, to derive exactly the same conclusions. This fact is indeed obvious: let a set of premisses P be given, and a sentence S that is derivable from P using a calculus K and not derivable using another calculus K'. If S is a logical consequence of P, then K' would not be complete; if S is not a logical consequence of P, then K would not be correct.

The general conclusion we have reached up to now can be summarized as follows: in a proper sense one should say that there is *one* logic (alethic logic), articulated into several domains, each domain being in turn expressible or "formalizable" by means of different logical calculi. Using a systems-theoretic approach this view can be expressed by saying that logic is a global system whose subsystems are the particular logics, while the different calculi do not constitute further subsystems, but only different descriptions of the functioning of a given subsystem. According to a different approach we could say that there is a general *concept* of logic (corresponding to the notion of alethic logic), that is "exemplified" by several particular alethic *logics* (sentential, first-order, second-order, class logic, etc.), that are formalizable by means of different logical calculi.

In both ways we can see how it is possible to maintain the uniqueness of logic (from a certain point of view) and the plurality of logics (from another point of view) in a consistent way. What we have seen, however, still remains at the margins of the real debate concerning the plurality of logics: we have only refuted the thesis that such a plurality is constituted by the mere existence of a plurality of logical calculi. The core of the debate, however, consists of accepting or rejecting the thesis that alethic logic is the *only* or the *genuine* logic. We shall now proceed to explore this issue.

7 The Broadening of the Domain of Logic

The path we have followed for determining the nature of logic has been that of analyzing the notion and the patterns of correct reasoning, as it is intuitively understood, and the most basic requirement we have found is that any correct reasoning must be truth-preserving. Indeed we have defined the notion of "logical consequence" with explicit reference to truth, then characterized a correct reasoning as that in which certain propositions are linked with their logical consequences, and finally have characterized logic as the idealization and explicitation of the links realized in correct reasoning. In this effort of idealization and explicitation the focus on the "links" (as we have already noted) led to disregard the meaning, and even the contingent truth, of the sentences actually occurring in a discourse, especially because these are often the source of fallacies in everyday reasonings. Yet human reasoning, being a part of the activity of thinking, never takes place in a vacuum, but is always unfolding within a context of meaning.

This, in particular, is reflected also in the intuitive notion of logical consequence, for which we usually require not only that premisses and consequences are equally true, but also that the premisses are "relevant" to the truth of the consequences. This fact was taken into account by traditional syllogistic, whose basic principle was: *quae conveniunt uni tertio conveniunt inter se* (things that are pertinent to a third thing are mutually pertinent). The two premisses of a syllogism must contain a common "medium" term, and at the same time be true: a double condition that grants that they are really intrinsically related, and for that reason really relevant to the conclusion (that consists in a sentence containing only the two "extreme" terms). Indeed the *venatio medii* (the "hunting" of the appropriate medium) was considered as the most challenging part of the construction of good arguments. In contemporary logic, the discussions on material and strict implications, and especially the construction of *relevant logics* constitute examples of a "broadening" of the horizon of purely alethic logic, which has led to logical systems that are not certainly "at variance" with alethic logic, but introduce stricter requirements for the "logical" inferences.[10]

That the consideration of the meaning is by no means just accessory or contingent in the formation of correct reasonings is clearly shown by the fact that several "immediate inferences" in our reasoning are simply the explicitation of certain meanings, from which more complex inferences can be constructed. For example, we consider an immediate "logical consequence" of affirming that a certain event was "necessary", to affirm that it was "possible" (but not vice versa); we consider a "logical consequence" of affirming that a given action is "obligatory" to affirm that its omission is "forbidden"; we consider a "logical consequence" of affirming that one "believes that p" that he does not "believe that non-p". If we consider these immediate inferences (and many others of a similar kind), we recognize in them the basic feature of the alethic notion of logical consequence (that is, it cannot happen that the premiss is true and the consequence is false), but we cannot say that the consequence is true *in virtue* of any linguistic link, just because there is no such link here (these are all examples of *immediate* inferences). The actual situation is that such inferences are, so to speak, simply the linguistic explicitation of a particular "conceptual domain", in which certain basic notions

[10] A good survey of the construction of relevant logics can be found in Belnap (1981).

"interdefine" each other in a way quite similar to that in which truth and falsity are mutually interdefined.

The three examples just presented obviously hint at immediate inferences belonging, respectively, to modal logic, deontic logic, epistemic logic, and have each a definite specificity, the specificity of a given "conceptual domain" that is possible (and useful) to render explicit, for example, through an appropriate axiomatization. The above examples could in fact be single items of such axiomatizations, whose function is comparable to that, let us say, of the axioms of the familiar alethic sentential logic. The specific nature of every conceptual domain, however, entails that differences be no less significant than similarities. For example: it is fully in keeping with the notion of truth that it has just one opposite, that is, falsity. Therefore alethic logic must consider only two "statuses" of a proposition (true and false), and the "bivalence principle" is appropriate to it. But if we consider the conceptual field of modality, it would be arbitrary to impose on it something like a bivalence principle, since to a state of affairs can be attributed not two, but three modalities (possible, impossible, necessary). Similarly, from a deontic point of view, an action may be not only obligatory or prohibited, but also permitted. These considerations indicate that a modal logic and a deontic logic should be "three-valued" rather than "two-valued", with the awareness, however, that these "values" must not be considered as "truth-values" (a confusion that has affected several controversies).[11]

The explicitation of these logics is delicate because they maintain a continuous interplay with alethic logic: in the first place, because the axioms of these logics are meant to be an explicitation of what modalities, deontic aspects, epistemic attitudes, *truly* are. In the second place because it is possible sometimes to translate in terms of true propositions the contents of axioms that do not refer to propositions (for example, instead of saying: "the state of affairs A is necessary", we can say: "the proposition S is necessarily true", where S is the linguistic description of A). Finally, axioms of a given logic may concern propositions, and yet not reflect all the features of a propositional logic. For example, we have given above the "epistemic" statement: "if A believes the sentence S, A does not believe the sentence non-S". The soundness of this statement relies: upon the epistemic fact that believing a proposition is equivalent to believing that it is true; upon the alethic fact that the negation of a true proposition is false; upon the alethic fact that one cannot believe at the same time that a proposition is true and false. All this, however, does not entail the validity of a principle of bivalence in epistemic logic. Indeed, while it is correct to say that "one never believes S and non-S",

[11] This feature is obviously compatible with the fact that, from a *formal* point of view, it is possible to take only one of such notions as primitive, and define the others by a suitable use of negations. This fact does not eliminate the other fact, that these notions are *conceptually* distinct.

it is not correct to say that "one always believes either S or non-S" (since one may well "suspend" belief in the case of some particular S).[12]

In the above examples our reasonings regard, so to speak, clear-cut cognitive situations, but in many cases we are obliged to formulate correct reasonings in less privileged cognitive situations, and we are indeed able to do so. For example, the clear-cut partition of modal logic according to which an event is either necessary, or possible, or impossible, is rather unsatisfactory for several purposes, in which we are interested in evaluating the "measure" of the possibility of an event: probability calculus is, from this point of view, a "logical" response to this legitimate desire, in the sense that it permits to give such a response not just on the basis of a guess, but of rigorous and eleborated reasonings.[13]

Already in traditional epistemology the term "probable" was used not for denoting some modal property concerning the occurrence of an event, but for indicating a particular state of mind with regard to the *truth* of a proposition: in that context it was usual to indicate as "ignorance" the lowest degree, with "doubt" the situation in which the mind has not taken a position yet, with "opinion" the state in which the mind is inclined to admit the truth of a certain judgment, with "certainty" the state in which the mind is in full possession of the truth. Opinions, being always affected by a certain degree of uncertainty, were said to be more or less "probable". Therefore probability was meant to admit of an indefinite spectrum of degrees, spanning from ignorance to certainty.[14] It is rather natural to translate this qualitative traditional view in terms of the modern notion of probability, by identifying ignorance with a probability $p = 0$, certainty with $p = 1$, and all the intermediate states with probabilities p_i with values between 0 and 1. Through such a translation it is also sensible to use the probability calculus to express the "logic" of such mental attitudes towards truth.

It would be mistaken, however, to consider such probabilistic logics as "at variance" with alethic logic, or as admitting a continuous spectrum of truth-values. For example, if we say: "it is 90% probable that Peter is at home in this moment", we cannot avoid that the event mentioned in the proposition (that is, the presence of Peter at home now) either occurs or does not occur (bivalence principle in ontological sense), and from that it follows that the proposition describing the event is either true or false (bivalence principle for

[12] Other well known examples are offered, for instance, by deontic logic, where it is clear that, in spite of its "structural analogy" with modal logic, this analogy is not total (cf. von Wright 1981).

[13] In the title of the already cited *Laws of Thought*, Boole already puts the calculus of probabilities on the same footing as logic.

[14] This classical doctrine is standard in all textbooks of the "scholastic" tradition, including quite recent ones. For example, it is presented in Verneaux (1959), pp. 93–98. In the 18th century it gave rise to several developments regarding the "probability of judgments", where the tools of the recently born probability calculus in its Pascalian version were applied.

declarative logic); the probability of 90% only indicates that we very strongly adhere to the truth of the said proposition (though not feeling completely certain about it).

The last considerations offer us a guideline for understanding the "logic" of the unfolding of different logics. If we go back to our original considerations, in which we have seen logic coming out of the needs of the investigation of truth, that is as a powerful *instrument* (this is the meaning of the Aristotelian *"organon"*) in the acquisition of *knowledge*, we can easily understand how the same Aristotle indicated the *ideal* conditions in which the use of such an instrument can give the best results: these conditions are those in which the reasoning starts from premisses that are "immediately true, better known than the conclusion and cause of it" (*Post. Anal.*, A, 2, 6). This is tantamount to saying that the ideal situation is that in which the reasoning can develop "under conditions of certainty". Humans, however, are obliged most of the time to develop their reasonings "under conditions of uncertainty", and the same Aristotle, in presenting the general aim of his syllogistic, precisely declared that his work "has as purpose that of finding a method for constructing syllogisms on whatever proposed problem, moving from probable premisses" (*Topics* A 1, 100a 22). This means that, even if we have at our disposal an instrument (such as alethic logic) that is "sure" and yields truth *if* applied to true premisses, we still remain with the task of evaluating to what extent the "probable" premisses of which Aristotle speaks are actually probable.

The methods and reasonings by means of which we try to establish the *degrees of confidence* in a given sentence (that is, in the *truth* of a given sentence) can legitimately be called a "logic", since they must necessarily move from certain true propositions, and lead us to other propositions we *intend* to be true, but whose truth is not absolutely granted (like in the case of standard alethic logic), such that we want at least to know "how confident" we can be that they are true. *inductive logic* is the most important example in this domain, and certain criticisms that have been levelled against it are essentially misplaced: its aim is often pictured as that of coming to *general* conclusions from a finite (though perhaps very large) set of *singular* true premisses, and it is too easy to note that such a transition is never absolutely granted. This is, however, a misrepresentation of the aim of inductive logic. Even in the case of enumerative induction, there is no pretention of attaining an absolutely true general conclusion, but only one of which we try to establish the "probability" in the traditional sense. That in this effort the probability calculus can be of great help is simply obvious, though certain precisions must be elaborated regarding the different meanings of probability that apply in this enterprise, and their relation with the notion of the "probability of a sentence".[15] One must be aware, however, that inductive logic does not reduce to this application of the probability calculus, nor does it have as main purpose that of providing some justification for generalizations. In fact,

[15] For this variety of meanings of probability cf. E. Agazzi (1988b).

its inductive methods are used in several sciences (and also in nonscientific contexts) to investigate some possible *causal* relations among events, on the basis of an available empirical evidence. In these cases we can speak of the "inductive support" provided by such evidence to a certain hypothesis, and it is possible to elaborate and formalize a "logic" of such a support that is not isomorphic with probability calculus.[16]

As a final example of a logic corresponding to the needs of reasoning under conditions of uncertainty we shall mention the circumstance that even the possibility that the premisses from which we start are, to some extent, mutually inconsistent does not imply the collapse of our correct thinking. *Paraconsistent* logics have actually shown how correct reasonings can be developed within contexts in which a certain degree of inconsistency is present (apart from more technical applications, this logic is rather close to the familiar situations of our daily reasonings, in which we are often far from sure that the system of our premisses is really consistent).

8 Logic and the Object of Discourse

Our considerations are a development of the remark stated above that reasonings cannot occur "in a vacuum". This is the obvious consequence that reasoning is a particular activity of thinking, and thinking cannot be "thinking of nothing"; in fact the activity of "thinking" is inseparable from the production of "thoughts" that are its very *actualization* and have that "mental reality" which we can equate with their sense. When we say that thinking cannot be thinking of nothing, however, we do not mean that thinking is

[16] A very instructive presentation of these different aspects is offered in Cohen (1981), where it is shown that the application of the "Pascalian" approach to the evaluation of the soundness of a scientific proposition (that is, essentially, the use of standard probability calculus for improving the acceptance of this proposition), that characterizes Carnap's inductive logic, is opposed to Popper's doctrine according to which a hypothesis is the more valuable the more *improbable* it is. In spite of being opposite, these doctrines are shown to be appropriate, respectively, for expressing the degree of appreciation that is given to a scientific proposition, in the first case, from the point of view of a "technological" purpose or, in the second case, from the point of view of a theoretical "explanatory" purpose. By deepening the analysis of the different systems of inductive logic elaborated in the past decades, Cohen shows certain shortcomings that can be eliminated in a "logic of inductive support" (that he has presented especially in Cohen 1970, 1977) which is not formalizable according to a "Pascalian" approach, but is formalizable using a generalization of a certain system of modal logic (the S4 system of C.I. Lewis). This discussion exemplifies very well how certain logical systems have been created in order to make explicit and controllable some rational arguments applied in the context of scientific research (and even according to different goals pursued in this research), and how they had to challenge several problems of a genuine formal logical nature.

thinking of thoughts or senses (since thoughts are not the "content" of thinking, but simply thinking in *act*). Hence, we must say that thought cannot be thought of nothing, and by this we underline that *intentional* nature of thought that has been analyzed several times by past and contemporary philosophical schools. It is at this stage that we can (and must) speak of an "object" of thought, and this is that to what thought *refers* (or its *referent*).

These clarifications are not just a pedantic digression, but a useful reminder for our discussion, since *truth* comes about precisely in this referential situation: it is the property of a thought whose content (sense) conforms to its intended object or referent. We are not obliged to find an agreement in specifying of what this "conformity" consists: it is enough to note that, in the same way that a thought necessarily is thought of *something*, also a proposition is necessarily true (or false) of *something* (that is, of its referent). In other words, a proposition is never true or false "in itself", but always relatively to a certain referent. What we have been saying about thoughts and propositions can be transposed to their linguistic expressions, that is, discourse and sentences. The conclusion is that an investigation of the truth conditions of a discourse cannot avoid taking into *serious* consideration also the particular nature of the *objects* of that discourse, since it is "in virtue" of such objects that a suitably constructed discourse can be true, and never "in virtue" of itself.[17]

The last affirmation seems to be contradicted by the existence of propositions that are usually considered as "true in themselves", or "always true": typically the propositions of logic, also called "logical laws". Yet even in such cases these propositions can be said to be true because, through a *reflective* act of thinking, we can take their *meaning* as the *object* of reflection, and recognize that "in virtue" of this meaning, they represent a universal scheme or "form" that will yield true propositions not "independently of any interpretation" of particular referents, but "in connection with whatever interpretation" of particular referents. For example, the sentence: "if all A are B and all B are C, then all A are C" can be said to be true in itself, or always true, only if a suitable *meaning* is attached to all its linguistic ingredients: sentential connectives must receive a certain meaning, not less than the quantifier "all", while A, B and C must receive the meaning either of names for classes or of names for properties; finally the meaning of "are" must be clarified: it can be taken as expressing essentially a class inclusion, or a predication and, in both cases, the "universal truth" of this logical law will be established "in virtue" of the intuitive meaning of such fundamental intellectual operations as the assignment of individuals to a class, or of the attribution of a property to individuals. In conclusion, the said sentence is *true* of a (rather complex) network of thinking operations whose description contains some "empty boxes" (the letters A,B,C): if these boxes are filled

[17] This problem has been deeply analyzed in the large literature on "truth-makers" which we cannot discuss in this paper.

with contents not of any kind, but of the kind prescribed by their *meaning*, the result will necessarily be a special true proposition about those particular contents.

9 Meaning, Sense, and Reference

Our considerations might sound rather peculiar to many readers, since we are maintaining that *meanings* can become the *object* of a certain kind of knowledge, and this seems to ignore the distinction between *sense* and *reference* that we have accepted in a preceding part of this paper. This is, however, just a superficial first impression. In fact, "sense" is normally understood as a mental representation, while "referents" are understood as concrete material "things" to which a certain sense can be properly related. This *distinction* (that in modern times was elaborated especially by Frege) is opportune, but it cannot be taken as a *separation*, and in order to stress the strict relation existing between sense and reference we explicitly speak of *meaning* as a composite unity of which sense and reference are the constitutive parts. This can be seen if we consider that any sense unavoidably "points to" or "intentionally refers to" an entity to which it applies, though this entity may happen not to exist concretely. This situation can be expressed by saying that a sense *encodes* (through the presence of a set of concepts) a specific class of entities, and these entities *exemplify* the sense.

In several particular cases sense can be constructed "artificially", and we proceed to investigate whether the class of referents it points to is empty or not. For example, in mathematics we can give a "definition" (explicit or axiomatic) by a coordination of certain concepts, and then look for possible "mathematical entities" that satisfy this definition. But even in common language we can construct expressions such as "golden mountain", or "the present king of France". It would be improper to say that they are "meaningless", because they can be *understood*, and hence have a *sense*, and it is in virtue of this sense that we can find out that they have no reference. This happens because this sense entails that possible referents should be endowed with certain properties defining a given class of objects (in the case of the second example, these properties are in part implicit, in part explicit: for example, the presence of "the" indicates that the class must contain just one individual, it is implicit that the referent be a concretely existing human person, and not just a character in a novel, the adjective "present" indicates a temporal coordinate, while "king" summarizes a set of political and legal prerogatives). If it happens that no individual *exemplifies* all these joint properties, we conclude that the expression denotes an empty class, and that no *referent* corresponds to this *sense*.

In the majority of the cases, however, the sense is not created artificially, but is obtained (by means of the intellectual operation that we call "abstraction") from the "intentional presence" of certain referents that we encounter in some kind of *experience.* Moreover, the sense is not established once and for all, but is continuously "controlled" (in its explicit formulation), through a comparison with the "intended referents". This is why sense is just a part of the meaning, and it cannot be completely disjoint from the referents that exemplify it.

An important remark is now fundamental for our discourse: humans are endowed with the capability of directing their intentional power not only to material things, but also to *abstract* objects, and in particular to meanings and senses (this peculiarity was already pointed out in the distinction of *intentio prima* and *intentio secunda* proposed by medieval philosophers). In this intellectual activity sense and meaning become *referents* of the intentional act, they are the *objects* of a second-level investigation, and it is possible to acquire genuine *knowledge* about them.

In this way we have given some necessary clarifications for supporting what we have said above, when we have maintained that the different "logics" must be understood as explorations and explicitations of the reasonings we perform within particular "conceptual domains" that correspond to certain fields of *meaning.* But now this clarification can be further developed. We have just maintained that the unity of sense and reference (that is, the delimitation of a certain field of meaning) occurs in the context of a particular kind of *experience.* Unfortunately, the concept of experience is usually affected by a too narrow empiricist characterization, according to which it means exclusively *sense-experience.* This narrow sense has been historically suggested by modern natural science and sanctioned by the authority of Kant. In common language, however, we normally use expressions such as "moral experience", "aesthetic experience", "religious experience", and so on. Are these expressions misleading? By no means: they indicate the basic fact that in our "contact" with reality we feel directly implicated, and induced to express *judgments* of specifically different kinds. For example, in front of a certain action we inevitably express the judgment that it is "good" or "evil" (and this judgment has nothing to do with an exact empirical "description" of this action); or we can judge a picture as "beautiful" or "ugly" (and, again, this does not correspond to any precise empirical description of the picture). Exactly as we "experience" a cherry "as" red (in a visual sense experience), we "experience" the action of helping a person in necessity "as" good in a moral experience, or a Schubert melody "as" beautiful in an aesthetic experience. (Even Kant, who could not use the term "experience" in the context of morality, having restricted its meaning to sense experience in the *Critique of Pure Reason*, speaks of the presence of the moral law in every human as a "Faktum der Vernunft", as a "fact" of Reason, in the *Critique of Practical Reason*: a very vague and imprecise expression that actually hints at that feature of "givenness" that, even according to him, characterizes the immediateness and irreducibility of any experience.)

We need not give additional examples and details. The above explanations may be sufficient for justifying the existence of different logics, and the justification is rather simple: whenever in a certain field of experience we are led to use reasonings or arguments for supporting *judgments*, the space for the explicitation of a specific "logic" is, in principle, open. It is a contingent fact that only a few of such logics have been actually constructed or sketched: this mainly depends on the "interest" we could have in rigorizing such reasonings, and this is only an historical fact. In certain fields, for example, we are still convinced that judgments are subjective, or can at most be supported by general and not stringent reasonings (like in the case of aesthetic judgments), and do not expect that a "logic" will be looked for. In other fields the interest for the explicitation of a logic has surfaced recently (like in the case of epistemic logic), and one cannot be a prophet in these matters.

There are domains, however, where the use of rigorous arguments is considered essential, and these are typically the sciences. Therefore, it is sensible that some special logics have been proposed (or will be proposed) for particular sciences, or scientific theories. But in order to see this in depth we need to follow further our investigation.

10 The Relevance of the Referents for Logic, or from Pure to Applied Logic

That logic cannot be reduced to a purely formalistic construction should be clear enough after all the considerations we have presented: being the explicitation of the forms of our correct reasonings, and reasonings being part of our thinking activity, logic cannot help being subjected to the variable "conditions" of thinking. Precisely this kind of consideration led Kant to develop a "transcendental logic" (in which rather the "a priori conditions" of thinking were stated). Our intention here is not that of discussing the sense and the possibility of such a logic; when we mention the "conditions" of thinking we want simply to stress that our thinking activity inevitably depends on, or is immersed in, a variable *cognitive* context, and even if we maintain, for example, that the most fundamental part of logic is that which makes explicit the truth-preserving linkages among propositions, we might be in difficulty in applying this fundamental part of logic in certain cases, owing to the *particular* content of the propositions.

This problematic situation was known already to traditional logic. For example, Aristotle discusses at length (in Chap. 9 of *De Interpretatione*) the possibility of assigning a truth-value to propositions such as "there will be tomorrow a sea-battle" (propositions referring to singular future contingent events), and comes to the conclusion that such propositions are neither true nor false (the issue has been taken up several times by posterior logicians under the classical title *"de futuris contingentibus"*). This Aristotelian position is very interesting, since the same Aristotle had explicitly defined a proposition as a discourse that is either true or false ("not every discourse is a proposition, but only that discourse in which truth or falsity reside", *De int.*, 4, 17a).

In spite of this he can consistently maintain that propositions regarding future contingent events are neither true nor false, because truth and falsity are properties pertaining to a *judgment*, in virtue of its way of expressing the "being" or "not being" of something, but it is not said that we can express such a judgment in all circumstances. There is, first, a cognitive aspect: future events cannot be directly *known* (they cannot be *referents* of an act of cognition); this does not totally exclude the possibility of saying something true about them, but only indirectly, when they can be the object of universal or necessary judgments. In the case of a future contingent event we cannot assign it (at present) neither existence nor non-existence, precisely because it is contingent, and contingency means the possibility of being and not being. This is an ontological consideration, that regards the *particular* nature of the referent of the discourse. From both points of view we come to the conclusion that a logic for the discourse about future contingents events must admit an intermediate value (or even a distribution of such values) between truth and falsity, that it might be equipped with some "temporal" parameters, and, briefly, be more or less different from usual logic: this owing to the *cognitive and ontological peculiarity of the intended referents* of the discourse.[18]

11 The Logic of Particular Sciences

This classical and relatively simple example offers a paradigm for a general situation: in every science the referents of the discourse are specific *objects* that are singled out and shaped by means of conceptual and operational tools, amounting at the same time to a characterization of their cognitive accessibility and ontological status. Therefore, it is altogether reasonable that the arguments we adopt in a given science be sensitive to such cognitive and ontological conditions, and can give rise, not in every case, but in certain cases, to special logics. A few well-known examples can be mentioned as a confirmation of this.

Let us consider as a first example elementary arithmetic, in whose modern rigorous presentations we find the *principle* of "*mathematical induction*". Even when this principle is formally expressed as a special "axiom", it must

[18] The above considerations obviously indicate the motivation that has led to the construction of *tense logics*. One must be aware, however, that the simple purpose of taking time into account does not directly produce a unique "type" of tense logic. For example, the logic elaborated by Prior (using temporal operators), and that proposed by Quine (using temporal quantifiers) are different not only from a technical point of view, but also because they reflect very different epistemological and even metaphysical tenets (cf. Haak 1978, pp. 156–162).

be recognized that it differs significantly *from* the other axioms, since it has a "logical" rather than a "numerical" nature in the following sense. While from the other axioms different properties of numbers can be deduced (for instance, the fact of being even, odd, prime, etc.), no numerical property can be deduced *from* this axiom, but it is *used* for *proving* that a certain property holds for *all* numbers in a distributive sense. Its logical nature was recognized, for example, by Poincaré, who interpreted it as summarizing an infinite chain of syllogisms.[19] A confirmation of this fact is offered by the various generalizations of this principle. In set theory (where transfinite numbers are admitted and defined) we normally use "transfinite induction" on ordinal numbers. In mathematical logic, several kinds of "proofs by induction" are used for proving a great many metatheorems concerning logical calculi (that can easily be presented as "generalized arithmetics"). It is not difficult to recognize that this principle is by no means "universal", but can be properly used only depending on the particular class of *referents* to which our reasonings are applied: only if this class is defined in a special way, that is, it is "constructively" defined, can we use a form of inductive proof (and also of "inductive definition") for treating the properties of its members. A class of entities is constructively obtained if its members are introduced in the following way: (a) certain "basic members" (or elements) are immediately assigned to the class; (b) a finite number of operational rules is indicated for obtaining or constructing additional members; (c) a "closure" condition is expressed, stating that only and all the members obtained in such a way are admitted in the class. A proof by induction consists then: (a) in proving that a certain property P holds for the basic elements; (b) in proving that, if P holds for certain members, then it also holds for the members obtained by applying the "constructive" rules to these members. In elementary arithmetic the basic element is the natural number 0 (or 1, according to preference); the constructive operation is just one, that is, the "successor" operation, and an inductive proof consists in showing that a certain property P holds for 0 (or 1), and in proving that, if it holds for n, it also holds for the successor of n. In the case of mathematial logic we can see, for example, that the class of sentences is constructed by defining as basic elements the "elementary sentences", by listing a finite number of operations (depending on the sentential connectives admitted) for obtaining new "composite" sentences from already accepted sentences, by introducing a closure condition. In order to "prove by induction" that all sentences have a certain property P, we must first show directly that all the elementary sentences have such a property, and then prove that, if certain sentences have that property, this is preserved in the sentences obtained through the application of the admitted operational rules.

[19] Poincaré has developed this view on several occasions, in the course of his often polemic discussions regarding the philosophical interpretation of mathematics. A short presentation is offered in the first chapter "On the nature of mathematical reasoning" of Poincaré (1902).

Since the class of terms or the class of proofs are also defined in a constructive way, proofs by induction can by produced on the construction of a term, or on the length of a proof, and so on.

The induction of which we have spoken is more properly called "logical induction", in order to distinguish it from the (very different) induction used in the empirical sciences, and has nothing to do with the so-called "inductive logic". This is why, at the beginning of twentieth century, certain scholars preferred to speak of a "proof by recursion" and, accordingly, one should say that such a kind of proof can be used when the referents of the discourse belong to a "recursively" defined class. Those authors clearly had in mind that such a procedure was of a "logical" nature, since it denoted a particular "way of reasoning or thinking". This is evident, for example, in the title of a famous paper of Skolem, where he speaks of the "rekurrierende Denkweise" (the "recursive way of thinking").[20]

We have devoted a rather detailed presentation to the use of the principle of logical induction because, on the one hand, it has all the features of a genuine *logical* principle, but, on the other hand, it is bound to the particular nature of the *referents* of the discourses in which it is used. Moreover, these discourses are *scientific*. For reasons of brevity we will not indulge in a similar detailed analysis of the two more examples we want to mention, also because they are well known and treated in a very abundant literature. Therefore, some hints will be sufficient.

Intuitionistic logic is known for not admitting the general validity of "classical" principles, such as the excluded middle and the double negation. This logic, however, was not created "artificially" or for the sake of paradox: it was a consistent implication of the "constructivist" point of view advocated by the intuitionistic school, that deeply affected the way of conceiving the ontological status of mathematical entities, the notion of mathematical existence, and the way of conceiving the correct cognitive attitude of the mathematician.

The second well-known example is *quantum logic*. Its development is bound to the fact that quantum "objects" are defined by means of conceptual and operational procedures that imply the impossibility in principle of simultaneously determining with a precision higher than a given threshold the value of conjugate magnitudes, with consequences regarding the purely statistical character of the predictions of the future state of a system, the outcome of experiments, etc. It has soon become clear that our knowledge of these objects could not avail itself of the traditional conceptual tools of the so-called "classical physics", and new tools had to be proposed. Some of them were theoretical principles of the new physics (such as the linear superposition, or the collapse of the wave packet); others were found in modifications of the probability calculus (using a non-Kolmogorovian version of it, capable of mirroring the actual behavior of probabilities in quantum mechanics).

[20] Cf. Skolem (1923).

Finally, another solution consisted of creating and formalizing a particular "logic" in which were mirrored the inferences that the scientist is obliged to adopt if he wants to adhere to the actual cognitive and ontological situation of his discourse and its referents.

12 Why Is It Logical to Admit Several Logics?

If we wanted to sketch a recapitulation of the whole of our reflections we could say that the plurality of logics is not an evidence imposed on us by the de facto situation of the existence of many logical *calculi*, for these are often only different "formulations" of a given logic. Such a plurality rather results from the distinction between *pure* and *applied* logic, since the s derives from the plurality of the applications of logic, which can entail certain integrations, restrictions or modifications of the rules of pure logic. This happens because the applied logics are nothing but idealizations of the correct reasonings humans adopt (or "apply") in the different contexts of their experience, when they try to provide a *justification* (by means of such reasonings) of certain *judgments* that are not immediately supported by the kind of evidence characteristic of a given type of experience.

It is usual, when a certain theory is axiomatized, to list its "non-logical" axioms, that is, those axioms that are meant to characterize the specificity of that theory, while the "logical" part is often reduced to the fact that a certain standardized *language* is used for the formulation of the axioms, and some known logical *calculus* is indicated as a tool for the proofs (even without explicitly listing its axioms and rules). What we are maintaining is that the "special" logics needed for making explicit the correct reasonings adopted in a given field should receive an intermediate position: their axioms should be put beside the "nonlogical" axioms of a theory (axioms that we could call "disciplinary"), as "specific logical" axioms (the discussion presented above regarding the special "position" of the axiom of induction in elementary arithmetic exemplifies concretely what we mean here).

These logics are not "at variance" with standard alethic logic, for the simple reason that they are only more restricted "in scope", and even when they appear to contain rules or principles in contrast to those of standard logic, a careful analysis shows that such exceptions are only the consequence of the impossibility of finding realized (in a special context of the discourse) the *conditions* for the application of the standard principles and rules. Standard logic, however, remains basic in two senses. First, because it is the framework within which all the other logics are expressed and formulated (it is, so to speak, the logic governing the metaconsiderations regarding those logics). Secondly, because even the special logics are nothing but an effort of fulfilling the task of logic as *such*, that is, to find explicit and controllable rules for checking the correctness of our reasonings, in the sense that they remain faithful to the links of "logical consequence". Therefore it is perfectly "logical" (in the sense that it conforms to the spirit and task of logic) to admit as many special logics as they are required for making explicit the conditions of "logical consequcnce" in the different types of discourse.

References

1. E. Agazzi (ed.): *Modern Logic. A Survey* (Reidel, Dordrecht 1981)
2. E. Agazzi (ed.): *Probability in the Sciences* (Kluwer, Dordrecht 1988a)
3. E. Agazzi: 'Probability: A Composite Concept'. (1988b) In: *Probability in the Sciences* ed. by E. Agazzi (1988a) pp. 3–26
4. N.D. Belnap: 'Modal and Relevant Logics: 1977'. In: *Modern Logic. A Survey* ed. by E. Agazzi (Reidel, Dordrecht 1981) pp. 131–151
5. R. Carnap: *Die logische Syntax der Sprache* (1934) Engl. trans.: 'The Logical Syntax of Language' (Routledge & Kegan Paul, London 1937)
6. L.J. Cohen: *The Implications of Induction* (Methuen, London 1970)
7. L.J. Cohen: *The Probable and the Provable* (Oxford University Press, Oxford 1977)
8. L.J. Cohen: 'Inductive Logic 1945–1977'. In: *Modern Logic. A Survey* ed. by E. Agazzi (Reidel, Dordrecht 1981) pp. 353–375
9. S. Haak: *Philosophy of Logic* (Cambridge University Press, Cambridge 1978)
10. F. Klein: *Vergleichende Betrachtungen über neuere geometrische Forschungen* (Deichert, Erlangen 1872)
11. J. Łukasiewicz: *Aristotles's Syllogistic from the Standpoint of Modern Formal Logic.* 2nd edn. (Oxford University Press, Oxford 1957)
12. H. Poincaré: *La science et l'hypothèse* (Flammarion, Paris 1902)
13. T. Skolem: 'Begründung der elementaren Arithmetik durch die rekurrierende Denkweise ohne Anwendung scheinbarer Veränderlichen mit unendlichem Ausdehnungsbereich'. Skrifter utgit av Videnskapsselkapet i Kristiania, I. Mathematisk-videnskabelig klasse **6** (1923) pp. 1–38
14. R. Verneaux: *Epistémologie générale.* (Beauchesne, Paris 1959)
15. G.H. von Wright: *An Essay on Modal Logic* (North Holland, Amsterdam 1951)
16. G.H. von Wright: 'Problems and Prospects of Deontic Logic. A Survey'. In: *Modern Logic. A Survey* ed. by E. Agazzi (Reidel, Dordrecht 1981) pp. 399–423

Does Metaphysics Need a Non-Classical Logic?

Francisco Miró Quesada

1 Trying to Clarify Concepts

To know if metaphysics needs a nonclassical logic, we must first have a clear insight into "metaphysics" and "nonclassical logic". Let us begin with metaphysics. From the beginning we encounter a mess of difficulties. If we focus our attention on the meaning of the term, we shall find that it is polysemic. Some of its significations are obscure, and not a few even absurd.

Among the definitions of metaphysics we find many conceptual differences. We find an idealistic metaphysics (Plato, Plotinus, Hegel, Schelling), a materialistic one (Marx, Engels), a critical one (Kant), realistic nonmaterialistic theories (Aristotle, Augustine, Thomas Aquinas and, generally, all Christian philosophers), thinkers that surpass, or try to surpass, the duality body-soul (Whitehead, Heidegger, Sartre, Ortega). An epistemologic metaphysics, that is, the theory of the most general conceptual presuppositions that must be used to develop physics (Bunge). A noneist metaphysics, and many more. Which one shall we select?

But, in spite of this proliferation, it is possible to discover common traits applicable to all the precedent definitions.

1. All metaphysical systems intend to provide a theory that encompasses the whole physical world, or the whole mental world, or both;
2. All metaphysical systems purport, consciously or unconsciously, to explicate the meaning of words, or to explain the presence of phenomena or the existence of facts (generally, although not always) in a defective way;
3. No empirically corroborable metaphysical systems can be found.

Let us take a look at the systems of Bergson, Heidegger, and Sartre. The core of Bergsonian metaphysics, is the idea of the "élan vital". Even if Bergson says that his conception of the "élan vital" does not have the intention to explain life evolution, he coined it to try to understand the prodigious proliferation of life. And trying to understand means two things: to explicate a meaning, or to explain why things are as they are. Bergson never says that the idea of the "élan vital" is to make clear the meaning of the word "life". The "élan vital" is employed to try to understand the marvelous variety of life in general.[1]

[1] Bergson: *L'évolution Créatrice*, p. 88, *Les deux sources de la moral et de la religion*, pp. 116, 117.

Sartre tries to explicate the meaning of the uncanny condition of human existence. And he finds the issue in the process of "néantisation". His theory of nothingness trys to find why human consciousness is asunder from the world of objects, and also the way we are related to other beings. His theory of the three dimensions of human existence, l'Être-en-soi, l'Être-pour-soi, and l'Être-pour-autrui explains with an intellectual rigor that is not generally found in existentialist philosophers, the fundamental ways of relationship established through human intercourse.[2]

But, what about logic? For a long time I tried to find the logic we ought to apply to metaphysical doctrines. Of course, I never found it. And I could not, for a most simple reason: because many theories of metaphysics employ different logics. For instance: Plato employs a fuzzy logic, because the ideas are models of things and some things are nearer than others to their corresponding idea. Aristotle created classical logic, and much of his metaphysics is developed through this logic. According to the modern conception, Aristotelian logic has a principle that is not found in classical logic: $(A \rightarrow B) \rightarrow \neg(A \rightarrow \neg B)$. But, putting aside this small formula, he surely did create classical logic. Plotinus employs two different logics: fuzzy and paraconsistent. Hegel employs dialectical logic. A.C.N. Da Costa creates paraconsistent logic. This system can be applied to inconsistent theories, (like Hegelian metaphysics) without trivializing them. Meinong creates the theory of objects, according to which there are definite nonexisting objects. And he suggested that the logic needed to study this strange world, must be independent vis-à-vis the traditional belief in the ontological aseptic nature of logic. This suggestion has led Richard Routley to create "noneist" or "neutral" logic.

To analyze all the mentioned types of metaphysics with their correspondent logical systems, would mean to write a gigantic text. We must, then, decide among many possibilities. We choose the following: Neoplatonism (Plotinus), Dialectics (Hegel), Empiricism (Stuart Mill, Hempel, Armstrong), and Noneism (Routley).

2 Plotinus

Plotinus' metaphysics is a doctrine of very deep concepts, and very beautiful literary style. It is also a system whose conceptual content is very wide. It includes a great deal of classical philosophy. But its more interesting part is, no doubt, his doctrine of hypostases. This doctrine can be considered as the "kernel" of his philosophy.

---

[2] Sartre: *L'Être et le Néant*, Troisième Partie, Chapitre III.

2.1 Hypostases

According to Plotinus there are three hypostases: The One ($E\nu$), the Spirit ($No\upsilon\varsigma$), and the Soul ($\Psi\upsilon\chi\eta$).[3] The Unity is the whole and it generates the Spirit. And the Spirit comes from the One. There is a dynamism, a kind of emergence of the Spirit, from the One, and of the Soul from the Spirit. So the relation of the three hypostases is transitive. The Spirit never becomes. The cause of thinking is different from the cause of entities. But they exist without negating each other. And the One is Spirit and Thought, Thinking and the thing thought, because there could not be any thought without Otherness and without Selfness. These processes, establish the following principles: Spirit, Entity, Selfness, Motion, and Restness. Motion on behalf of Selfness, Otherness on behalf of Thinking and Thought, because when Otherness is exluded, it becomes the One that must be silent. Moreover, the objects of Thought must have a reciprocal Otherness, and also Selfness as far as it is one with itself. But it is also a Commonness, for all of them. So, as far as they are different, the Otherness is established. The Plurality of Principles that are given in this way, constitute Number, Quantitiveness, Besideness and, Propriety. And, from all of them, come the Principles that are mentioned a few lines above.[4]

It is obvious that the three hypostases are described utilizing vague concepts. The concepts about the emergence of the Intellect from the Unity, and the emergency of the Soul from the Intellect, are quite imprecise, but highly suggestive. We think that the philosophy of Plotinus is a paradigm of what we call "literary philosophy". We employ the epithet "literary" in a positive way. There are items which cannot be clearly apprehended. The most profound views of being and of emergence from being, escape our possibilities of precise knowledge.

But vagueness is not opposed to knowledge, and conceptual precision does not mean the only possibility of apprehending the truth. There are many propositions that are precise but untrue. For instance $2 + 2 = 5$. There is vague and true knowledge. For instance, it is true that two objects may have different essences. But the concept of "essence" is vague. How many distinctive traits must an essence have to be different from another one? One, two, ten, ... one hundred? The reply is that it depends on the level of abstraction in which the object is situated. But what precisely is the meaning of the abstraction level? The concepts we have referred to, are vague, and cannot be analyzed with the resources of classical set theory and its corresponding classical logic. But they are quite easily approached through fuzzy logics supra.[5]

---

[3] Plotinus **10**, pp. 209, 210, 214

[4] Plotinus **10**, pp. 217 and following; **49**, pp. 118 and following

[5] We say "classical logic" because there is only one classical logic. But we say "fuzzy logics" because there exist many different, although related, fuzzy logics.

3 Logic

3.1 Provisos

To understand the concept of "fuzzy logic", we must first lay out the concept of "fuzzy set". Fuzzy sets (crisp sets) are quite different from classical sets. There are no crisp sets in the real Universe. Mathematical objects do not exist. Physical objects are never precise. We never see a precise sphere, we only see spheroidal objects that approach spherical shape. We never see any real object, even if it is apparently an object with a crisp form. At first view, a billiard ball appears as a perfect sphere. But if we look at it through a very powerful magnifying glass, we shall discover little irregularities on the surface. If we do not, we can have recourse to a strong microscope, and even to an electronic one.[6]

So our world is fuzzy and, apparently, we cannot indict Plotinus for having committed the crime of conceptual fuzzyness. We say "apparently" because there are different levels of fuzzyness. For instance, a common chair appears to us, with utmost clarity, a piece of furniture. In ordinary life, we never link it with fuzzy perceptions. Quite the contrary is the central conception of the "hypostases" in Plotinian philosophy.

We may qualify fuzzy concepts with respect to their fuzzyness. For instance: very little, little, regular, rather big, big, very big. The fuzzyness of the hypostatic concepts is very big indeed.

4 Hegel

4.1 Metaphysics and Logic

If one system of philosophy is metaphysical it is, no doubt, the Hegelian. And if a system is so directly related to logic as the Hegelian, it is impossible to distinguish the one from the other. It is true that the Hegelian logic is rather peculiar. But who is the logician brave enough to throw the first stone? The simplest system of logic, the classical, is irrelevant. Is irrelevance not peculiar? If it is not, then, "the moon is made of cheese", implies "Napoleon was a great general". One must be crazy to accept such enormity. For this reason, a logician, or a person that pretends to be a logician, must be tolerant. This mélange of metaphysics with logic shows that, if we want to understand Hegelian logic, or at least to make serious efforts to understand it, we must focus our attention on his "Logic", the most important of his books, the least read of them and, believe it or not, the easiest to read.

[6] It is difficult to imagine the way a microscope is employed to find the irregularities on the surface of a billiard ball. Nevertheless, we think it could be possible using some very sophisticated apparatus.

According to the great German philosopher, the Leitmotiv of philosophy is the "Idea". The "Idea" has an internal dynamism that develops through itself, without ceasing to be the same. In the Phenomenology of the Spirit, Hegel shows the way this development manifests itself through history. But, in his Logic, the dynamism is purely conceptual. The development of the Idea, follows a way that is determined through different steps whose manifestation has a determined order. Although every step follows from the preceding ones following a variable order, there is a general principle that impinges in all the steps: the "dialectical triad" thesis, antithesis, and synthesis. The point of departure is Being. Being without qualification is Nothingness. Being is the thesis, and Nothingness is the antithesis. The Synthesis is "Movement", because it consists in passing from Being to Nothingness. But this passing is not a nullifying of Being and Nothingness. Quite the contrary, both continue the dialectical march towards the last step, that is none other than the "Idea". And this Idea is exactly the same as the Idea that was the point of departure.[7] All the dynamic process is a logical process that develops in a "logical time". The process is difficult to understand because the triadic steps are, in many places, arbitrary. But there are steps that are quite understandable.

4.2 Holism

One of the greatest difficulties to follow the Hegelian speculations, is the holism that is present in all his texts. If we try to understand the Logik, we must read, with a few exceptions, all the text. But being a holist is no sin at all. It is a philosohical position adopted by Quine, Davidson, Rorty and others, generally, although not always, pragmatists. Why must Hegel be banned from holism?

However, there are some texts that belong to the few exceptions that, with due effort, are understandable. One of them is the following: paraphrasing his argumentation, he says that the famous Kantian example: $5 + 7 = 12$, is a proof that the equality of $5 + 7 = 12$ has two parts: the first is plural, the second is single. And this relationship of plurality with singlehood clearly shows that there are a priori synthetic propositions.[8]

It is worthwhile to pay attention to Hegel's information about mathematics. Most analytical philosophers consider that he was completely ignorant concerning formal sciences, but it is quite the contrary. In several parts of the Logik he quotes several mathematical texts from great mathematicians. He says, rightly, that the concept of infinitesimal is contradictory. He says, correctly, that Euler utilizes this concept when he describes the Newtonian method of infinitesimals, and he quotes two of his books. He also cites, to support his analysis of the concept of infinity, three important mathematicians, Landen, Barrows and, the most important of them, Pierre de Fermat.

[7] Hegel: *Wissenschaft der Logik* Zweiter Teil, pp. 498 and following

[8] Hegel: *Wissenschaft der Logik* Erster Teil, pp. 156 and following; Zweiter Teil, p. 447

He also speaks about Newton, and quotes his book Principia Mathematica Philosophiae naturalis. And he writes about all the mentioned texts, in a way that shows he had read them. The way he interprets the texts written by the quoted authors is, in my opinion, extravagant. But, nevertheless it is a proof of the vast information he had concerning the matters he spoke about. Another text, understandable enough, is the following.

> The corresponding statement, namely, that there is not any contradiction present around, must not preoccupy us at all. An abolute determination of the essence must be found in all experience, in any reality, in any concept... Current experience speaks from itself, that there is, at least, a set of contradictory things, contradictory tendencies, whose contradiction is not the result of an external reflection but is to be found in itself. Moreover, it cannot be considered only as a normativity that may be present here, or there, but it is the negative in its essential determination, the principle of all self movement, that is not other thing than its self interpretation (Hegel, Zweiter Teil, pp. 58f)

Evidently the logic that must be applied to develop his theories is dialectic logic. But there is a problem with this because, as we know, there are many dialectical logics. And in Hegelian texts, it is possible to find more than one. The great difficulty of dialectical logic is that, to develop it, the deductive rules must be "rationally" changed according to the moves that are being made. I think that the dialectical system created by Batens is the one that better complies with this condition.

5 Empiricism

5.1 Empiricist Metaphysics

That any kind of empiricist system is, in some way or another, related to metaphysics is rarely noticed. But it is much rarer to be conscious that empiricism is a doctrine that, in its most profound foundation, is a realistic metaphysics. Of course, a convinced empiricist would be outraged by this qualification. And he would say that an empiricist system does not comply with the third condition we have set out to be a metaphysical system, namely, to be nonempirically corroborable. Now, according to empiricist philosophy, there are two kinds of propositions: those that directly describe sensible impressions, and those that are derived through induction from the former. The derived propositions must be corroborated by means of observation. But any sensorial corroboration presupposes that there are objects that have several sensorial aspects. When we try to corroborate an inductive proposition, we do not refer to sensations, but to our perceptions. When we describe, for instance, a cube whose sides are green and red, we do not say that the cube we are looking at is only a combination of sensations of green and red colors. We are saying that the cube is an object that we perceive, whose faces are green and red. And a perception is always of real objects.[9] When an empiricist has to generalize an inductive process, he must make numerous observations. And to make them, he must apply an existential quantifier. Let us suppose that to state a general physical law, he has made one thousand observations: $O_1, O_2, \ldots O_{1000}$. In each of them he must have enunciated that an object o_i $(1 \leq i \leq 1000)$, has the property P. $P(x)$ is the general form of the statements. So, a condition that must be complied by every observation is $\exists x P(x)$. But an existential quantifier, remits to real objects. Qualities do not exist. Colors are qualitites of real objects, but they are not real. What is real, is the colored object that is being observed. As we see, empiricism cannot be developed without a realistic basis.[10]

What is the logic used by empiricist philosophers? This question simply answered is that: most of them believe that logic is derived through induction from sensorious experience.[11] But they are unaware of the logic they really employ. Generally, they employ two kinds of logic: classical and some type of epistemic logic.

6 Noneism

6.1 Noneism Metaphysics

Noneism was created by Richard Routly.[12] According to him the universe is a realm that comprises nonexistent objects. In this universe are included, beside real objects, sets that are contradictory or even absurd, and objects that we can think about, or imagine, but that are beyond the limits of reality. For instance, propositions like "the Squond is square" and "the Squond is round", in which Squond is an object that is square and round, are quite legitimate. Of course the Squond does not exist but – and this is the important point – it can be thought. Examples of less aggressive insolence, are more than abundant. For instance "the fairy Wandolina is very gentle" or "Zeus is the most powerful of the Olympian gods" are expressions fully understandable. The noneist system complies the three conditions to be metaphysical theory.

[9] There are cases in which perceptions are delusory, for instance, in cases of grave schizophrenia. But a delusory perception is always the perception of an object. Delusory perceptions always have the characteristic traits of real perceptions: to be the apprehension of an object.

[10] The ideas we have used to show that empiricism cannot do without realistic presuppositions have been inspired by the book *Exploring Meinong's Jungle, and beyond* of Richard Routley.

[11] Mill, pp. 202, 208

[12] The following exposition I based on Routley's book.

1. It encompasses the whole world, physical and mental,
2. it offers explications of meaning, and explanations of facts, and
3. it is not empirically corroborable.

That it contains objects with certain properties that are described by unverifiable propositions, is essential in an noneist universe. We have just shown two of them. This unverifiability is quite different from the one we find in physical theories. In these theories, when a new fact is discovered, its corresponding proposition must be deduced from the premises that are valid in the system.[13] This means that it must be explained, and to explain a proposition P, means that P must be logically deduced from previous theorems of the system (every axiom is a theorem).[14] This deduction is ordinarily done utilizing classical logic. Some physicists pretend that the only way to deduce quantum propositions, is by means of a "quantum logic". But this is a far-fetched methodology, because the reasons they expose to justify their claim, are based on the fact that the DeMorgan laws are not valid. But these laws are invalid, because they are identified with the addition of certain states that have nothing to do with logic.

I dare to qualify noneism as a metaphysical theory, because it complies with conditions 1, 2, and 3. That it complies with the first, is evident. Condition 2 is also complied with because noneist theory explains many facts. For instance, how is it possible that a child of five or six years, understands perfectly well a fairy tale. The only explanation of this miracoulous fact is that, as the tale advances, the child is apprehending objects that do not exist. This apprehension must not be confounded with imaginative objects. The objects the child imagines do not exist, but they are viewed as subjectdependable. On the other hand, the objects we apprehend listening to the story are not subjectdependent. The child does not invent them, he (or she) sees the images that the story is suggesting. Different children can "see" different images, but the images that a child has will have a certain structure, and the structure will be the same for all children (and, of course, for all adults). Moreover, if a child "sees" that color C_i is adjacent to color C_k, the other child will "see" other colors D_i and D_k but in the same order, that is D_i and D_k, will be adjacent.

The third condition is also complied by the system: nonverifiability with respect to sensory objects, and this nonverifiability is absolute. It is assured by definition. We qualify it as "metaphysical", applying a frequent proce-

[13] The word "system" is used in its usual meaning. This meaning is quite different from the way it is employed in physics, in which it means an object.

[14] We say "axioms" because, in mathematical physics, the propositions from which all others must be deduced, play the role of axioms. With the exception of constructive mathematics, some axioms of geometry (for instance, the axioms of "betweenness", and finite set theory, all the axioms are hypothetical propositions. The truth probability of a theory increases through corroborations of its explanations, or through the verification of its predictions.

dure in pure mathematics. If we define a complicated concept having definite properties, say, $\{A, B, C, D\}$, then, on behalf of easiness, we devise another concept that includes the first one, but has a note E, (there can be more than one) present in $\{A, B, C, D, E\}$. A classical example is the concept of zero. We can develop set theory without employing it. But as the theory advances, it becomes frightfully complicated.

6.2 Noneist Universe

Noneist philosophy is based on two principal ideas: the Ontological Commitment, OP, and the Characterization Postulate, CP.[15]

Since 1965, Richard Routley began to publish several texts which were "bahnbrechend". In 1980 he published a book "Exploring Meinong's Jungle, and Beyond" in which he develops a ripe and systematic theory of his philosophical thought. Although his exposition is clear from a technical point of view, he does not make any pedagogical concessions. It is a difficult book, even for the sophisticated, because many of its principal thesis are so revolutionary that, to be grasped, one must apply much intellectual energy.

The core of the matter is the mental power to apprehend objects that do not exist.[16] Classical logic is founded on what Routley calls the "Ontological Commitment".[17] And this assumption is blatantly manifested in the principle (or inference rule) of the existential quantifier. According to this principle, if there is an existent object, then it is possible to apply the existential generalization. For instance, if F is the predicat "nice", if a is a lamb, and the universe of interpretation is the classical set theory, the expression $F(a)$ means that the lamb a is nice. Then, by the principle (or the rule of existential generalization we can put $\exists x F(x)$. But suppose that the object we are thinking about is not a lamb, but a fairy. Then it is impossible to apply the principle of existential generalization, for the very simple reason that fairies do not exist. The same argument can be applied to an infinity of objects.

Most authors, like Russell, Quine, and others, consider that inexistent objects are always the same: the null-set. But this belief cannot be accepted because there are nonexisting objects that can be perfectly well differentiated. The traits that distinguish the Olympian god Zeus, are completely different from the traits that determine the Hipogriff. To cope with this proliferation of objects Routley has coined the Characterization Postulate, CP in what follows (and without "the").

The objects we apprehend, existent or nonexistent, have traits that permit us to recognize them. There are three ways to achieve this recognition:

---

[15] Routley, 1980, pp. 423–425.

[16] Routley: *Exploring Meinong's Jungle, and Beyond*, pp. 423–425, 440.

[17] idem pp. 423–425, 617–618.

1. The possibility of recognizing existent objects.
2. The possibility of recognizing nonexistent objects whose properties do not depend on our personal choice.
3. The possibility of recognizing objects which are freely created by ourselves.

It is very simple to recognize real objects by their traits. For instance, a chair, a book, a boat, etc. The quantity of traits we must know to recognize them is context dependent. For instance, a woman accompanied by her son, is going to a party. Before going outside, she becomes aware that she has forgotten the purse she had selected, and says to her son: "Please John, bring me the red purse I want to go with." John goes to the closet where his mother keeps her purses. But among them there are four red purses. In this case it is impossible to know which purse she was thinking about. But among the purses there is one that is red and is, also, the biggest one. In this case it suffices to say "John, bring me my biggest purse." So, we see that real objects can be, most times, easily characterized. Of course, there are exceptions. For instance, to find the traits that characterize a quark is very difficult. Theoretical physicists had a lot of trouble finding them. Also, cosmologists have tried, up to the present times, to determine the characteristic traits of the hypothetical galaxy which constitute the center of the swarm of galaxies among which is our own Galaxy (Milky Way).

The second possibility is surprising, because it encompasses mathematics, literature, and poetry. The mathematical truth $2+3=5$ and the personality of Sherlock Holmes do not depend on our caprice. We can, of course, add a few, or even many traits, to the ones that characterize the personage, but there is a set of traits we are obliged to admit, because without them Sherlock Holmes, such as he is described in Conan Doyle's stories, would nolonger be the same. We call "nuclear characterization" the set of qualities that determine the traits that cannot be eliminated from an object, nor added to it, because, otherwise, due to CP, it would be another object.

The realm of objects, integrated by a multitude of items, is infinitely wider than the realm of real objects, because it embraces every kind of item, belonging to reality, and belonging to irreality. In the second case, we can apprehend items that belong to imagination, and items that appertain to contradictory sets, or even to absurd sets, like the "squond". The realm conceived by Routley is, by far, the richest of all systems ever conceived by man. But it is important to become aware that sounds like "bla-bla-bla", are not objects at all. Universality means that we can apprehend any kind of object, but this capacity has nothing to do with senseless noises.

7 Noneist Logic

7.1 Quantification

The principal objection that can be made to traditional logic is the narrowness of its quantifiers. Universal quantification is doubly limited. Relative to physical universes it cannot be appliable to nonexistent objects. And, in relation with mathematical universes, it cannot be applied to contradictory and to absurd objects (remember the "squond").

To overcome this limitation, noneist logic makes a daring reinterpretation of the meaning of "quantification". If "$F(x)$" means that the object x has property F, then "$(x)F(x)$" states that every object, relative to the chosen universe, has property F. And "$\exists xF(x)$" means that there is an object that has property F. Quantification has, like classical logic, universal generalization, universal instantiation, existential generalization and existential instantiation. Due to the fact that noneist quantification is indifferent to the nature of the universe in which it is being applied, Routley calls it "neutral logic".[18]

We expect that the four examples we have presented concerning the relationship of logic to metaphysics, are more than sufficient to give an affirmative answer to the interrogation expressed in the title of this paper: "Does Metaphysics Need a Non-classical Logic?" Yes, it does!

References

1. Aristotle: *Prior Analytics* (Clarendon Press, Oxford 1928a)
2. Aristotle: *Posterior Analytics* (Clarendon Press, Oxford 1928b)
3. D.M. Armstrong: *Perception and the Physical World* (Routledge & Kegan Paul, London 1961)
4. D. Batens: 'Dynamical Dialectical Logics'. In: *Paraconsistent Logic. Essays on the Inconsistent* ed. by G. Priest et al. (Philosophia Verlag, München 1989)
5. H. Bergson: *L'Évolution Créatrice* (Presses Universitaires de France, Paris 1981)
6. H. Bergson: *Les Deux Sources de la Morale et de la Religion* (Presses Universitaires de France, Paris 1982)
7. M. Bunge: *Foundations of Physics* (Springer, New York 1967)
8. M. Bunge: *Treatise on Basic Philosophy* (Reidel, Dordrecht 1977)
9. M. Bunge: *Epistemología* (Editiorial Ariel, Barcelona, Caracas, México 1980)
10. N.C.A. Da Costa: *Sistemas Formais Inconsistentes* (Universidade Federal do Parana, Doctoral Dissertation 1963)
11. N.C.A. Da Costa: 'On the Theory of Inconsistent Formal Systems'. Notre Dame Journal of Formal Logic **15** (1974) pp. 497–510
12. F. Engels: *Dialektik der Natur* (Dietz Verlag, Berlin 1959)
13. C. Englewood: *The Roots of Reference* (Open Court, Illinois 1973)

[18] Routley, pp. 79–95, 174–180

14. G.W.F. Hegel: *The Phenomenology of Mind* (Macmillan, New York 1931)
15. G.W.F. Hegel: *Wissenschaft der Logik* (Meiner, Leipzig 1951)
16. C.G. Hempel: *Philosophy of the Natural Science* (Prentice-Hall, New Jersey 1966)
17. I. Kant: *Kritik der reinen Vernunft* (Reclam, Leipzig 1924)
18. B. Kosko: *Il Fuzzy-Pensiero* (Baldini & Castoldi, Torino 1997)
19. R. Kruse et al.: *Foundations of Fuzzy Systems* (John Wiley & Sons, New York 1994)
20. J. Maritain: *El orden de los conceptos* (Club de Lectores, Buenos Aires 1994)
21. K. Marx: *La Ideología Alemana, con la Colaboración de Engels* (Ediciones Populares, Lima 1985)
22. K. Marx: *Die Heilige Familie* (Dietz Verlag, Berlin 1995)
23. A. Meinong: *Über Gegenstandstheorie* (Meiner, Hamburg 1988)
24. S. Mill: *A System of Logic, Ratiocinative and Inductive* (University of Toronto, Toronto 1981)
25. F. Miró Quesada: *Lógica I-Filosofía de las Matemáticas* (Editorial San Marcos, Lima 1980)
26. F. Miró Quesada: 'Paraconsistent Logic: Some Philosophical Issues'. In: *Paraconsistent Logic. Essays on the Inconsistent* ed. by G. Priest et al. (Philosophia Verlag, München, Hamdem, Wien 1989)
27. F. Miró Quesada: 'Logic, Mathematics, Ontology'. In: *Philosophy of Mathematics Today* ed. by E. Agazzi & G. Darvas (Kluwer Academic Publishers, Dordrecht 1997)
28. J. Ortega y Gasset: *En torno a Galileo* (Revista de Occidente, Obras Completas, Madrid 1933)
29. H.T. Nguyen, E.A. Walker: *A First Course in Fuzzy Logic* (CRC Press Boca Ratón, Boston 1997)
30. Plato: *Parmenides* (The Loeb Classical Library, Harvard University Press, London)
31. Plato: *La République* (Societé d'Éditions 'Les Belles Lettres', Paris 1966)
32. Plotinus: *Plotins Schriften* (Meiner, Hamburg 1956–1960)
33. G. Priest: 'Classical Logic *aufgehoben*'. In: *Paraconsistent Logic. Essays on the Inconsistent* ed. by G. Priest et al. (Philosophia Verlag, München, Wien)
34. W. v. O. Quine: *From a Logical Point of View* (Harvard University Press, Cambridge Mass. 1961)
35. W. v. O. Quine: *The Ways of Paradox* (Random House, New York 1966)
36. W. v. O. Quine: *Philosophy of Logic* (Prentice-Hall, New Jersey 1970)
37. R. Routley: *Exploring Meinong's Jungle and Beyond* (Philosophy Department, Research School of Social Sciences, Canberra 1980)
38. R. Routley: *Relevant Logics and their Rivals* (Ridgeview Publishing, Ohio 1982)
39. R. Routley: 'Two Systems of Paraconsistent Logic'. In: *Paraconsistent Logic. Essays on the Inconsistent* ed. by G. Priest et al. (Philosophia Verlag, München, Hamdem, Wien 1989)
40. Saint Augstín: *La Cité de Dieu* (J.J. Dubochet, Le Chevalier et comp., Éditeurs, Paris 1845)
41. Santo Tomas: *Suma contra Gentiles* (Biblioteca de Autores Cristianos, Madrid 1967)
42. J.-P. Sartre: *L'Être et le Néant* (Presses Universitaires, Paris 1943)

43. F.W.J. Schelling: *Sistema del Idealismo Trascendental* (Anthropos, Editorial del Hombre, Barcelona 1988)
44. A.N. Whitehead: *Process and Reality* (The Macmillan Company, Cambridge 1936)
45. L. Zadeh: 'Fuzzy Sets'. In: *Information and Control* **8** (1965) pp. 338–356
46. L. Zadeh: *Inference in Fuzzy Logic* (Computer Science Division, University of California, Berkeley 1980)

Logic and the Philosophical Interpretation of Science

Dudley Shapere

1 The Promise of the Vienna Circle

The Vienna Circle began gathering in the early 1920s with a profound objective. Developments of fundamental importance in science, particularly the advent or relativity and quantum theories, had occurred which clearly upset much of what has become known as classical physics. But these new theories seemed to carry much more in the way of implications. For not only the classical scientific theories, but also the classical philosophies which attempted to interpret science, its methods, and its goals, seemed to have been refuted by these new scientific revelations. It was the explicit original aim of the Vienna Circle and likeminded thinkers elsewhere, and of the logical empiricist movement they ultimately spawned, to develop, through close examination of the new scientific developments and their implications, a new view of the nature of knowledge, the methods of reasoning by which the search for that knowledge should proceed, and the goals at which it aimed. It was a laudable mission indeed, and one sort of result was some extremely valuable (if ultimately controversial) insights into the nature and methods of the new sciences, typified by the work of Carnap, Reichenbach, and others on space and time and the interpretation of quantum mechanics.[1]

But in those heady days, there was also another major development which impressed the philosophically-focused members of the Circle and their allies elsewhere. This came from the formal sciences, in two related developments: first, the development of the new logic, the achievement, following Frege, of Whitehead and Russell's *Principia Mathematica*, and second, the axiomatic view of mathematical systems urged by Hilbert as an essential step in establishing the consistency of such systems through examining them from a 'metamathematical' perspective.

It was not long before a number of the philosophers of the Vienna Circle and their sympathizers had merged these formal developments into a methodology with which to carry out their program of investigating the nature of

[1] For example, H. Reichenbach, *The Philosophy of Space and Time* (Dover, New York 1958); *The Direction of Time* (University of California Press, Berkeley 1956); *Philosophic Foundations of Quantum Mechanics* (University of California Press, Berkeley 1948); R. Carnap: "Der Raum" *Kant-Studien, Ergänzungshefte* **56**, 1922.

scientific knowledge and, through such studies, the nature of the reasoning involved in science.[2] The motivation here, too, was clear. The new logic seemed powerful enough to express all deductive arguments, mathematical or otherwise; indeed, it appeared to be, as the Circle's patron saint Wittgenstein had implied, the study of all possibilities.[3] If so, then logic should provide the key to understanding what it is to be science, all possible science, anything that could properly count as science, whether past, present, or future. And understanding science in this sense would be gained through analyzing, logically, the terms we use in *talking about* science: the metascientific concepts, like 'is a law', 'is a theory', 'is a piece of evidence', 'confirms', 'explains', and so forth. The sum total of the separate analyses of the metascientific terms would be an understanding of science, its components and the methods by which it proceeds. The methodology involved was thus the new Frege-Russell-Whitehead logic, as gradually modernized and extended in the decades following publication of *Principia Mathematica*, together with the distinction between object-level and metalevel of analysis that had proved so useful in evading paradoxes and had been enshrined by Hilbert as an important distinction relevant to investigating mathematical systems. The thesis was that these new logical and metalogical analyses would illuminate the understanding of science, and in particular the new twentieth century developments in relativity and quantum theory that had inspired the Circle to gather. I will refer to this program as the *logistic program*, and to the associated thesis asserting the achievability of this goal as *the logistic thesis*.[4]

My first concern in this paper will be to remind us of the well-known failure of the logistic program in nearly all its detailed efforts. The fact that that failure was so general will raise the question of whether the program was doomed from the outset, in its very conception. I will argue that it was: that it misconceived, at a basic level, the character of the scientific enterprise, and, by focusing on the formal analysis of metascientific concepts and by conceiving scientific reasoning as primarily logical, the logistic program blinded itself

---

[2] The views of individual thinkers associated with the Vienna Circle did, of course, vary considerably. My focus in this paper is not on those individual variations, but on the views which became most widely influential: the views of Carnap and, later, Hempel, which came to dominate broader programs in the philosophy of science in the 1930s, 1940s, and 1950s. (Reichenbach, who stemmed from Berlin rather than Vienna, was also highly influential, especially after he emigrated to the United States. But his influence lay more with specific doctrines than with broad programs and methods.)

[3] L. Wittgenstein, *Tractatus Logico-Philosophicus* (Routledge & Kegan Paul, London 1949). In this work, Wittgenstein understood the new logic as basically truth-functional (propositional calculus). Though logical empiricists would recognize that quantification cannot be fully reduced to propositional calculus, they still agreed with Wittgenstein that logic is the study of all possibilities.

[4] The term "logistic" is often used in other ways which, however, are close enough to the present usages to warrant its adoption here.

to matters of central importance for understanding the 'nature' of science. In particular, it quickly diverted attention away from the original concerns of the Vienna Circle, the close examination of the radically-new scientific developments in such areas as relativity and quantum mechanics. These defects had influence well beyond the demise of logical empiricism, continuing to dominate much (though certainly not all) of philosophical studies of science for the succeeding decades of the twentieth century, even contaminating the views of many major critics of logical empiricism.

The question, "Does science need a new logic?" can be interpreted in at least two ways. On the one hand, it can be understood as a question of what Carnap would have called the 'object-level': Do any *specific* areas of science today require a new logic in order to solve *specific* problems arising in those areas?' (One thinks immediately of the problems of interpretation of quantum mechanics, and the proposal that those problems might be illuminated or even solved by a new 'quantum logic'.) On the other hand, the question can have a 'metalevel' focus, namely, 'Does science (in the *general* sense) require a new logic, or at least a more persistent and competent application of the logic we have, in order to understand its *general* character and procedures?' This paper will focus on the latter interpretation of the question, the importance of which will be found to lie in the fact, first, that it has dominated many twentieth-century attempts to understand what science is and does, and secondly, that profound error permeated both its program and the thesis governing it. However, although the central focus is on this general problem, at the end of the paper I will remark, though only briefly, on the problem of specific needs for a new logic within specific areas.

2 Two Clues to the Roots of the Logistic Failure

By the 1960s it had become clear, even to some of its most influential advocates, that this logistic program for analyzing science had failed in all its specific aspects. Even among avowed logical empiricists, there was wide agreement that attempts to provide analyses of each item on its list of metascientific concepts were unsatisfactory or uncompelling. Nor were attempts to clarify the nature of scientific reasoning, either by considering it to be an application of the new deductive logic, or by developing a new inductive logic, any less so. (Other proposals for new sorts of logic, for instance, modal logic, also played roles which I must ignore here. Also, in this paper I must omit explicit discussion of the logical empiricist emphasis on language and meaning, even though the objections there are both closely-related and equally severe, and the influences on much of later twentieth-century philosophy equally pernicious.) Since, according to the logistic program, an understanding of the nature of science in general was to result from the sum total of such analyses and clarifications, the failure of that program was widely admitted to have been, for all intents and purposes, complete.

I will not attempt to review those objections here – they are, after all, familiar. Rather, I will ask whether the very ubiquity of failure, the fact that it is found everywhere in the logistic program, raises the serious question, already mentioned, of whether the failures were due merely to lack of sufficient effort and imagination in applying the tools of logic to analyzing individual metascientific concepts, or whether the universality of failure was a symptom that something more fundamental was awry with the entire logistic program – that it was misguided from the outset, in its very thesis that logic is the key to understanding science and its forms of reasoning. Two clues from the history of the endeavor lead to the conclusion that the failure was not of the first kind, a lack of effort and imagination in the various subtasks of the logistic program, but of the second, more fundamental sort, that the program was misconceived from its beginning.

The first clue comes from the fact that, under the influence of the logistic program, the overwhelming majority of papers in the standard philosophy of science journals became logical exercises. The next result was that, for much of the duration of the logical empiricist movement (that is, approximately until the 1960s), the parent motivation of that movement, the Vienna Circle's concern with understanding the recent revolutionary changes that had taken place in physics, was largely lost. That it was lost is not surprising when we reflect more closely on the logistic program's claims. Its thesis can be understood more fully by considering it as consisting of three subtheses.

1. There is a general nature of science, a 'meaning' specifying what counts as scientific, that can be ascertained in advance of and independently of science itself.
2. The way to ascertain this general nature is to delimit a class of metascientific terms (ones used in talking about science) and to use the tools of the new logic to analyze what is involved in each. The sum total of such analyses would specify what counts as scientific.
3. The same tools are to be used to analyze what counts as legitimate scientific reasoning. Indeed, to most philosophers of science who styled themselves as logical empiricists (and also to Karl Popper, who did not), all reasoning, scientific included, is specified by formal rules, deductive where possible, or inductive where questions of confirmation are concerned, or by other sorts for other purposes. (Popper of course denied the need or even the possibility of an inductive logic of confirmation, and steadfastly affirmed that deductive logic is the only logic available.)

It was all too easy to interpret these three subtheses as forcing attention away from any specific scientific theories: after all, scientific theories come and go, the subtheses imply, while the general nature of science, its logical structure and methods, remain the same. Indeed, to focus too heavily on the particular scientific theories of a certain epoch would cause errors, by confusing their idiosyncratic features with the logic common to and definitory of all science. And so the focus of the movement was soon turned away from the revolutionary developments in quantum theory and relativity, to purely logical exercises supposedly applicable to all science, not just to the recent upheavals.[5] Content was irrelevant, indeed dangerously misleading, to analyses whose central focus must be on logical form. Ironically, the logistic thesis incorporated a deep Platonic-Aristotelian quest for essences: a view that Plato has Socrates express in the *Meno*, that if philosophers want to give us an understanding of what bees or virtue are, they must not offer us a swarm of bees or virtues in all their particulars, but rather the common nature of bee-ness or of virtue. The logistic program was all too similar: if one wants to know what science is, one must *not* examine the particulars of specific scientific theories; rather, one must try to get at what all science has in common, the common nature of anything that has, does, or ever will count as science. The only difference from Plato's bees lay in the tools by which such common natures are to be discovered and defined, and the characteristics of the common natures themselves. We must begin at least to suspect that perhaps the logistic program and its central thesis and subtheses were flawed, and that the fundamental failure lay in its ignoring of the specifics of particular scientific theories.

A second clue to determining whether the logistic program was flawed in this fundamental way comes, in yet another irony, from the thinkers who, in the 1960s, saw themselves as critics of the logical empiricist approach to understanding science, and who saw themselves as presenting a new program which would correct the deficiencies inherent in that movement. For reasons to be explained shortly, I will call them the 'postclassical'[6] philosophers of science, and in explaining their general program I will focus on the influential ideas of Thomas Kuhn's 1962 book, *The Structure of Scientific Revolutions*, though most of what I say here also applies to his fellow postclassicalists, notably Toulmin, Hanson, and Feyerabend. As the logical empiricists found inspiration in the logic and metamathematics of their era, so for Kuhn and his cohorts there was a new inspiration: the newly-professionalized field of the history of science. That post-World War II subject had already revealed that scientific change was immensely complex, involving alterations not only in substantive scientific claims, but often also in its concepts, methods, standards of scientific legitimacy, and explanatory goals. The opening sentence of Kuhn's book amounted to a manifesto: 'History, if viewed as a repository for more than anecdote or chronology, could produce a decisive transformation in the image of science by which we are now possessed.'[7]

---

[5] I emphasize once more that not all adherents of the movement were so diverted. But the majority of workers were, as an examination of the contents of journals shows.

[6] "Early postclassical" might be more appropriate, but will not be used here.

[7] T.S. Kuhn: *The Structure of Scientific Revolutions* (University of Chicago Press, Chicago 1970) p. 1.

No longer was the essence of science, and of its component metascientific concepts – 'theory', 'explanation', and the rest – to be sought, for there is no such thing; everything of importance about science lies in the particulars of its state at a particular epoch or in a particular tradition, and can vary from one such epoch or tradition to another. Yet here again lay the seed of philosophical disaster. For in viewing everything about science as determined by its tradition, or rather by the 'paradigm' defining and governing it, this postclassical movement condemned itself to fall into relativism. It was an extreme relativism, in which not only substantive theories, but also what count as reasons, legitimate theories, explanations, and even the meanings of words, could differ in an 'incommensurable' way from what so counts in another paradigm-tradition. Being thus paradigm-determined, what two traditions called legitimate explanations would exhibit no similarities whatever, and no paradigm-transcendent reasons whatever were left available by which the relative merits of two paradigm traditions could be compared and assessed.[8]

What had begun as a focus on history, and how that history could illuminate the transition from one scientific tradition to the next, ended in total inability, on doctrinal grounds, to account for such events as the rise of relativity and quantum mechanics and other major transitions in the history of science. Once more we are confronted with one of the many deep ironies of this entire twentieth-century episode. For the aim with which the Vienna Circle began, to understand these two great early twentieth-century scientific revolutions – an aim which was abandoned by its offspring, logical empiricism – was here revived by its most severe critics. (No wonder Kuhn's book was published under the rubric of the International Encyclopedia of Unified Science, instituted as an official organ of logical empiricism by Otto Neurath!) Yet those critics also abandoned that worthy aim, their own intoxication coming not from the overimbibing of logic and mathematics, but now from a conception of history that led them to deny that such revolutions could be explained at all. Their historical focus entailed a focus on the content of science at a specific time, rather than on the eternal essence of 'science'. Yet they too, were led by the very conception of their program to deny that anything having to do with scientific reasoning could be said about particular major scientific transitions. Only within a tradition could anything be said about reasoning, explanation, legitimate methodology, and so forth, and that differed radically, incommensurably, from what could be said about those subjects in a different tradition. There was no such thing as 'the en-

[8] Under the pressure of such criticisms, Kuhn and his followers retreated in various stages from the strong position suggested (rather ambiguously) in *Structure*. Nevertheless, analogous criticisms continue to haunt the various retreats, as I argue elsewhere (see, for example, D. Shapere: "Incommensurability, Reasons, and Radical Change in Science" forthcoming in proceedings of a conference held in Hannover, Germany, devoted to "Incommensurability and Related Matters".

terprise of science' in which all of the traditions were engaged; each tradition does its own thing, independently of and incommensurably with, any other tradition. Thus, for the new postclassical philosophers of science, major scientific change, scientific revolution, was left as mysterious, even unanalyzable, now, in the absence of anything that could count as a paradigm-transcendent reason, as it had been for their logical empiricist predecessors, who had denied that reasoning (a 'logic of discovery') was involved in arriving at new scientific ideas.

3 Nature of and Contrasts Between the Two Programs

Despite this commonality of their initial motivation and the character of the failure, the logical empiricist and postclassical movements embodied profoundly different conceptions of how we are to go about understanding science. Logical empiricism was based on two basic departures from traditional versions of (classical) empiricism. First, it abandoned the view that empiricial science consists of the application of rules by which, from a collection of data, a theory or explanation can be inferred. (Mill's methods of experimental inquiry typify this view.) The abandonment was not immediate, for early logical empiricists, like Carnap, tried to show how the world of theory could be 'logically constructed' from the pure givens of sense-perception.[9]

Ultimately, however, even Carnap rejected such views in favor of a 'hypothetico-deductive' conception of science in which antecedent hypotheses were required from the outset of inquiry, not as a result of inquiry but as requirements antecedent to empirical inquiry: given empirical data were still a necessary ingredient in science, but their role was in the testing of hypotheses, not in their derivation: they could not serve as the base from which hypotheses were to be derived. This train of events explains not only the departures from Millean empiricism; it also explains the continuity of logical empiricism with that earlier empiricism: for knowledge is still conceived as resting on experience, even though it is no longer understood as being derived therefrom. Thus, this first element, of continuity with departure, explains the term 'empirical' in the name of the logical empiricist movement. I have discussed the failings in this aspect of the latter movement elsewhere,[10] and it is relevant here only in order to bring out the full context of the

[9] R. Carnap: *The Logical Structure of the World and Pseudoproblems in Philosophy* (University of California Press, Berkeley 1967).

[10] D. Shapere: "The Methods of Science". In: *The Creation of Ideas in Physics* (Kluwer, Dordrecht 1995) ed. by J. Leplin, p. 13–27; "Testability and Empiricism". In: *The Reality of the Unobservable* ed. by E. Agazzi & M. Pauri (Kluwer, Dordrecht 2000); "Building on What We Have Learned" to appear in proceedings of the 1997 meeting of the Académie Internationale de Philosophie des Sciences; *The Rational Dynamics of Science*, Chap. III, forthcoming.

developments under discussion. The second departure is more directly relevant. It is as follows. Whereas traditional empiricists like Hume relied on the then-prevailing associationist psychology to explain the joining of simpler ideas into more complex ones, the logical empiricists insisted that the focus should be on *reasons* for accepting or rejecting ideas (*i.e.*, now, hypotheses or theories) rather than on psychological forces. As we have seen, they equated these reasons with logical reasoning, more exactly with the new logic of Frege, Russell, and Whitehead, and with any extensions thereof that might be found necessary and feasible. This second departure explains the term 'logical' in the name of the movement. Together, these two departures, and the associated continuities with earlier empiricism, define *classical empiricism*, as I understand it, in terms of an advocacy of the view that all our ideas, and those we count as knowledge in particular, depend in some way on sense-experience.[11] What I referred to above as 'postclassical philosophies of science' reversed this dependency in a fundamental way. In them, even what counts as an experience, is so permeated, shaped, by this background theory or paradigm that an independent 'given' in experience cannot even be separately distinguished. In this respect, the postclassical program, and the radically relativistic thesis that was associated with it, is at heart deeply antiempirical.

4 Roots of the Difficulties, and Their Solution

Let us now return to the discussion of the two clues which I said indicate the roots of the failures of twentieth-century philosophies of science and would point the way to a diagnosis of the failures and a prescription for resolving those difficulties. Recall that (together, of course, with the repeated failure of all attempts to interpret scientific metascientific concepts in terms of the new deductive logic or any extension thereof), the two clues were as follows:

[11] The use of new methods of observation employing instruments, going beyond sense-perception and relegating it to a peripheral role in the quest for knowledge, suggests a nonclassical version of empiricism, in which what counts as an "interaction" with nature is specified by what is learned. Based on a paper (D. Shapere: "The Concept of Observation in Science and Philosophy". In: *Philosophy of Science* XLIX/4 1982 pp. 485–525) I have developed such an *interactional empiricism*, in which we learn, and learn how to learn, through interacting with nature. This view recognizes scientific change is so deep-rooted as to preclude any antecedently-specified "nature" of science; all aspects of science, including what used to be considered "metascientific" ones, are in principle subject to change in the light of what we learn. In these latter respects, this interactional view has affinities also with the "postclassical" philosophies, though avoiding their relativism. A few further aspects of this view are discussed in the remainder of this paper.

First, the concern of the logical empiricist program with the *specific* content of science, and with the identification of that general nature with the 'logical form' of metascientific concepts, and the consequent abandonment of concern with the content of actual science such as the development of quantum and relativity theories.

Second, the concern of the postclassical program with the *specific* content of science, and with the changes in that specific content occurring over the history of a scientific subject or tradition. The original central concept of this historical concern was that of a global perspective ('paradigm', 'high-level background theory') defining and governing a particular tradition, and this in turn entailed the dependency of content and reasoning on that global perspective. Thus was generated the relativism which forsook any possibility of explaining the rationality of major scientific innovations such as the development of quantum and relativity theories.

These clues bring out clearly both the flaws and appeal of the two opposing programs. The participants of the Vienna Circle meetings saw the need to understand the reasoning and implications of the revolutionary developments in physics in the first three decades of the twentieth century. But the movement they produced, logical empiricism, became so infatuated with the equally new and powerful formal logic and metamathematics that, by the very content-independent nature of their program, they failed to exploit any lessons from those developments. The flaws and errors of the postclassical philosophers of science were exactly the reverse. Whereas the logical empiricists saw the need to focus on reasoning (while equating it with logic), their postclassical critics saw the need to attend to the content of science, but they interpreted that content in terms of an extreme historical relativism in which there can be no rational connection whatever between one paradigm-tradition and another, and so lost the potential to account for the reasoning involved in major events in the history of science. In sum, both the logistic and postclassical historicist programs began with the recognition of the need to understand radical changes in scientific views – but both ended by denying the possibility of such understanding.

In retrospect, the appeals of the two programs are these. For logical empiricism, there was the focus on reasoning in science, though what was conceived as reasoning was restricted to formal logic. For the postclassicists, there was the insistence on examining the content of science, though that insistence was corroded by the historical relativism into which the movement descended. Can the two appealing motives be merged into a coherent view, while avoiding the associated flaws? I believe they can. We *do* need to examine the content of science, and not merely its essence or logical form – the lesson the postclassical program saw, though in a distorted way. But we also need to show how reasons do at least give guidance in even the most revolutionary of scientific innovations – the original aim, at least, of the logistic program, though distorted by overconcentration on the form of reasoning rather than the content. The secret of how to merge these is known intuitively to every working scientist worth his or her salt: it is that the content of the science that is accepted at any given epoch provides the reasons guiding, and sometimes driving, further inquiry; it is the content of science that provides the background information, the reasons, in terms of which new problems can be stated, new methods introduced and applied, and new hypotheses formulated and tested, and either rejected or else accepted and deposited in the pool of background information which can then be used as reasons to lead to further development of ideas. We apply prior knowledge in this sense to particular problems, selecting the relevant background information piece by relevant piece (not as a monolithic global determinant), from the pool of what has been learned, and apply it to do a variety of jobs in the particular problem-situation at hand. In short, we build new a inquiry on what we have learned, or have the best reasons to think we have learned; this is not only how we learn, but also how we learn how to learn. Except to philosophers, the need to have prior knowledge in new problem-situations, and to use it as reasons guiding further action or thought, is a universally-known secret. Yet neither the logical empiricists, with their disavowal of any 'logic of discovery', nor their postclassical critics, with their monolithic paradigms and high-level background theories that govern an epoch's science wholesale, ever managed to grasp this secret, at least in any fully-realized way.

How does a scientific proposition come to be admitted to the pool of background information, to serve as guidance in further inquiry? Here, there is time only to sketch the outlines of an answer which I have developed more fully elsewhere. In the sixteenth through eighteenth centuries, a piecemeal approach began to play a central role in inquiry, in which separate subject-matters ('domains'), like motion, gases, heat, electricity, and light were examined and explained in isolation from considering any other subject-matter. Things could conceivably have turned out that such an approach would not work: as holistic philosophers throughout history have insisted, the relations between different domains might be too tight to permit such piecemeal study. But as a matter of contingent fact, the approach managed to achieve what it promised to do: explanatory theories were developed which accounted for such areas of investigation as those mentioned above – motion, gases, magnetism, heat, electricity, and light.

These successes led to the elevation of the piecemeal way of inquiring about such domains to the level of a criterion of how to go about inquiry in new areas. In the nineteenth and twentieth centuries, another criterion was gradually added, extending the piecemeal approach still further. It was found that explanations of many domains which had been previously studied in isolation from one another could cohere with one another, and even unite what had previously been understood separately. One such unification had already been achieved in the seventeenth century by Newton, in his fusing of terrestrial and celestial physics. But the following two centuries saw successive unifications, some of the most famous of which are the unification of electricity and magnetism (most notably by Faraday), of electromagnetism and light (Maxwell), of electricity and chemistry (the various theories culminating in the valency theory of chemical combination), of matter and light (quantum mechanics), the partial joining of the electromagnetic and weak interactions (electroweak unification, by Salam and Weinberg), the juxtaposition (to say no more), based on mathematical similarity, of the strong and electroweak forces (the Standard Model of elementary particles and forces, based on the separate theories, both being gauge theories), the application of the Standard Model of the very small, the quantum realm, to the Big Bang theory of the origin of the universe, and, outside of physics and astronomy, the unification of species brought about by Darwin's theory of evolution, and the synthesis of that theory with Mendelian genetics (synthetic theory of evolution), and the rapidly-growing unification of the history of the origin and composition of the solar system with the interpretation of our earth's history that begins to convert the term '*geo*logy' into an oxymoron. My apologies if I have omitted anyone's particular favorite from my very short list. In any case, whatever the psychological motivations individual scientists may have had to seek unity, there was now *reason* to consider unification to be a goal of science: namely, it had proved possible to get unifications, to fulfil what the attempt promised to do; as Pauli might have put it had he lived, at least some things that God had (in appearance at least) put asunder had proved possible for man to join after all. The results of unification could therefore be added to the attempts to explain isolated domains and incorporated into the pool of background information guiding further research, at least until specific doubt should arise leading to question their qualification for playing this guiding role. We not only learn about the universe in which we live; we also learn how to think, to reason, about it.

This piecemeal approach to inquiry dates to very early in evolutionary history. Just as Einstein had to have a deep grasp of classical mechanics and electromagnetic theory to see a problem of their compatibility and resolve it, so also our hunter-gatherer ancestors had to know what stones could be used to make good tools, and where to find such tools, what plants were edible, and where they could be found and when they could be harvested. In contrast to Einstein, for them such background information was necessary to survive, and was applied wherever relevant in daily practice. But this contrast does not detract from the common need for prior knowledge in confronting and resolving problems. In this sense, science as I have described it is a rational descendant of a far earlier practical and local need for knowledge, but with the practicality and locality no longer directing the quest for and application of knowledge.[12]

---

[12] D. Shapere: "The Ultimate Furniture of the Universe", forthcoming in proceedings of a conference held at San Sebastian, Spain, on "Concepts of Nature from Greek Thought to Quantum Mechanics".

5 What Does It Mean to Say That Science Is Content-Driven?

Science is content-guided, by what it has learned already. Its reasons for proceeding in a certain direction, sometimes to gain new knowledge, consist not of logical rules, but rather of the knowledge we have already acquired (or at least think we have acquired). We have to go on the basis of something. As long as we can be critical of what we take to be background information, and alter or reject it when necessary, what better basis could we have than what we have learned, including what we have learned about how to learn? But this accumulated learning, the pool of background information, is not found by logic. In sophisticated science, the prior knowledge we apply to gain new knowledge is itself gained by investigation which itself was gained by applying still earlier prior knowledge, using it and criticizing it, all the way back to the deepest levels of evolutionary history. In making this remark about the dependence of inquiry for rational guidance on the results of prior inquiry, I do not mean to belittle the role of logic in science. Far from it: of course logical inference plays an important role in science, especially in the form of mathematics.[13] But logic is far from all there is to reasoning in science, and it is not even the central part that needs to be understood: if logic really is concerned with all possible worlds, then it can tell us little or nothing about why we have come to believe certain things about our particular world, our particular line of the universal truth table, the particular state-description that holds for this universe. The reasons leading to what we have come to believe about our particular universe must be sought in the successive critical applications of what we have come to believe to new problems, new investigations, new forms of inquiry.

Today, science must be content-guided; perhaps someday it will be more content-driven than it can be now, in the sense that its background information would be more nearly sufficient to raise and solve all problems of interest. Science would then approach complete autonomy, self-sufficiency, in its reasoning. But the pool of background information available to us today is still insufficient to drive us (as opposed to giving guidance) always in the right directions to solve the problems facing us in science, much less to drive the well-formed formulation of all the problems that there must be. For science to be truly autonomous in its reasoning, we would have to know everything. In the case of individual theories, like quantum mechanics, we do not really understand whether there is a clear difference between what the theory itself

[13] I do not mean to imply that the roles of mathematics in science are strictly limited to logical inference. On the contrary, mathematics has proved the key to enabling the human mind to conceive ideas that go beyond the limited thought-processes having to do with practical matters on an everyday, local level. This idea is elaborated in my "On the Introduction of New Ideas in Science" in: *The Creation of Ideas in Physics* (Kluwer, Dordrecht 1995) pp. 189–222, and the essay referred to in footnote 12.

says (and what we use to guide further inquiry) and an 'interpretation' superimposed (no pun intended) on the theory. We must for the present admit that we can only be guided by prior knowledge, which may be inadequate for the tasks to which it is applied.

One of the most important lessons of the twentieth century's quest for understanding the nature, goals, and methods of science is that the human mind, a product of evolution selected for practical purposes of survival in a very local, rather middle-sized environment, cannot have much faith in its ability to anticipate nature or how to go about learning about it, explaining it, or talking and thinking about it. The universe of the very large, of the very small, of very high velocities and high and low extremes of energy, turns out not to be intelligible in terms of the categories of thought developed through all those millions of years - an ontology of immediate separate objects and events affecting us and each other, on a very restricted local level, both spatially and temporally. This vague and primitive ontology, which I have elsewhere[14] called the Primal View of Nature, was to take a wide variety of interpretational directions in classical philosophy and science, and, in one or another of those interpretations, shaped what various philosophers and scientists thought of as explanatory. But today it is to be hoped that we have learned that the Primal View in its multifarious forms is inadequate, and that we must learn new ways of understanding nature. Perhaps it may even turn out that we can come to understand how we are to understand only by introducing a new logic. But that is another question, and, typically of scientific investigation, its outcome cannot be predicted in advance, but must await the results of careful inquiry, guided, as always, as much as possible by the content of what we have learned.

References

1. R. Carnap: 'Der Raum'. Kantstudien, Erg. Heft no. 56, Berlin 1922
2. T. Kuhn: *The Structure of Scientific Theories* (University of Chicago Press, Chicago 1970)
3. H. Reichenbach: *The Philosophy of Space and Time* (Dover, New York 1958)
4. H. Reichenbach: *Philosophic Foundations of Quantum Mechanics* (University of California Press, Berkeley 1948)
5. H. Reichenbach: *The Direction of Time* (University of California Press, Berkeley 1956)
6. D. Shapere: 'Building on What We Have Learned: The Relations Between Science and Technology'. In: *Advances in the Philosophy of Technology* ed. by E. Agazzi, H. Lenk (Society for Philosophy and Technology, Newark, Delaware 1999) pp. 239–255
7. D. Shapere: 'On the Introduction of New Ideas in Science'. In: *The Creation of Ideas in Physics* ed. by J. Leplin (Kluwer, Dordrecht 1995) pp. 189–222

[14] see footnote 12

8. D. Shapere: 'Testability and Empiricism'. In: *The Reality of the Unobservable* ed. by E. Agazzi, M. Pauri (Kluwer, Dordrecht 2000) pp. 153–164
9. D. Shapere: 'The Methods of Science'. In: *The Creation of Ideas in Science* ed. by J. Leplin (Kluwer, Dordrecht 1995) pp. 13–27
10. D. Shapere: *The Rational Dynamics of Science.* forthcoming
11. D. Shapere: *The Ultimate Furniture of the Universe.* Forthcoming in: Proceedings of Conference on 'The Concept of Nature from Greek Thought to Quantum Mechanics' held at San Sebastian, Spain, October 1–5, 1998
12. L. Wittgenstein: *Tractatus Logico-Philosophicus* (Routledge & Kegan Paul, London 1922)

How Set Theory Impinges on Logic

Jesús Mosterín

1 The Set-Theoretical Universe

Reality often cannot be grasped and understood in its unfathomable richness and mind-blowing complexity. Think only of the trivial case of the shape of the earth. Every time the wind blows, a bird flies, a tree drops a leaf, every time it rains, a car moves or we get a haircut, the form of the earth changes. No available or conceivable geometry can describe the ever-changing form of the surface of our planet. Sometimes the best we can do is to apply the method of theoretical science: to pick up a mathematical structure from the set-theoretical universe, a structure that has some formal similarities with some features of the real-world situation we are interested in, and to use that structure as a model of that parcel of the world. In the case of the earth, the structure can be an Euclidean sphere, or a sphere flattened at the poles, or an ellipsoid, but of course these structures do not represent the car and the hair, and so are realistic only up to a point.

The largest part of scientific activity results in data, in contributions to history (in a broad sense). Only exceptionally does scientific activity result in abstract schemata, in formulas, and in theories. In history there is truth and falsity, but we are not sure whether it makes sense to apply these same categories to an abstract theory. We pick up a mathematical structure and construct a theory. We still have to determine its scope of application or validity, and the range of its realizations. If it is consistent, it will have at least mathematical realizations, and that we can know *a priori*. But the range of its real-world applications is a matter for empirical research to ascertain.

Mathematical (or theoretical) science departs drastically from the usual ways of representing and understanding. In order to realize how utterly different its method of representation is from ordinary language, it suffices to ponder how remote the set-theoretical universe (the reservoir of mathematical structures) is from anything in ordinary language or everyday experience.

Mathematicians create the set-theoretical universe from almost nothing, from just the empty set, by means of successive (and never-ending) iterations of the operations of the power set and the union. The ordinal numbers include the natural numbers and extend the possibility of carrying out iterations into the transfinite. The usual ordinal numbers β such that (for some ordinal α) $\beta = \alpha + 1$ are the successor ordinals. The limit ordinals are the ordinals that are not successor ordinals. Von Neumann proved a general recursion theorem that allows us to define ordinal functions by transfinite recursion over all the ordinals. So he was able to give a precise definition of the cumulative hierarchy of sets by means of the ordinal function V_α:

- $V_0 = \emptyset$
- $V_{\alpha+1} = \wp(V_\alpha)$
- $V_\lambda = \cup_{\beta<\lambda} V_\beta$

Assuming the axiom of foundation (i.e., that all sets are in the cumulative hierarchy), von Neumann defined the set-theoretical universe V as the union of all the V_β (for any ordinal β): $V = \cup_{\beta\in\Omega} V_\beta$. This definition has become canonical. The set-theoretical universe is usually figured (see Fig. 1) as an inverted cone, whose (inverted) vertex is the empty set and that proceeds upward by iterations of the operations of power set and union, indexed by the ordinals. To each ordinal α corresponds a new slice of sets (the sets with rank α. The union of all those slices is the set-theoretical universe.

All the mathematical structures used in science for modeling the real world (the natural and real numbers, the Euclidean and non-Euclidean spaces, the probability spaces, the vector spaces, the Hilbert spaces, the differential manifolds, the tensor fields, etc.), all of them appear somewhere in this inverted cone. The rank of a structure is the least ordinal β such that the structure is in the slice V_β of the set-theoretical universe.

Concerning the inverted cone of the set-theoretical universe, we can ask (1) how wide it is, and (2) how tall it is. Different axioms and hypothesis

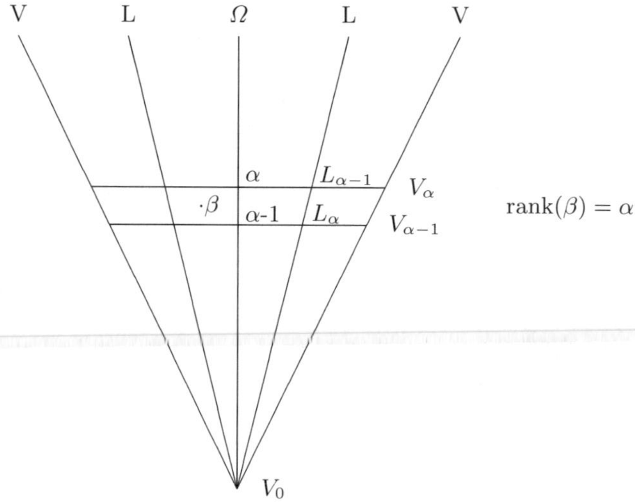

Fig. 1. Scheme of the set-theoretical universe

determine the width and the depth (or height) of the cone. The Continuum Hypothesis and Gödel's axiom of constructibility ($V = L$) concern the width of the cone. If we accept the Generalized Continuum Hypothesis (GCH), the cone is narrower. If we reject it, it becomes wider. The axiom of constructibility makes for an especially narrow cone. On the other hand, large cardinal axioms concern the height of the cone. The axiom of infinity, the existence of $\aleph_1$, of inaccessible cardinals, of Mahlo cardinals, of weakly compact cardinals, of measurable cardinals, of Woodin cardinals, and so on, make for deeper and deeper (or higher and higher) cones.

Set theory is deep and full of open questions. One could be tempted to think that logic is a harmless pursuit, independent of set-theoretical assumptions, but it is not.

2 Second-Order Logic and Set Theory

At first sight, second-order logic looks like the most natural framework for formulating such mathematical theories as natural number arithmetic, Euclidean geometry, mathematical analysis, and even ZFC set theory, all of which are categorical in second-order language. Only in second-order logic are all these theories complete, and only in second-order logic can their corresponding mathematical structures be characterized uniquely, up to isomorphism. Unfortunately, it is impossible to mine all these alleged riches from second-order logic. There cannot be any complete calculus for deducing all the theorems of a second-order theory from its axioms, and second-order logic itself is indeterminate. The set of its valid formulas cannot be known. If it were known, it would settle all the many open questions of set theory. As a matter of fact, we need to settle all the open questions of set theory before second-order logic can be made determinate.

A sentence of pure second-order logic is a closed formula of second-order logic that does not contain any symbol besides logical constants and quantified variables. It does not contain any specifically mathematical or set-theoretical symbol. Corresponding to each open question in set theory there is a sentence (a closed formula) of pure second-order logic, such that the formula is logically valid if and only if the set-theoretical question has an affirmative answer.

3 The Axiom of Choice

No one really doubts the axioms of ZF. The first doubts were expressed in relation to the axiom of choice (AC), that, in one of its versions, says that there is a universal choice function, i.e., a function that assigns to each nonempty set in the universe one of its members. It is equivalent to many different mathematical statements, like the well-ordering theorem. The well-ordering theorem was conjectured by Cantor in 1883. It was a necessary keystone of the Cantorian construction of set theory. Only if every set can be well ordered can we be sure that every set has an ordinal as its order type and an aleph as its cardinality. Cantor tried to prove it, but failed. It was first proved by Zermelo in 1904, but only under the assumption of the axiom of choice, to which it is equivalent.

The following sentence of pure second-order logic (interpreted on any universe) says that the universe can be well-ordered.

$$\exists W[\forall xyu(Wxy \wedge Wyu \Rightarrow Wxu) \wedge \neg Wxx \wedge \forall xy(Wxy \vee Wyx \vee x = y)$$
$$\wedge \forall Z(\exists xZx \Rightarrow \exists u(Zu \wedge \forall x(Zx \Rightarrow Wux \vee u = x)))].$$

This formula is satisfied by a structure iff the universe of the structure can be well-ordered (i.e., is well-ordered by some relation).

4 The Continuum Hypothesis

The other conjecture Cantor unsuccessfully tried to prove was the continuum hypothesis. With the benefit of hindsight, we now know that it could not be proved, as it is independent of the rest of the axioms of set theory.

In order to be able to formulate a formula of pure second-order logic equivalent to the continuum hypothesis in a compact way, we have to introduce some abbreviations or definitions. Notice that these abbreviatory devices are fully dispensable. Any formula formulated with the so-defined symbols is just an abbreviation of another and longer formula of pure second-order logic. Here are some definitions:

Y is injectable into Z (smaller than or equal to Z): $Y \preceq Z$

$$Y \preceq Z \Leftrightarrow_{df} \exists W[\forall xyz(Wxy \wedge Wxz \Rightarrow y = z) \wedge$$
$$\wedge(Wyx \wedge Wzx \Rightarrow y = z) \wedge \forall u(Yu \Rightarrow \exists xWux \wedge Zx)]$$

Y is bijectable onto Z (equinumerous with Z):$Y \sim Z$

$$Y \sim Z \Leftrightarrow_{df} Y \preceq Z \wedge Z \preceq Y$$

Y is smaller than Z : $Y \prec Z$

$$Y \prec Z \Leftrightarrow_{df} Y \succ Z \wedge \neg Y \sim Z$$

Z is infinite: $Inf(Z)$ [Dedekind]

$$Inf(Z) \Leftrightarrow_{df} \exists Y[\forall x(Yx \Rightarrow Zx) \wedge \exists x(Zx \wedge \neg Yx) \wedge Z \sim Y]$$

Y is the power set of X : $\wp(X, Y)$

$$\wp(X,Y) \Leftrightarrow_{df} \exists W[\forall u(Xu \Leftrightarrow \exists z Wuz) \wedge \forall u(Yu \Leftrightarrow \exists z Wzu) \wedge \\ \wedge \forall Z(\forall u(Zu \Rightarrow Xu) \Rightarrow \exists v \forall x(Zx \Leftrightarrow Wxv)) \wedge \\ \wedge \forall uv(\forall x(Wxu \Leftrightarrow Wxv) \Rightarrow u = v)]$$

[Here, the first formulas mean $Dom(W) = X$ and $Range(W) = Y$. W mimics the membership relation between elements of X and subsets of X (elements of Y).]

The Generalized Continuum Hypothesis (GCH) says that, for any infinite set A, there are no intermediate cardinalities between $\mid A \mid$ and $\mid \wp(A) \mid$. It is equivalent to the validity of the pure second-order logic formula abbreviated below,

$$\text{GCH} \Leftrightarrow_{df} \forall XYZ(Inf(X) \wedge \wp(X,Y) \wedge Z \prec Y \Rightarrow Z \preceq X)$$

In ZFC set theory we can neither prove nor disprove GCH, because it is independent of the rest of the axioms of standard set theory, as proved by Kurt Gödel and Paul Cohen. So, GCH is a logical truth iff the generalized continuum hypothesis is a set-theoretical truth, but we do not know whether it is (or we want it to be) true, and so neither do we know whether the corresponding formula is a logical truth.

5 Inaccessible Cardinals

A cardinal κ is inaccessible iff

1. κ is an uncountable cardinal (i.e., $\kappa > \aleph_0$).
2. κ is a strong limit cardinal (i.e., the power set (or power of 2) construction on inferiors does not lead us to κ: for any cardinal $\lambda : \lambda < \kappa$ implies $2^\lambda < \kappa$).
3. κ is regular (i.e., κ is not the supremum of a set of fewer than κ smaller ordinals).

Regularity of cardinals can also be defined in terms of cofinality. The cofinality of $\alpha, cf(\alpha)$, is the minimum ordinal β such that there is a function $f : \alpha \rightarrow \beta$ with $range(f)$ cofinal in α. The cofinality of a limit ordinal $\lambda, cf(\lambda)$, is the smallest cardinal κ such that λ is the supremum of κ smaller cardinals. A cardinal κ is regular iff $cf(\kappa) = \kappa$.

If κ is an inaccessible cardinal, then κ is a fixed point of the aleph function: $\aleph_\kappa = \kappa$. If κ is an inaccessible cardinal, then all of ZFC axioms are true in V_κ (i.e., V_κ is a model of ZFC). A consequence of this last fact (via Gödel's 2nd incompleteness theorem) is that the existence of inaccessible cardinals is unprovable in ZFC. The relative consistency is also unprovable.

In ZFC we can neither prove nor disprove the existence of inaccessible cardinals. So the assertion that there are inaccessible cardinals adds new strength to the theory, it is an axiom that adds new depth (or height) to the set-theoretical universe.

There are inaccessible cardinals iff there are sets of inaccessible cardinality. The predicate $Inacc(Z)$ says that the set Z is of inaccessible cardinality. It can be defined in pure second-order logic (remember that the symbols Inf and $\wp$ are just abbreviations):

$$Inacc(Z) \Leftrightarrow_{df}$$
$$Inf(Z) \wedge \exists Y(Y \prec Z \wedge Inf(Y)) \wedge \forall X(X \prec Z \Rightarrow \exists Y(\wp(X,Y) \wedge Y \prec Z) \wedge$$
$$\wedge \forall XYW[X \prec Z \wedge \forall u(Xu \Leftrightarrow \exists xWux) \wedge \forall u(Yu \Leftrightarrow \exists xWxu) \wedge$$
$$\wedge \forall y(Xy \Rightarrow \forall V(\forall x(Vx \Leftrightarrow Wyx) \Rightarrow V \prec Z)) \Rightarrow Y \prec Z].$$

This formula says that Z is infinite and uncountable, that the power set construction on inferiors does not lead us to Z (strong limit), and that the range of any function that applies inferiors to the elements of inferiors does not lead us to Z (which is equivalent to regularity). So, the formula defines a set of inaccessible cardinality.
$\exists X Inacc(X)$ is satisfiable iff there is an inaccessible cardinal.
$\neg\exists X Inacc(X)$ is a logical truth iff there are no inaccessible cardinals.
Of course, still stronger axioms have been proposed, asserting the existence of larger and larger cardinals (like Mahlo cardinals, weakly compact cardinals, measurable cardinals, Woodin cardinals or supercompact cardinals). Each of these axioms is independent of the previous ones and implies them. But (with some ingenuity and much space) we could reformulate all these axioms as sentences of pure second-order logic.

6 First-Order Logic and Set Theory

Everyone agrees that second-order logic is set-theory in disguise, but first-order logic is often supposed not to be contaminated by any set-theoretical decisions. Is this so?

The set of logically valid sentences of first-order logic is the set of all sentences that are satisfied (or true) in all structures (or in all interpretations on all domains). Depending on which and how many sets there are, there will be more or less logically valid sentences. The more sets there are, the less formulas will be satisfied in all of them, i.e. the less formulas are logically valid.

First, a trivial point, usually we do not admit in our semantics (or model theory) structures with an empty universe. Because of this, formulas like the following are logically valid:

$$\exists x(Px \vee \neg Px) \exists xx = x \forall xZx \Rightarrow \exists xZx$$

If we admit structures with the empty set as universe, then these formulas become invalid, obviously.

Some first-order formulas (called infinite schemata by Quine) are only satisfied by infinite structures, like the following:

$$\forall x \exists y Rxy \wedge \forall x \neg Rxx \wedge \forall xyz(Rxy \wedge Ryz \Rightarrow Rxz)$$

$$\forall x \exists y \forall z(Rxy \wedge \neg Rxx \wedge (Ryz \Rightarrow Rxz))$$

[the same, in prenex normal form]

$$\forall xy(f(x) = f(y) \Rightarrow x = y) \wedge \exists y \forall x f(x) \neq y$$

The negations of these formulas, like

$$\neg \forall x \exists y \forall z(Rxy \wedge \neg Rxx \wedge (Ryz \Rightarrow Rxz))$$

are logically true if all sets are finite, but are not valid if there are infinite sets.

Many invalid (assuming infinite sets) first-order formulas become valid formulas if we restrict our set-theoretical background to finite sets. The formula

$$\begin{aligned}&\forall xyz(Rxy \wedge Ryz \Rightarrow Rxz) \wedge \forall x \neg Rxx \wedge \forall xy(Rxy \vee Ryx \vee x = y) \Rightarrow \\ &\Rightarrow \exists y \forall x(Rxy \vee x = y)\end{aligned}$$

says that if R is a lineal order, then R is not dense. Of course, no finite ordering is dense. If there are only finite sets, this formula is logically true. If there are infinite sets, this formula is not logically valid, as many infinite orderings (like the rationals or the reals) are dense. The first-order formula

$$\begin{aligned}&\forall xyz(Rxy \wedge Ryz \Rightarrow Rxz) \wedge \forall x \neg Rxx \wedge \forall xy(Rxy \vee Ryx \vee x = y) \Rightarrow \\ &\Rightarrow [\forall xy(Sxy \Rightarrow Rxy) \wedge \exists xy Sxy \Rightarrow \exists z \forall xy(Sxy \vee Syx \Rightarrow Szx \vee z = x)]\end{aligned}$$

says that if R is a lineal order and S is an nonempty suborder of R, then S has a minimum. Its validity would imply that any lineal ordering is a well-ordering. Indeed, every finite lineal ordering is a well ordering, and so the formula is valid if all sets are finite. But many infinite lineal orderings (like the integers or the rational or the reals) are not well-orderings. So, the formula is not valid if there are infinite sets.

In contrast with second-order validity (which is not well defined), first-order validity is well determined, both if we accept infinite sets or if we keep to finite ones. But in both cases we get different notions of validity, different sets of valid sentences.

If we reject infinite sets, and admit only finite sets, we get many more valid first-order formulas. As there are fewer sets, and so fewer structures, there are more formulas satisfied by the fewer structures. All the formulas usually considered valid continue to be valid, but many new ones become valid, like the formulas previously considered.

We have seen that second-order logic depends massively on set theory, but also first-order logic has some dependency. In the polemic about logicism in the first decades of the twentieth century, the status of the axiom of infinity played a crucial role. The polemic closed with the agreement that the axiom of infinity belonged to set theory and had nothing to do with logic. But, as we have just seen, it has much to do with logic, even with first-order logic. Whether we accept it or not in the metatheory as background of our definitions, we get quite different (extensionally different) first-order logics.

The standard notion of first-order logic is the one that accepts infinite sets in the set-theoretic background. This logic is semantically complete, as Gödel first proved in 1930. This means that the set of the valid sentences of first-order logic is recursively enumerable, or, in other words, that it can be generated by the successive application of the rules of a deductive calculus. This is also equivalent to saying that the calculus allows us to deduce all the consequences of a given set of premisses. All the proofs (Gödel's, Henkin's, etc.) of the semantic completeness of first-order logic proceed by the construction of certain infinite sets (sets of terms, in Henkin's case), and lose all acceptability if we reject infinite sets. Furthermore, if we reject infinite sets, we reduce so drastically the amount of available structures that the number of valid formulas increases considerably. The set of valid formulas becomes much larger and complex, so much so that it ceases to be recursively enumerable, as proved by Trakhtenbrot in 1950. So, first-order logic is only semantically complete in so far as we countenance infinite sets in the metatheoretic background.

The other way around, if we replace standard second-order logic (in which the second-order variables vary over the full power set of the universe of the structure) by Henkin second-order logic (in which variables can vary on any particular subset of the power set of the universe), then there are many more structures available than in the standard case, and so there are fewer formulas satisfied in all of them, i.e., there are fewer valid formulas. The set of valid formulas becomes so much smaller and less complex, that it even becomes recursively enumerable.

References

1. P. Cohen: 'The Independence of the Continuum Hypothesis'. In: I and II. Proceedings of the National Academy of Sciences USA, **50** (1963–64) pp. 1143–1148 and **51** pp. 105–110
2. K. Gödel: *The Consistency of the Axiom of Choice and of the Generalized Continuum Hypothesis with the Axioms of Set Theory* (Princeton University Press, Princeton 1940)
3. L. Henkin: 'Completeness in the Theory of Types'. Journal of Symbolic Logic **15** (1950) pp. 81–91
4. I. Jané: 'A Critical Appraisal of Second-Order Logic'. History and Philosophy of Logic **14** (1993) pp. 67–86

5. A. Kanamori: 'The Mathematical Development of Set Theory from Cantor to Gödel'. The Bulletin of Symbolic Logic **2** (1996) pp. 1–71
6. J. Mosterín: 'La frontera entre lógica y teoría de conjuntos'. In: *Symposium Quine* ed. by J.J. Acero and T. Calvo (Universidad de Granada Press, Granada 1987) pp. 209–227
7. B.A. Trakhtenbrot: 'The Impossibility of an Algorithm for the Decision Problem in Finite Classes'. [in Russian] Dokl. Akad. Nauk SSSR, **70** (1950) pp. 569–572
8. J. Von Neumann: 'Über die Definition durch transfinite Induktion und verwandte Fragen der allgemeinen Mengenlehre'. Mathematische Annalen **99** (1928) pp. 373–391

Geometries and Arithmetics

Graham Priest

1 Introduction: Geometry and Arithmetic

Arithmetic, geometry, and logic are the three great *a priori* sciences of Kant's *Critique of Pure Reason*. According to Kant, the mind has certain cognitive structures which, when imposed on our "raw sensations", produce our experiences. The first two, space and time, are dealt with in the Transcendental Aesthetic. The third, the categories, is dealt with in the Transcendental Analytic. In the case of all three, a certain body of truths holds good in virtue of these *a priori* structures; and these constitute the three corresponding sciences; geometry in the case of space, arithmetic in the case of time, and logic in the case of the categories. As the difference in location in the *Critique* indicates, the sciences are not entirely on a par: geometry and arithmetic are synthetic; logic is analytic. Nonetheless, each, as a science, is certain and, essentially, complete. This gives us Euclidean geometry, (standard) arithmetic, and Aristotelian logic.

There are few now who would agree with the Kantian picture of these three sciences—at least in its entirety. It has disintegrated, not just under the pressure of philosophical criticism, but under the pressure of developments in science itself. The science with the clearest modern status is, perhaps, geometry. There are many geometries; and which one is to be applied to actual space is an *a posteriori* matter. Following Frege, many twentieth-century philosophers have taken arithmetic and logic to be both *a priori* and analytic.[1] The purpose of this paper is to argue that arithmetic, at least, is in exactly the same camp as geometry.

The status of arithmetic has been a contentious issue during the twentieth century, and I am hardly the first to argue for the position I have just stated. Quine's celebrated arguments in 'Two Dogmas of Empiricism' result in the same conclusion. I think that Quine's argument does establish the a posteriority of arithmetic.[2] However, it is not my aim to go over this ground here,

[1] For example, Wright (1983) shows that standard arithmetic may be derived in second-order logic augmented by "Hume's Principle": if X and Y are in one-to-one correspondence then the number of Xs is the same as the number of Ys. This certainly looks as though it could be analytic.

[2] Though, as an attack on analyticity, I think the argument fails. See Priest (1979).

nor do I think that the central philosophical argument I shall employ is very new. (It is simply a sophisticated version of the old saw: if we kept counting two rabbits and two rabbits and getting five, arithmetic would have been refuted.) But I shall build the argument on the existence of relatively recent work on inconsistent arithmetics. I hope this will give it a novel twist.

I will start by reviewing the modern situation concerning geometry. I will then discuss what alternative arithmetics look like. Next, I will argue that the applicability of an arithmetic is an *a posteriori* matter. I will end with a few comments concerning logic.

2 Non-Euclidean Geometry

Until the nineteenth century, 'geometry' just meant Euclidean geometry; but in the first part of that century some different geometries were developed. Initially, these were obtained by Lobachevski and Bolyai simply by negating one of the postulates of Euclidean geometry in order to try to find a *reductio* proof of it. But under Riemann, the subject developed into one of great generality and sophistication. In particular, he developed a highly elegant theory concerning the curvature of spaces in various geometries. Whether non-Euclidean geometries were to be called geometries *in strictu sensu* might have been a moot point; after all, they did not describe the structure of physical space. But they were at least theories about objects called 'points', 'lines', etc., whose behaviour bore important analogies to that of the corresponding objects in Euclidean geometries. Moreover, Riemann realised that it might well be an empirical question as to which geometry should be applied in physics.[3]

Within another 50 years, and even more shocking to Kantian sensibilities, Riemann had been vindicated. The General Theory of Relativity postulated a connection between mass and the curvature of space (or space-time), which implied that space may have non-zero curvature, and so be non-Euclidean. Predictions of this theory were borne out by subsequent experimentation, and the Theory is now generally accepted.

How to understand the status of physical geometry as it emerged from this affair is still philosophically contentious. The simplest interpretation is a realist one.[4] Geometry in physics is a theory about how certain things in physical space, i.e., points, lines, etc., behave; and a non-Euclidean geometry gets it right. The alternative to realism is nonrealism, of which there are many kinds. One is reductionism: talk of geometric points and lines is to be translated without loss into talk of relationships between physical objects. This is the view most famously associated with Leibniz,[5] but has found few

[3] On the history of non-Euclidean geometry, see Gray (1994); and on Riemann in particular, see Bell (1953), vol. 2, chap. 26.

[4] This is endorsed by Nerlich (1976).

[5] And also Aquinas, *Summa Theologia* I, *quaest.* 46.

modern adherents. Another kind of nonrealism is instrumentalism.[6] Geometry has no descriptive content, literal or reductive. It is merely auxiliary machinery for the rest of physics. As such, we may choose whatever such machinery makes life easiest elsewhere; and a non-Euclidean geometry does just that. One notable version of instrumentalism is that according to which, once we have chosen a geometry, its claims become true by convention. Such conventionalism is often associated with the name of Poincaré.[7]

Whichever of these spins one puts on the ball, the changes in geometry have forced us to draw a crucial distinction. We must distinguish between geometries as pure mathematical structures and geometries as applied theories. As pure mathematical structures, there are many geometries. Each is perfectly well-defined proof-theoretically or model-theoretically. What holds in it may be *a priori*. By contrast, which pure geometry to apply to the cosmos as a physical geometry is neither *a priori* nor certain, but is to be determined by the usual criteria of physical science.

3 Nonstandard Arithmetics

Having spelled out the situation for geometry, let me now address the question of whether the situation is the same for arithmetic. There are two issues to be addressed here: whether there are alternative arithmetics, as there are alternatively geometries; and whether the question of which one to apply is *a posteriori*. I will take the questions in that order.

Normal arithmetic is the set of sentences of the usual first-order language that are true in the standard model, the natural numbers, $0, 1, 2, \ldots$, as subject to the usual arithmetic operations. We may take an alternative arithmetic simply to be one that is inconsistent with this. In other words, we form an alternative arithmetic by throwing in something false in the standard model. Naturally, if consistency is to be preserved, other things must be thrown out. There are two possibilities here.

The first is that we retain all the axioms of Peano Arithmetic, but add the negation of something independent of Peano Arithmetic but true in the standard model. We then have a theory that has a classical nonstandard model.[8] As a rival arithmetic, such theories are a little disappointing, however. For, as is well known, any model of such a theory must have an initial section that is isomorphic to the standard model. In a sense, then, such theories are not rivals to standard arithmetic, but extensions thereof.

The second, and more radical, way of obtaining a nonstandard arithmetic is to add something inconsistent with the Peano axioms, and jettison some of

---

[6] This is endorsed in Hinckfuss (1996).

[7] See Poincaré (1952).

[8] In this model there may even be solutions to diophantine equations that have no solution in the standard model – by the solution to Hilbert's tenth problem – though these solutions will have no name in the standard language of arithmetic.

these. (This is the analogue of how non-Euclidean geometries were initially produced.) In principle, this could produce many different systems, but I know of only one to be found in the literature. This jettisons the axiom which says that numbers always have a successor, and adds its negation, producing a finite arithmetic.[9] Although there are such systems, then, there is no well-worked out theory of their general structure.

A more radical way still of producing a nonstandard arithmetic, for which there is now a general theory, is to drop the consistency requirement. We may then add the negation of something true in the standard model *and jettison nothing.* This situation is novel enough to warrant an extended introduction.

4 Solving Equations

A driving force behind the development of mathematics can be seen as the extension of the number system in such a way as to provide solutions to equations that have no solution. Thus, for example, the equation $x+3 = 2$ has no solution in the natural numbers. Negative numbers began to be used for this purpose around the fifteenth century. Or consider the equation $x^2 = -1$, which has no solution in the domain of real numbers. This occasioned the introduction of complex numbers a little later. In each case, the old number system was embedded in a new number system in which hitherto insoluble equations found roots.

Now consider Boolean equations. A Boolean expression is a term constructed from some of an infinite number of variables, p, q, r... by means of the functors $\wedge$, $\vee$, and $^-$ (complementation). A Boolean equation is simply an equation between two Boolean expressions. The simplest interpretation for this language is the two-element Boolean algebra, $\mathcal{B}_2$, whose Hasse diagram is:

$$\begin{array}{c}\top\\ \updownarrow\\ \bot\end{array}$$

($\wedge$ is interpreted as meet, $\vee$ as join and $^-$ as order-inversion). Within this interpretation many Boolean equations have solutions. For example, the equation $p \vee \overline{p} = q$ is solved by $q = \top$ and $p = \top$ (or $\bot$). But many equations have no solutions, e.g., $p = \overline{p}$. It is natural, then, to extend the algebra to one in which all equations have solutions. The simplest such one is the algebra $\mathcal{D}_3$, whose Hasse diagram is as follows:

[9] See van Bendegem (1987). Goodstein (1965) gives an arithmetic where a number can have more than one successor, though it would be more accurate to describe this as an arithmetic in which there is more than one successor function.

$$\begin{array}{c}\top\\ \updownarrow\\ \mu\\ \updownarrow\\ \bot\end{array}$$

(where operations are interpreted in the same way; in particular, μ is a fixed-point for $^-$). In this structure the equation $p = \overline{p}$ is solved by $p = \mu$. More generally, it is not difficult to check that if every variable is assigned μ, any Boolean expression evaluates to μ. Hence, every Boolean equation has a solution. (The one just given is rather trivial; but in general, there will be others.)

The construction here is a special case of a more general one. The three-element algebra just given is a De Morgan algebra with a fixed point for negation.[10] By the same argument as before, every Boolean equation has a solution in an algebra of this kind. The general result is that every Boolean algebra can be embedded in such an algebra.[11] By Stone's theorem, every Boolean algebra can be embedded in a power-set algebra. Hence, it suffices to prove the result for power-set algebras. Let A be any set, and let $\langle \wp(A), \cap, \cup, ^- \rangle$ be its power-set algebra. Let $S = \{-1, 0, 1\}$ and let $F = S^A$.[12] If f,$g \in F$, define the functions $f \wedge g$, $f \vee g$ and $\overline{f}$ as follows:

$$\begin{aligned} f \wedge g(x) &= \min\{f(x), g(x)\} \\ f \vee g(x) &= \max\{f(x), g(x)\} \\ \overline{f}(x) &= -f(x). \end{aligned}$$

It is straightforward to check that $\langle F, \wedge, \vee, ^- \rangle$ is a De Morgan algebra. ($f \leq g$ iff for all $x \in A$, $f(x) \leq g(x)$.) Also, if f_i is the constant function with value i, then $\overline{f_0} = f_0$. Finally, it is easy to check that if $B \subseteq A$, and $\hat{B}$ is the following function:

$$\begin{aligned} \hat{B}(x) &= 1 \quad \text{if } x \in B \\ &= -1 \text{ if } x \in B - A \end{aligned}$$

then the map $B \mapsto \hat{B}$ is an embedding, as required. (Note also that if, for some a, $A = \{a\}$, so that $\wp(A)$ is $\mathcal{B}_2$, then F is $\mathcal{D}_3$, with $\top = f_1$, $\mu = f_0$ and $\bot = f_{-1}$.)

The result is analogous to one of the fundamental theorems of algebra, that every field can be extended to one over which all equations have solutions (i.e., all nonconstant polynomials have roots), an algebraic closure. (The algebraic closure of the reals is, of course, the field of complex numbers.) The above proof shows that any Boolean algebra can be closed in a similar way.[13]

---

[10] A De Morgan algebra is a structure $\langle D, \wedge, \vee, ^- \rangle$, where $\langle D, \wedge, \vee \rangle$ is a distributive lattice, and for all a, $b \in D$, $a = \overline{\overline{a}}$ and $a \leq b \Rightarrow \overline{b} \leq \overline{a}$.

[11] The following proof is due to Greg Restall.

[12] In general, S could be any subset of the reals containing 0 and closed under $-$.

[13] Strictly speaking, the result is not quite the analogue, since the algebra obtained is not a Boolean algebra. The exact algebraic analogue is that every De Morgan

5 Models of the Inconsistent

Algebras have many applications. The relevant one in the present context, is to provide a structure of truth-values for a language. The fact that every Boolean expression can be evaluated to μ in $\mathfrak{D}_3$ means that if we take these as the semantic values of a propositional language, and interpret connectives in the obvious way, then, provided we take the designated values, Δ, to be those $\geq \mu$, every sentence is satisfiable; as, more generally, is every set of sentences. These semantics characterise the paraconsistent logic LP,[14] and the value μ may be thought of as *both true and false.* Such a value might seem a rather odd one. One might be tempted to call it an imaginary truth value, for the same reason that $\sqrt{-1}$ was called imaginary. But there is nothing really imaginary about imaginary numbers. Mathematically speaking they are just as *bona fide* as real numbers. Indeed, they even have applications in physics. Leave quantum mechanics out of this; even in classical physics, magnitudes such as impedance are given by complex numbers. Similarly, mathematically speaking, there is nothing imaginary about μ. And like imaginary numbers, μ may even have important applications. For example, one application for μ might be to take it to be the truth value of paradoxical self-referential sentences.

Propositional LP can be extended to a first-order logic in an obvious fashion. Let L be a first-order language. An LP interpretation for L is a pair, $\langle D, d\rangle$, where D is a nonempty domain, d maps each constant into D, each n-place function symbol into an n-place function on D, and each n-place predicate to an n-place function from D into $\mathfrak{D}_3$. For identity, we require that $d(=)(x, y) \in \Delta$ $(= \{\top, \mu\})$ iff $x = y$. The semantic values of formulas are assigned in the obvious way (the quantifiers being treated as the analogues of conjunction and disjunction). An inference is valid iff there is no interpretation for which all the premises are in Δ, but the conclusion is not.

It is clear that classical interpretations are special cases of LP interpretations. It follows that LP is a sub-logic of classical logic. It is a proper sub-logic, since in it $\alpha \wedge \neg\alpha \nvDash \beta$. However, it is to be noted that the logical truths of LP are exactly those of classical logic.[15] As is to be expected, every set of first-order sentences has a model. There is always the trivial model, with a one-element domain, such that for every predicate, P, $d(P)$ is the con-

algebra (and *a fortiori* every Boolean algebra) can be extended to a De Morgan algebra in which all equations have solutions. And, in fact, the techniques of the proof can be extended to show this. Every De Morgan algebra is isomorphic to a field of polarities (Dunn (1986), p. 189). The members of such a field are pairs of the form $\langle X, Y\rangle$ where each of X and Y is a subset of some underlying set, A. The closure is now $S^A \times S^A$, operations being defined in the natural way. The rest of the proof is then much as before.

[14] See, e.g., Priest (1987), chap. 5.

[15] For a proof, see Priest (1987), chap. 5.

stant function with value μ. But, in general, there will be many non-trivial models too. One useful way of constructing these is as follows.

Let $\mathfrak{I} = \langle D, d \rangle$ be any interpretation. Let $\sim$ be any equivalence relation on D, that is also a congruence relation on the denotations of the function symbols in the language (i.e., if g is such a denotation, and $d_i \sim e_i$ for all $1 \leq i \leq n$, then $g(d_1, ..., d_n) \sim g(e_1, ..., e_n)$). If $d \in D$ let $[d]$ be the equivalence class of d under $\sim$. Define the *collapsed interpretation*, $\mathfrak{I}_\sim = \langle D_\sim, d_\sim \rangle$, as follows. $D_\sim = \{[d]; d \in D\}$; if c is a constant, $d_\sim(c) = [d(c)]$; if f is an n-place function symbol, $d_\sim(f)([d_1], ..., [d_n]) = [d(f)(d_1, ..., d_n)]$ (this is well defined, since $\sim$ is a congruence relation); and if P is an n-place predicate:

$$\begin{aligned} d_\sim(P)([d_1], ...[d_n]) &= \top \text{ if } \forall e_1 \sim d_1 ... \forall e_n \sim d_n, d(P)(e_1, ...e_n) = \top \\ &= \bot \text{ if } \forall e_1 \sim d_1 ... \forall e_n \sim d_n, d(P)(e_1, ...e_n) = \bot \\ &= \mu \text{ otherwise.} \end{aligned}$$

(It is easy to check that $d_\sim(=)$ is as required for an LP interpretation.) In effect, the collapsed model identifies members of an equivalence class to produce a composite object. The predicates true or false of the composite object are exactly those true or false of all the objects that compose it. In particular, then, if two members of the class have inconsistent properties the equivalence class is an inconsistent object.

It is now not difficult to prove the *Collapsing Lemma*: if the value of α is in Δ in $\mathfrak{I}$, then it is in Δ in $\mathfrak{I}_\sim$. First we show that for any term, t, $d_\sim(t) = d([t])$. Applying this fact secures the atomic case. The result is then proved by induction.[16] The Collapsing Lemma assures us that if an interpretation is a model of some set of sentences, then any interpretation obtained by collapsing it will also be a model. This allows us to construct non-trivial models of inconsistent theories. Let us see how by returning to arithmetic.

6 Inconsistent Arithmetic

Let L be the usual language of first-order arithmetic, with function symbols for successor, addition and multiplication, and one binary predicate, =. Let N be the natural numbers; let $\mathfrak{N}$ be the standard model of L; and let A be the set of sentences true in $\mathfrak{N}$. Any model of A, I will call a *model of arithmetic*.

Example one. Let $n, p \in N$ and $p > 0$. Define a relation, $\sim$, on N, thus:

$$x \sim y \text{ iff } (x, y < n \text{ and } x = y) \text{ or } (x, y \geq n \text{ and } x = y \pmod{p})$$

It is easy to check that $\sim$ is a congruence relation on N. Let $\mathfrak{N}_n^p$ be the model obtained by collapsing with respect to it. The Collapsing Lemma assures us that it is a model of arithmetic. It is finite; it has an initial tail of length n

[16] See Priest (1997).

that behaves consistently. The other numbers form a cycle of period p. The successor graph can be depicted as follows:

$$\begin{array}{ccc} 0 \to 1 \to \ldots \to & n & \to n+1 \\ & \uparrow & \downarrow \\ & n+p-1 \leftarrow & \cdots \end{array}$$

Example two. Let $\mathcal{M}$ be any non-standard classical interpretation of A. Define the relation $\sim$ as follows:

$$x \sim y \text{ iff } (x,\ y \in N \text{ and } x = y) \text{ or } (x,\ y \text{ are non-standard})$$

Again, it is easy to check that $\sim$ is an equivalence relation that is also a congruence on the arithmetic operators. Let $\mathcal{N}_\Omega$ be the model obtained by collapsing with respect to it. $\mathcal{N}_\Omega$ contains the standard interpretation, plus an inconsistent "point at infinity". The successor graph can be depicted as follows:

$$\begin{array}{r} 0 \to 1 \to \ldots\ \Omega \\ \circlearrowleft \end{array}$$

Just as the algebra $\mathcal{D}_3$ allows the solution of Boolean equations that do not have a solution in a Boolean algebra, the inconsistent arithmetics allow solutions to arithmetic equations that do not have them in the standard model. (Any equation that has a solution in $\mathcal{N}$, has a solution in any model of arithmetic.) For example, $x = x + p$ has solutions in any $\mathcal{N}^p_n$, and if s and t are any non-constant polynomials in x, then $s = t$ has a solution in $\mathcal{N}^1_n$ (namely, n) and $\mathcal{N}_\Omega$ (namely, Ω).[17] This fact will be important later.

7 The General Structure of Models

The foregoing provides all that we need to pursue the philosophical issues raised, but let me digress for a section to describe the general structure of LP models of arithmetic (which include, of course, all the classical models).[18]

Let $\mathcal{M} = \langle M, d\rangle$ be any model of A, and let us define the ordering on M in the usual way: $x \leq y$ iff $\exists z\ x + z = y$. If $i \in M$, define $N(i)$ (the *nucleus* of i) to be $\{x \in M; i \leq x \leq i\}$. (In a classical model, $N(i) = \{i\}$, but this need not be the case in an inconsistent model. For example, in $\mathcal{N}^p_n$ $\{x; n \leq x\}$ is a nucleus.) Any two members of a nucleus define the same nucleus. Now, if N_1 and N_2 are nuclei, define $N_1 \preceq N_2$ iff for some $i \in N_1$ and $j \in N_2$, $i \leq j$. $\preceq$ is a linear ordering.

If $i \in M$, *i has period $p \in M$* iff $i + p = i$. (In a classical model every number has period 0 and only 0. But again, this need not be the case as $\mathcal{N}^p_n$ demonstrates.) All members of a nucleus have the same periods. We may thus

[17] On solutions of equations in inconsistent models, see, further, Mortensen (1995).
[18] I shall not give proofs. These can be found in Priest (2000a).

speak of the periods of the nucleus itself. If a nucleus has a period distinct from 0, I will call it *proper*. If $N(i) \preceq N(j)$ then any period of $N(i)$ is a period of $N(j)$. Hence any nucleus after a proper nucleus is a proper nucleus.

If i is in a proper nucleus, its successor, i' is in the nucleus. i may have more than one predecessor (as $\mathcal{N}_n^p$ demonstrates). However, i has a unique predecessor, $'i$, in that nucleus. Thus, if i is in a proper nucleus, its *chromosome*:

$$\ldots,''i,'i,i,i',i''\ldots$$

is contained in the nucleus. A chromosome is either a finite cycle or is isomorphic to the integers. Moreover, all the chromosomes in any one nucleus are either finite cycles of some minimum non-zero period (in which case the minimum non-zero periods of all subsequent nuclei are divisors of it—not necessarily proper), or are isomorphic to the integers. (Both sorts of nuclei are possible.)

Thus, the general structure of a model is a linear sequence of nuclei. There are three segments (any of which may be empty). The first segment contains only improper nuclei. The second segment contains proper nuclei with linear chromosomes. The final segment contains proper nuclei with cyclical chromosomes of finite period. A period of any nucleus is a period of any subsequent nucleus; and in particular, if a nucleus in the third segment has minimum non-zero period, p, the minimum non-zero period of any subsequent nucleus is a divisor of p. Thus, we might depict the general structure of a model as follows (where m is a multiple of n):

$$0,1,\ldots \boxed{\begin{array}{c}\ldots a \rightarrow a'\ldots \\ \ldots b \rightarrow b'\ldots \\ \vdots\end{array}} \ldots \boxed{\begin{array}{ccc}c_0\ldots c_i & & \\ \uparrow & \downarrow & \ldots \\ c_{m-1}\ldots c_i' & & \end{array}} \ldots \boxed{\begin{array}{ccc}d_0\ldots d_i & & \\ \uparrow & \downarrow & \ldots \\ d_{n-1}\ldots d_i' & & \end{array}} \ldots .$$

8 Empirical Applications

Let us now return to the philosophical issues. Each model of arithmetic (or its theory)—save the standard one—is an alternative pure arithmetic, in the same sense that there are alternative pure geometries: abstract mathematical structures dealing with objects (numbers or points) that behave in a way recognisably similar to the corresponding objects of the standard theories. Moreover, as we have seen, there are arithmetics which do not even have initial sections isomorphic to the natural numbers (whilst still verifying all of standard arithmetic). Thus, the first point of similarity with geometry is established.[19]

[19] It might, at this point, be suggested that *real* arithmetic is second order, and that this is categorical. Hence there is only one arithmetic. The reply is that we can still obtain non-standard arithmetics by varying the Peano postulates.

To establish the second point, it must be shown that which of these arithmetics one is to apply, is an *a posteriori* matter. And this may well be doubted. Could there be applied non-standard arithmetic in the same way that there is applied non-Euclidean geometry?

I take the answer to be 'yes'. As I said in the introduction, there are general Quinean arguments why this is the case, though I do not intend to discuss these here. Instead, I will attempt to establish the point by telling a story to show how we might come to replace a standard application of ordinary arithmetic with a different one.[20] I do not want to claim that the situation described is a possibility in any real sense: it is merely a thought experiment. But if it succeeds it will show the required conceptual possibility. And this suffices to show that applied arithmetic is *a posteriori*, in the same way that applied geometry is.

There are a few such stories already in the literature, notably, those of Gasking (1940). Gasking was criticised by Castañeda (1959), who argued that if we use a different arithmetic and preserve our standard practices of counting, we end up with either a simple change of terminology or inconsistency. I will not discuss whether Castañeda's arguments work, here. I intend to finesse them by describing a situation that might motivate an inconsistent arithmetic, specifically, one of the form $\mathcal{N}_n^1$.[21]

Let us suppose that we come to predict a collision between an enormous star and a huge planet. Using a standard technique, we compute their masses as x_1 and y_1, respectively. Since masses of this kind are, to within experimental error, the sum of the masses of the baryons (protons and neutrons) in them, it will be convenient to take a unit of measurement according to which a baryon has mass 1. In effect, therefore, these figures measure the numbers of baryons in the masses. After the collision, we measure the mass of the resulting body, and obtain the figure z, where z is much less than $x_1 + y_1$. Naturally, our results are subject to experimental error. But the difference is so large that it cannot possibly be explained by this. We check our instruments, suspecting a fault, but cannot find one; we check our computations for an error, but cannot find one. We have a puzzle. Some days later, we

---

Moreover, the Collapsing Lemma applies to *second-order* *LP* as well, and this can be used to give non-standard second-order paraconsistent arithmetics.

[20] There are certainly other possible examples. For example, even if arithmetic is not the form of the intuition of time, as Kant thought, one might tell a story where circumstances suggested the possibility of calibrating time with a non-standard arithmetic. In virtue of the cycles in inconsistent arithmetics, stories of time-travel would appear to be particularly fruitful here.

[21] Some of the philosophical issues surrounding inconsistent arithmetics, were aired in Priest (1994). There, it is argued that our ordinary arithmetic might be an inconsistent one. A critique of this is to be found in Denyer (1995), with replies in Priest (1996). Denyer's criticisms are not relevant here. For present purposes, I accept that our ordinary arithmetic is the usual one. My concern is with inconsistent arithmetics as revisionary.

have the chance to record another collision. We record the masses before the collision. This time they are x_2 and y_2. Again, after the collision, the mass appears to be z (the same as before), less than $x_2 + y_2$. The first result was no aberration. We have an anomaly.

We investigate various ways of solving the anomaly. We might revise the theories on which our measuring devices depend, but there is no obvious way of doing this. We could say that some baryons disappeared in the collision; alternatively, we could suppose that under certain conditions the mass of a baryon decreases. But either of these options seems to amount to a rejection of the law of conservation of mass(-energy), which would seem to be a rather unattractive course of action.

Then someone, call them Einquine, fixes on the fact that the resultant masses of the two collisions were the same in both cases, z. This is odd. If mass has gone missing, why should this produce the same result in both cases? An idea occurs to Einquine. Maybe our arithmetic for counting baryons is wrong.[22] Maybe the appropriate arithmetic is $\mathcal{N}^1_z$. For in this arithmetic we may have $x_1 + y_1 = x_2 + y_2 = z$, and our observations are explained without having to assume that the mass of baryons has changed, or that any are lost in the collisions! Einquine hypothesizes that z is a fundamental constant of the universe, just like the speed of light, or Planck's constant.

While she is thus hypothesising, reports of the collisions start to come in from other parts of the galaxy. (The human race had colonised other planets some centuries before.) These reports all give the masses of the two new objects as the same, but all are different from each other. Some even measure them as greater that the sum of their parts. Einquine is about to give up her hypothesis, when she realises that this result is quite compatible with it. Even if the observer measures the mass as z', provided only that $z' > z$ then $z = z'$ in $\mathcal{N}^1_n$, and their results are the same!

But this does leave a problem. Why do observers consistently record result that differ from each other? Analysing the data, Einquine sees that values of z (hers included), are related to the distance of the observer from the collision, d, by the (classical) equation $z = z_0 + kd$ (where z_0 and k are constants). In virtue of this, she revises her estimate of the fundamental constant to z_0, and hypothesizes that the effect of an inconsistent mass of baryons on a measuring device is a function of its distance from the mass. Further observational reports bear this hypothesis out; and Einquine starts to consider the mechanism involved in the distance-effect.

We could continue the story indefinitely, but it has gone far enough. For familiar reasons, there are likely to be theories other than Einquine's that could be offered for the data. Some of them might preserve orthodox arithmetic by jettisoning conservation laws, or by keeping these but varying some physical auxiliary hypotheses. Others might modify arithmetic in some other,

---

[22] We already know that different sorts of fundamental particles satisfy different sorts of statistics.

but consistent, way (which would be just as good for present purposes). An obvious suggestion here is that we might use, instead, a finite consistent arithmetic, with maximum number z_0, which is its own successor. (For all we have seen of the example so far, this would do just as well.) However, there might well be reasons that lead us to prefer the inconsistent arithmetic. Notably, this arithmetic gives us the full resources of standard arithmetic, whilst the finite arithmetic does not. For example, in the inconsistent arithmetic it is true that for any prime number there is a greater prime, which is false in the finite arithmetic. This extra strength might cause us to prefer it. Alternatively, the difference might even occasion different empirical predictions, which verify the inconsistent arithmetic. Indeed, whatever arithmetical theories we consider applying, these theories may become more or less plausible in the light of further experimentation, etc.

Just as with geometry, the question of the status of the applied arithmetic is another matter. One might tell a realist story about this. Collections of physical objects have a certain physical property, namely size; and sizes, together with the operations on them, have the same objective structure as the numbers and corresponding operations in the non-standard arithmetic. Hence, facts about the mathematical structure transfer directly to the physical structure, and this is why it works. Alternatively, one could tell instrumentalist stories of various obvious kinds. For example, Gasking (1940), runs a conventionalist line on arithmetic that is similar to Poincaré's on geometry. But it is not necessary to go into this here: the main point is made. It is quite possible that we might vary our arithmetic for empirical reasons, even to an inconsistent one. There can, therefore, be alternative applied arithmetics, just as there are alternative applied geometries.[23]

9 Non-Standard Logics in Science

It may fairly be pointed out that the revision of arithmetic envisaged in the previous section is more radical than was the adoption of a non-Euclidean geometry. For the arithmetic revision requires the adoption of a non-standard logic, whilst the geometric revision did not.

It might be argued that this shows such revision to be impossible after all. For logic, it might be thought, cannot be revised. Such an argument, though would be incorrect. Logic can be revised: it has been revised. The adoption of Frege/Russell logic early in the 20th century did just that. To give but one example, traditional logic counts the following inference as valid—it is an instance of the syllogism *D*arapti—whilst modern quantification theory does not:

[23] The revision of arithmetic I have just described is a local one, in that it is only the counting of *baryons* that is changed. It would be interesting to speculate on what might happen which could motivate a global change, i.e., a move to a situation where everything is counted in the new way.

All men are mortal.
All men are liars.
———————————
Some mortals are liars.

There is therefore nothing sacrosanct about logical theories.[24]

The observation, rather, cuts in the other direction. Science uses mathematics like a multi-national company uses the resources of an under-developed country. It steps in and helps itself to anything that it wants to use. Thus, scientists have always been mathematical opportunists. Whether the mathematics involved infinitesimals, imaginary numbers, bizarre functions (like the Dirac δ-function), or whatever, it mattered not whether this had an accepted logical foundation. If the mathematics gave the right answers, science employed it, and everything else went along for the ride. Thus, if an example of the kind described in the last section were ever to arise, it would *force* a revision of logic. (If it had not already been revised: there are good and quite independent reasons for revising our logical theory anyway.) Thus, none of the Kantian trilogy—geometry, arithmetic and logic—is immune from revision under the pressure of scientific advance.[25]

References

1. E.T. Bell: *Men of Mathematics* (Pelican Books, London 1953)
2. P. Benacerraf, H. Putnam (eds.): *Philosophy of Mathematics; Selected Readings* (Blackwell, Oxford 1964)
3. J.-P. van Bendegem: *Finite, Empirical Mathematics; Outline of a Model* (Rijksuniversiteit Gent, Gent 1987)
4. H.-N. Castañeda: 'Arithmetic and Reality'. Australasian Journal of Philosophy **37** (1959) pp. 92–107; reprinted in Benacerraf and Putnam (1964), pp. 404–71
5. N. Denyer: 'Priest's Paraconsistent Arithmetic'. Mind **104** (1995) pp. 567–75
6. J.M. Dunn: 'Relevance Logic and Entailment'. In: *Handbook of Philosophical Logic.* III/3. ed. by D. Gabbay, F. Guenthner. (Kluwer, Dordrecht 1986)
7. D.A.T. Gasking: 'Mathematics and the World'. Australasian Journal of Philosophy **18** (1940) pp. 97–116; reprinted in Benacerraf and Putnam (1964) pp. 390–403
8. R.L. Goodstein: 'Multiple Successor Arithmetics'. In: *Formal Systems and Recursive Functions* ed. by J. Crossley, M. Dummett. (North Holland Publishing Company, Amsterdam 1965) pp. 265–271
9. J.J. Gray: 'Euclidean and Non-Euclidean Geometry'. In: *Companion Encyclopedia of the History and Philosophy of Mathematical Sciences.* sect. 7.4; ed. by I. Grattan-Guiness (Routledge, London 1994)
10. I. Hinckfuss: 'Instrumentalist Theories and Space and Time'. In: *Natural Kinds, Laws of Nature, and Scientific Methodology* ed. by P. Riggs (Kluwer, Dordrecht 1996)

[24] The question of the revision of logical theories is discussed at much greater length in Priest (2000b), which paper partly overlaps with this one.

[25] An earlier version of this paper was given at the European meeting of the Association of Symbolic Logic, held at Donostia/San Sebastián, July 1996.

11. C. Mortensen: *Inconsistent Mathematics* (Kluwer, Dordrecht 1995)
12. G. Nerlich: *The Shape of Space* (Cambridge University Press, Cambridge 1976)
13. H. Poincaré: *Science and Hypothesis* (Dover Publications, New York 1952)
14. G. Priest: 'Two Dogmas of Quineanism'. Philosophical Quarterly **29** (1979) pp. 289–301
15. G. Priest: *In Contradiction* (Martinus Nijhoff, The Hague 1987)
16. G. Priest: 'Is Arithmetic Consistent?' Mind **103** (1994) pp. 321–31
17. G. Priest: 'On Inconsistent Arithmetics: Reply to Denyer'. Mind **105** (1996) pp. 649–59
18. G. Priest: 'Inconsistent Models of Arithmetic; I Finite Models'. Journal of Philosophical Logic **26** (1997) pp. 223–35
19. G. Priest: 'Inconsistent Models of Arithmetic: Part II, the General Case'. Journal of Symbolic Logic **65** (2000) pp. 1519–1529
20. G. Priest: 'Non-Euclidean Geometry and Non-Standard Logic: a Tribute to Łukasiewicz'. In: *Proceedings of the Conference Łukasiewicz in Dublin* ed. by M. Bagrahmian; to appear.
21. C. Wright: *Frege's Conception of Numbers as Objects* (Aberdeen University Press, Aberdeen 1983)

Remarks on Criteria of Truth and Models in Science

Giuseppe Del Re

1 Premise

The aim of this paper is a contribution to discussions on the role of logic in the sciences of Nature from the side of the scientists. The remarks given below are justified by the consideration that the logicians are expected to provide procedures and rules, or at least critical analyses capable of helping scientists to make sound argumentations, especially to distinguish between mere conjectures and reliable explanations. Now, in the author's opinion, there are aspects of the nature of scientific truths which were ignored during the golden age of physicalism and now call for attention. Among these aspects there are the role of analogiesand the distinction between reasonable and true statements.

2 Logic for Science

2.1 What We Mean by Logic Here

Science attains knowledge by sensible (experimental) operations and mental operations. We are interested here in mental operations, particularly the procedures by which inferences are made in science.

With no pretence to originality[1], we shall take logic to be, in a broad sense, the metascience which studies the *mental* operations involved in the attainment of a truth judgement, in particular

1. the *procedures* by which we try to obtain knowledge from premises and data;
2. the *criteria and rules* by which we decide that the knowledge obtained is valid – i.e., that the *propositions* of which it consists are "true".

[1] The disputes about logic and its nature are beautifully presented by John Passmore: *A Hundred Years of Philosophy* (Penguin, London 1968), especially in chaps. 6–8.

This definition is akin to that of argumentative logic,[2] but, as mentioned, it is not intended to compete with those of Frege, Quine, Kripke, and others; it is only offered to let the reader know what the author means here by logic. It allows, at any rate, for the existence of alternative logics. More importantly, it applies to cases in which one should really speak of verisimilitude rather than truth, for the validity of a statement is identified with its *reasonableness*. If logic (in a broad sense) is to be concerned not only with rigid, though possibly probabilistic rules, but with reasonable arguments leading to moral certainty, then the scientists might profit from knowing if and how its rules apply in this connection.

More light from the specialists of logic would be welcome particularly in the existence judgments of science. It is often taken for granted that the end result of scientific thinking is either a reliable description of facts or a set of propositions in the form 'if A then B', where A and B are events in the sensible world. A probabilistic logic will be somewhat broader by allowing for propositions such as 'if A then either B or C', and possibly use quantifiers to give propositions a richer content. But science, nowadays, is also issuing a large number of *existence* statements, e.g., 'pulsar NP0532 exists and is in the Crab Nebula' or simply 'there are invisible stellar objects with a peculiar radiowave emission in regular bursts, to be called pulsars.' Existence statements such as these, since they too are results of mental operations, require an *ad hoc* critical analysis.

3 Truth in Science

3.1 Metaphors, Models, and the Direct-Access Level

Even in the frame of strong realism – the belief that there are things and beings existing independently of us and intelligible to us – the process of understanding should consist, as Latin *comprehendere* and *intelligere* signify, in garnering, classifying, and storing information with related data already present in the mind. In other words, understanding is information processing ending in the hierarchical storage of the output according to the built-in structure of the human mind, though keeping a record of whatever hierarchy comes from out there'.[3]

---

[2] Cf., e.g., the article by Chaim Perelman: 'Logica'. In: *Enciclopedia del Novecento* (Istituto dell'Enciclopedia Italiana, Roma 1975) III pp. 1046–1054, and refs. therein.

[3] Kant's notion that the *Ding an sich* is unknowable is probably obsolete, for, especially after the achievements of computer science, it now appears difficult to doubt that information processing in our mind obeys the laws governing communication in sensible reality. Even 'cognition scientists' now seem to admit that the real problem lies in consciousness; cf. David Chalmers: *The Conscious Mind: In Search of a Fundamental Theory* (Oxford University Press, Oxford 1996).

A second point is that understanding proceeds via 'mental constructs,' and there are grounds for the conjecture that those constructs are essentially images[4] – perhaps what scholasticism calls 'phantasmata.' An important point in favour of this conjecture is a well-known but often neglected fact. As Ernest Weekley, a British grammarian, put it:[5]

> Every expression that we employ, apart from those that are connected with the most rudimentary objects and actions, is a metaphor, though the original meaning is dulled by constant use.
> Thus, in the above sentence, *expressions* means 'what is squeezed out', to *employ* is 'to twine in' like a basket maker, *to connect* is to weave together, *rudimentary* means 'in the rough state', and *an object* is 'something thrown in our way'.

Images are essential especially for science, which has the difficult task of studying the properties of huge objects such as the remote star Betelgeuse and extremely tiny objects such as an electron. That Betelgeuse should be a ball of fire can be somehow deduced from its appeareance, especially after the Hubble telescope has succeeded in resolving it as a disk. But what we know of it is actually a small two-dimensional picture obtained by a telescope, and the claim that it is a ball of nuclear matter with such and such properties is justified by analogieswith our ordinary experience. Concerning the electron, suffice it to mention that its 'spin' is always presented with comments to the effect that it is not really a rotation of the electron on its own axis; but the very word betrays its origin – the image of the electron in the mind of the physicist as a tiny (some add, *brilliant* green) ball.

Perhaps the most instructive visual model of this kind is the stick-and-ball model of a molecule, on which the concept of molecular structure rests.[6] In the course of less than one and a half centuries, this model has allowed the chemists to understand and prepare molecules of all kinds, from benzene derivatives to DNA. Indeed, the discovery of the genetic code was entirely based on it, and was obtained by a path completely independent of physics, though made easier by physical methods of analysis and detection. Now, the stick-and-ball model, as its name says, is a structure in ordinary space at the ordinary scale. Of course, the chemists know that it cannot be treated as a mere magnification of a molecule: but it provides the image the chemists have in mind when they study a chemical substance. In other words, the

[4] Cf., e.g., the papers by D. Ingvar and by V.B. Mountcastle: *The Principles of Design and Operation of the Brain* In: J.C. Eccles, O. Creutzfeld (eds) (Pont. Acad. of Sciences, Vatican City 1990), p. 433ff, p. 523ff.

[5] Quoted by Owen Barfield: 'The Rediscovery of Meaning'. In: *Adventures of the Mind* (second series) ed. by Richard Thruelsen, John Kobler (Random House/Vintage, New York 1961).

[6] G. Del Re: 'The Ontological Status of Molecular Structure'. Hyle, An International Journal for the Philosophy of Chemistry (Karlsruhe 1998) **4**, pp. 1–23.

arguments of the chemists are essentially referred, albeit with corrections and reservations, to objects belonging to the realm of objects that man knows by direct sensible experience, particularly by sight and touch.

We shall call this realm 'man's direct-access level of reality'.[7] Our contention is that most explanatory arguments of science, when referred to objects not belonging to this level, are arguments by analogy involving images associated with models in the direct-access level.

3.2 Science and Arguments by Analogy

Visual models have become especially important in science ever since the time of Galileo. The historical impact of the dispute between Galileo and the Roman theologians has made their posterity forget that the telescope and the microscope contributed to the great jump in scientific knowledge no less than the principle that facts should be the ultimate criteria of truth. Perhaps the only one amongst the greatest scientists of the seventeenth century to grasp the metaphysical significance of the new tools was Blaise Pascal. In his famous thought on 'the two infinities' he clearly called attention to levels of reality associated with size.[8]

Let us illustrate in more detail the following claim that, as soon as we are dealing with objects that do not belong to our direct access level, our minds appear to work via analogieswith suitable objects (models) belonging to our direct-access level. Even a mircoscopic mite is considered an insect only because it appears to have, *under the microscope*, the characteristics of macroscopic acari, and to respond like them to changes in its environment. At the astronomical scale, as mentioned, the same happens; indeed most 'observations' actually consist of photographs, radiotelescope data, and the like, so that theoretical arguments play an essential role in the derivation of statements about the 'reality' to which they refer. These theoretical arguments are based on models, in the above sense of objects belonging to the direct-access level of reality serving as terms of analogy.

The importance of this consideration can be especially seen in the case of submicroscopic objects, where its implications are extremely serious. The best known case is particle–wave dualism. This arises from the fact that the physicist wants to *describe* a particle in order to set up experiments. In view of such a description they need direct-access models: and it turns out that two such models are necessary, one being a wave train in a pool, the other a tiny sphere. The famous wave-corpuscle dualism may thus be seen as just the statement that a single direct-access model is not satisfactory for analogical arguments.

[7] The specification 'man's' is needed because other living beings, e.g., bacteria, may have a different direct-access level.

[8] Pascal, Pensées, Lafuma 199.72.

It would thus seem that the problem for a logic to be suitable for quantum particles – a problem already well studied, particularly by Weingartner[9] – should be restated in a way different from the current one: a quantum logic should be a logic for analogical argumentation, allowing for multiple analogies.

To show more concretely what we mean, let us consider the electron. Adapting classical examples of analogical argumentation, one could formulate the particle–wave dualism as follows:

a macroscopic wave is delocalized;
the electron can be described as a wave;
if this holds for all processes involving an electron,
then the electron is delocalized;

a macroscopic particle is localized;
the electron can be described as a particle;
if this holds for all processes involving an electron,
then the electron is localized.

As is well known, it was a great merit of Niels Bohr and the other great physicists of his time that they placed emphasis on the 'if' clauses above, and showed that the apparently incompatible conclusions were both 'true' as long as observability could be considered a criterion of truth.
The existence argument, in the case of the electron, runs as follows:

the negative electric charge, as defined in electrostatics, exists;
all phenomena involving negative electric charges can be explained
by assuming that the electric charge is carried by submicrosopic corpuscles;
these carriers, called electrons, are far smaller than any object
belonging to our direct access level of reality;
therefore, we can grasp their nature only by analogies.
Indeed, they are so remote from anything belonging to our ordinary
experience that they can only be described by two antithetic models.

3.3 Rational or Reasonable?

The term 'reasonable' as applied to a judgement is still, as far as the author knows, an open problem. For example, the well-known philosophical dictionary of A. Lalande says that this word implies essentially conformity to common sense and to generally accepted value judgements, an idea of moderation and right measure. At the same time, it gives (correctly) *vernünftig*

[9] Cf. his paper in this book.

as the German translation. There seems to be an ambiguity here. This ambiguity can be best shown by asking whether 'this is reasonable' is equivalent to 'this stands to reason' or 'there are reasons for this.' Clearly, in the former case reason means right logical thinking, in the latter 'reason' means 'grounds', German *Grund.* The ambiguity is in the Latin word *ratio*, and this proves that the two meanings must have much in common. On the other hand, *Vernunft* and *Grund* are not the same thing; an analysis of this point and related ones is to be found in the beautiful study by Heidegger on the *Der Satz vom Grund.*

We shall adopt here the following definition: a judgment is called reasonable if there are sufficient grounds to accept it without further discussion – i.e., to consider it true.[10]

We suggest here that, to the extent to which they are essentially based on analogies, the propositions of science that are normally treated as true are actually reasonable in the sense just stated, particularly when they are existence judgments concerning entities not belonging to man's direct-access level of reality.

Two aspects of reasonableness in the sense adopted here are probably where scientists studying submicroscopic objects most need the logicians' help: axiom quality and coherence.

The problem of axiom quality arises from the fact that the principles of the natural sciences must be axioms in the etymological sense, they should be propositions 'worthy of confidence'. Are there rules which can guide a scientist in the assessment of this quality? The analysis by Henry Poincaré of the principle of energy conservation (Meyer's law) is, in our opinion, a masterful example of the kind of analysis we are thinking of. The concluding remarks of that analysis give an idea of the logical problems involved:[11]

> ...je ne veux retenir de toute cette discussion qu'une impression, c'est que la loi de Meyer est une forme assez souple pour qu'on y puisse faire rentrer presque tout ce que l'on veut. Je ne veux pas dire par là qu'elle ne correspond à aucune réalité objective ni qu'elle se réduise à une simple tautologie, puisque, dans chaque cas particulier, et pourvu qu'on ne veuille pas la pousser jusqu'à l'absolu, elle a un sens parfaitement clair. Cette souplesse est une raison de croire à sa longue durée, et comme, d'autre part, elle [la loi de Meyer] ne disparaîtra que pour se fondre dans une harmonie supérieure, nous pouvons tra

[10] The problem of telling when a sientific statement is 'true' has been a major issue for the philosophers of science of the twentieth century, and it is superfluous to give references here. We confine ourselves to citing the study by Morton L. Schagrin: 'On Being Unreasonable,' *Philosophy of Science* March 1973, pp. 1–9.

[11] Henri Poincaré: *La science et l'hypothèse (1902)* (Champs-Flammarion, Paris 1968) pp. 148–149. The English translation is that published by Dover (New York 1952, no translator's name), with the word 'subtle' replaced by 'flexible', which is the correct equivalent of French 'souple.'

> vailler avec confiance en nous appuyant sur elle, certains d'avance que notre travail ne sera pas perdu. – *I wish to retain here only one impression of the whole of this discussion, and that is, that Meyer's law is a form flexible enough for us to be able to put into it almost anything we like. I do not mean that it corresponds to no objective reality, nor that it is reduced to mere tautology; since, in each particular case, and provided we do not wish to extend it to the absolute, it has a perfectly clear meaning. This flexibility is a reason for believing that it will last long; and as, on the other hand, it will only disappear to be blended in a higher harmony, we may work with confidence and utilise it, certain beforehand that our work will not be lost.*

All this can be summarized by saying that the principle of conservation of energy is 'reasonable,' i. e. that there are good grounds for believing that it corresponds to a feature of reality. But what exactly is the argument? Poincaré did not know, and, as far as we know, the ontological status of such a fundamental principle as conservation of energy is no better understood today. It seems to us that a rigorous study of its logical status could be of great help.

The question of coherence is another open question. When applied to an argumentation, the adjective 'coherent' is usually considered a synonym of 'rational', or 'logical'. This is probably a simplification, especially when analogies are involved; in fact, more emphasis on coherence (seen as 'systematic consistency') might be illuminating in connection with the procedures of contemporary science. Besides, the modern concept of coherence should probably conform to that which has been elaborated by physical optics. In terms of photons, coherent light is light consisting of photons which have the same frequency, phase, and direction. More generally, scientists tend to speak of coherence whenever there is close cooperation of parts. This means that in nature coherence is not a set of cause-effect relations. In practice, it may be treated as a kind of regularity that corresponds to a relation in space as well as in time that is possibly due to a common efficient cause, but is not in itself a system of cause-effect relations in the current sense. We could say that two independent chains of cause-effect pairs of events (processes) are coherent if there is a nonrandom relation between pairs of events belonging each to one of the two processes. Now this notion of coherence is expected by the scientists to apply to their (analogical) argumentations and to provide support for claims of truth, especially of existence judgements. Truth as coherence was widely discussed about a century ago, particularly by Bernard Bosanquet. Our remarks differ from those of Bosanquet and others in that we are not proposing solutions to any problem in philosophy or logic. We are only claiming that, in the practice of scientific research, truth is reached as moral certainty supported at the same time by chains of implications and proofs of consistency. This claim is not in itself a novelty; where, in our opinion, it is actual and worth further attention is the fact that the analogical nature of scientific statements concerning objects on levels of reality other than man's direct-access level does not seem to have been clearly realized. The question is, can the logicians establish the rules and limitations of current argumentations of science on what there is and on what really happens in those levels of reality?

Significant? Not Significant? The Dilemma of Statistical Induction in Scientific Research

Italo Scardovi

1. In 1876, Charles Darwin terminated a series of experiments that had lasted eleven years. During the experiments, he had compared groups of self-fertilised plants and groups of cross-fertilised plants in successive generations of different species. These lengthy experiments, whose original aim was the detection of possible differences in fertility between the two groups, produced an unexpected result unexplainable by the dominating inheritance theory. The result was "blending inheritance": in an adult state, the cross-fertilised plants were longer, on average, than the self-fertilised plants.

The differences in the averages were present in all the observed species; however, the individual variability within the groups caused partial overlaps, towards the extremes: the cross-fertilised plants showed greater linear development than the self-fertilised plants, but the stems of the latter were longer than those of the former. In other words, there was a transvariation between the distributions, which called for particular caution in the interpretation of the respective average values.

In the book dedicated to the discussion of these results (The Effects of Cross- and Self-fertilisation in the Vegetable Kingdom, 1876), Darwin questioned the validity of his experiments, in which he had paid careful attention to ensure the greatest possible uniformity in the groups of plants compared in the experiments, allowing that they differed in the one relevant factor for the experiment: the reproductive mechanism. Darwin realised that he had to rescue that data from the most deceptive of logical commitments for an experimental result expressed by natural variability: the suspicion (the smaller the groups observed, the more consistent the suspicion) that the difference between the mean values could not be "significant", or in other words, that it could not be distinguished from what is known in statistics as a "sampling error" (a random error that tends to diminish as the groups involved are enlarged).

Darwin knew better than anyone that the quantitative features of any form of life are the expression of an intrinsic variability: the same synchronic variability in which he had ingeniously identified a necessary condition for the evolution of the species. Origin of Species, first published in 1859 and then in its 6th edition, had given sufficient and well-discussed proofs of this. In the silence of his monastery Gregor Mendel (1865), unknown to (almost everyone) had discovered the secret combinatorial algorithms of that variability, of its genesis.

2. The Darwinian experiences reflect the classic paradigm of the experimental sciences: the comparison between a situation in which a given factor acts and a situation where it does not act. From a naturalistic point of view, the history of science gives us several examples of this paradigm, starting from research on "spontaneous generation"; from the observations of Francesco Redi (1665) to those of Lazzaro Spallanzani (1765), of Louis Pasteur (1862) and John Tyndall (1876). All these were comparisons between experimental situations that differed in the respective presence and absence of one relevant circumstance; classic experiences in which we can immediately recognise the canon that in John Stuart Mill's System of Logic (1843) is given the name "method of agreement and difference"; "always-never"-type experiences aimed at resolving a "true-false" alternative. Francesco Redi, a Galilean naturalist, compared two groups of twenty-four flasks each one containing different organic tissues. The two groups were different in one important aspect: all the flasks of one group were open, whereas the flasks of the other were closed by a thin layer of gauze. The appearance of fly larvae in the open ones and the complete absence in the closed flasks brought Redi to refuse the generatio *ex putri* theory and to consider Harvey's assumption *omne vivum ex ovo*.

The fragments in the individual flasks belonged to different species, according to the criterion of the variety of the examples, but they could, however, be reduced to a common name, thus representing repeated manifestations by of the same reality: organic tissue in decomposition. The plurality of the observations in the two groups therefore served to give inductive strength to the inference, but it was to be reconsidered according to the subjective and inter-subjective context of the "confirmation". The same can be said, passing from insects to infusorian protozoa, of the experiments performed a century later by Lazzaro Spallanzani: nineteen "boiled jars" compared to nineteen nonboiled jars. Spallanzani's experiment was another "always-never" experiment and the plurality of the jars, within each group, served to confirm the proof of an invariant property.

If not all Redi's closed flasks had been shown to be free from insects and not all Spallanzani's boiled jars had revealed to be each free from infusorians, the "agreement" within each group, as essential as the "difference" between the two groups, would have been lacking. Redi-type experiments concern properties common to all the single observations and one could say, in Virgil's words, "ab uno disce omnes". Virtually – David Hume would say "rationally" – when comparing just two flasks, one open and one closed, the same difference between the two experimental situations would have emerged and thus would have suggested the refusal of the theory of spontaneous generation.

The experimental condition of Darwin's botanical observations is different, as the empiric context is different: quantitative and variable. The plurality of the plants within the observed groups was not the mere repetition of the same characteristic; it was the expression of the intrinsic variability of the quantitative feature observed, i.e., the length of the stem, which varied from plant to plant. Here, the plurality of the observations is essential for statistical properties to come, as properties of the whole of the phenomenon and should, be reconsidered first of all within the context of the "discovery", to use Reichenbach's term. The repetition of the observations, i.e., the plurality of the plants, therefore plays an objective, heuristic role: it pertains to the very nature of the phenomena and the comparison is made between the summarised quantity of the lengths observed within the groups. Once again, the canon of difference, but with a different meaning. Averages summarise and conceal, just as any statistical value, the graduation of the values that make them up. This is the fundamental difference between Darwin-type and Redi-type experimental situations.

Claude Bernard had also stumbled across individual variability some time before, in founding a "medicine experimental" (1865); and had encountered considerable difficulties in reducing his quantitative physiology to the deterministic assertions of a whole tradition. Darwin, on the contrary, showed an anticipatory methodological consciousness of the problems and the principles of statistical inference. He knew that the groups under observation, or "statistical samples", as we now call them, were but limited subsets of vaster reference groups and he instinctively realised that individual variability acts on these subdivisions, making them more unstable the smaller they are. Hence his caution in drawing general conclusions from the differences observed in his samples.

These may seem, and today to a great extent they are, obvious considerations. However, without clarifying certain premises regarding phenomena and concepts, there is a risk of not being able to distinguish problems that are fundamentally different. Of not distinguishing, for example, the probability of an induction (the well-known problem of Hume) from the induction of a probability (the problem of statistic inference); of not distinguishing between the principle uncertainty of any nondeductive inference and the inductive strength of the statistic properties of a whole.

3. Going back to Darwin's doubt on the probative worthiness of his experimental results, tricked by the transvariant variability, that he solves by posing a question to his cousin, Francis Galton, an outstanding member of the "biometric movement". With what criterion, this is the question, can one judge "significant" the difference between the average length of the self-fertilised plants and the length of the cross-fertilised plants, when one knows that two groups of plants, however they may be taken, will always have different mean values for one quantitative characteristic or another? "Significant"

is the word that was soon chosen in research language to mean "significant of the presence of at least one factor not due to random sampling".

This is the protoscheme of all the problems to which the "statistical theory of significance" caused, among the biometricians, by Darwin-type questions would have given a first methodological solution. Today, scientists are well aware that any statistical property of a variable characteristic observed in a sample is always subject to fluctuations due to the randomness of sampling, i.e., to the effects of so-called "sampling error". This is the origin of the question, to which Darwin's dilemma offered a rational paradigm: a true-false alternative, where "true" is hidden within great numbers and "false" can be confuted, albeit without certainty.

An important methodological answer is R.A. Fisher's theory. The suspicion that the properties observed in a sample cannot be generalised beyond its context is expressed by Fisher with the "null hypothesis", and it is with respect to this that one poses the significant-not significant dilemma, one of the two propositions is true according to the classical principle of the dilemma *tertium non datur*. However, in the context of the phenomena expressed by individual variability, the alternative cannot be resolved if not in terms of probability.

It was Fisher who established, as early as 1925, formal criteria for testing statistical significance according to a procedure that can be reinterpreted as the "hypothetical-deductive canon". The scheme remains that of the classic hypothetical assumption, "Ifthen", but the following is not univocal, but plurivocal. The "antecedent" is the null-hypothesis: it assumes that the difference between two compared samples is pure randomness and as such, cannot be generalised; and the "consequent" is the space belonging to all the results compatible with this hypothesis: an exhaustive and disjunctive plurality over which the probability of the certain event is divided. The rule is to consider the null-hypothesis refuted if the observed result (usually the difference between two means) is found within the interval of values to which it is attributed, in the variable representative of the plurality of the consequences, a probability inferior to a given conventional limit. An inconclusive confutation, inspired by a principle, which is still a dilemma: "either a rare event will take place, or the null-hypothesis is not true".

4. In experimentation, it has become a bad habit to make standardised and automatic use of this methodology, thus distorting its original purpose. For Fisher, the theory of significance fulfilled the necessity of not underestimating the risk of drawing unfounded conclusions and it aimed at adding a further element of judgement to the many that a researcher uses on his results; obviously, it is not the only element and not necessarily the conclusive one. Once the null-hypothesis has been dismissed in probability, temporarily overcoming one of the horns of the dilemma, the heuristic dialogue between experience and reason, reality and fantasy starts. It goes without say that the scientist will employ other elements of judgment, in the first

place, "prior knowledge", according to the inferential theory that replaces the Bayes-Laplace theorem. However, prior knowledge can have an illuminating value and, in general, it does in the science that Kuhn calls "normal", i.e., the science that explains a paradigm, not the so-called "revolutionary" science, that overturns a paradigm. This is the case of Darwin's examination: "blending inheritance" was certainly rather unenlightening prior knowledge.

Over the years, an important set of rules has been drawn up, concerning the techniques of the theory of significance, that went far beyond the original scientific purposes. Drafted by Jerzy Neyman, it has been translated into a formal, coherent apparatus that is useful in all repeated sampling procedures; for example, in the systematic checking of serialised industrial production considered to provide rules for "inductive behavior" that in the long run, are capable of ensuring a number of incorrect conclusions that do not exceed a calculable limit. This logic that inspires a preference strategy between possible decisions, opening to sample examination a whole world of actions and choices.

However, strategy is one thing, and knowledge something else; statistical induction for convenience is one thing and statistical induction as a knowing act albeit probable, is another; it is one thing to assume a decision in the alternative between different degrees of the expected utility, it is quite another to test a hypothesis in scientific research. Two different ways of dismissing the dilemma of statistic inference, both of them coherent with their respective dominions of validity. The prevalence, even from a cultural point of view, of economic pragmatism, so typical of the times we live in, has lead us to the confusion of different ends and different instruments, although Fisher warns – in his many writings on the subject, also in Statistical Methods and Scientific Induction (1955) – not to reduce the proof of a hypothesis to an automatic procedure of acceptance and not to forget that the analysis of significance always implies critical attention to the experimental situation in question.

Unfortunately, we have ended up confusing, in the same methodology, the proof of a hypothesis and a batch of goods, mistaking science for convenience, truth for utility. It is indeed a question of convenience that imposes the refusal of the null-hypothesis, according to Neyman's canon, with the mere aim of advantaging a predefined alternative hypothesis. As though Darwin, refusing the hypothesis of "sampling error" should in itself claim victory for "blending inheritance", from which he wisely steered clear.

The methodology inspired by the Bayes-Laplace theorem, taken as an algorithm of inductive inference, contrasts with Neyman's criterion that one cannot choose between two alternative hypotheses without considering beforehand their two degrees of preliminary admissibility, or the "prior" probability, of the hypotheses themselves. And this is true. However, the Neymanian school replies that in a repetitive sample, or rather in the long run, the "prior" probabilities tend to lose their importance. And this is also true. However, that long run is not Darwin's experience and, in general, of the scientific research on quantitative and variable phenomena. Bayesian inference becomes indispensable when the empiric observations are very scarce or indeed concern a single event. However, the single event is not typical of scientific research; it is typical of strategic decision.

These brief notions, I believe, are sufficient, without going into formal stration, to take into account the various schools of thought on the dilemma of statistical inference. Each one has its own internal coherence regarding contexts and aims. And they should therefore not be confused. They all hide a trap: an uncritical use of particular techniques, as though they were rules to substitute reason.

References

1. T. Bayes: 'An Essay Towards Solving a Problem in the Doctrine of Chances'. Philosophical Transactions (London, December 1763)
2. C. Bernard: *Introduction à la médecine expérimentale* (J.B. Baillière, Paris 1865)
3. C.R. Darwin: *The Effects of Cross and Self Fertilization in the Vegetable Kingdom* (Murray, London 1876)
4. R.A. Fisher: *Statistical Methods for Research Workers* (Oliver and Boyd, Edinburgh 1925)
5. R.A. Fisher: *The Design of Experiments* (Oliver and Boyd, Edinburgh 1935)
6. R.A. Fisher: *Statistical Methods and Scientific Inference* (Oliver and Boyd, Edinburgh 1956)
7. P.S. de Laplace: 'Mémoire sur la probabilité des causes par les événements'. Académie des Sciences Paris (1774) VI

Part II

Alternative Proposals

Outline of a Paraconsistent Category Theory*

Newton C.A. da Costa, Otávio Bueno, and Analice Volkov**

Summary. In this paper, we show how a certain kind of paraconsistent category theory can be developed. We also want to contribute to a better knowledge of Oubiña's and Bosch's work in category theory.

The aim of this paper is two-fold: (1) To contribute to a better knowledge of the method of the Argentinean mathematicians Lia Oubiña and Jorge Bosch to formulate category theory independently of set theory. This method suggests a new ontology of mathematical objects, and has a profound philosophical significance (the underlying logic of the resulting category theory is classical first-order predicate calculus with equality). (2) To show in outline how the Oubiña-Bosch theory can be modified to give rise to a strong paraconsistent category theory; strong enough to be taken as the basis for a paraconsistent mathematics which encompasses all classical mathematical results.

1 The Theory OB_0

In this section, we outline the Oubiña-Bosch theory of categories, which will be denoted by OB_0. The details of OB_0 can be found in Bosch [1964], Oubiña [1966], and [1969], and Volkov [1997]. What follows in this section is essentially Oubiña [1969] (and Volkov [1997]).

The underlying logic of OB_0 is the classical predicate calculus with equality, as formalised in Kleene [1952], but implication, equivalence, and conjunction are symbolised by $\rightarrow$, $\leftrightarrow$, and $\wedge$, respectively, and function symbols are allowed.

The primitive symbols of OB_0 are the following: (a) predicate symbols: $\subseteq$ (binary), comp (ternary), and funct (unary); (b) function symbols: o (ternary), I (unary), and P (unary).

* Thanks to Steven French for comments on an earlier version of this work.

** It is with great regret that we report that Analice Volkov passed away a few months after finishing her contribution to this paper. It was a sad and unfortunate loss to all of us.

Intuitively, the objects to which the theory OB_0 refers to are categories. After some developments, the notion of set is introduced by definition: a set is a special kind of category.

The postulates (axioms, axiom schemes, and rules of inference) are those of the classical predicate calculus (with equality and function symbols; see Kleene [1952]), plus the postulates P1–P23 introduced below.

1.1 Subcategories

$x \subseteq y$ is read x is a subcategory of y.

P 1 $(x \subseteq y \wedge y \subseteq x \leftrightarrow x = y) \wedge (x \subseteq y \wedge y \subseteq z \to x \subseteq z)$

Definition 1. $x \subset y =_{def.} x \subseteq y \wedge x \neq y$. *(If $x \subset y$, we say that x is a proper subcategory of y.)*

1.2 Union of Subcategories

P 2 $x \subseteq z \wedge y \subseteq z \to \exists v(v \subseteq z \wedge x \subseteq v \wedge y \subseteq v \wedge \forall u((u \subseteq z \wedge x \subseteq u \wedge y \subseteq u) \to (v \subseteq u)))$

Definition 2. *The category v of postulate P_3 is unique. It is called the union of x and y in z, and it is denoted by $x \cup_z y$.*

Theorem 1.
- $x \cup_z y = y \cup_z x$
- $x \subseteq y \wedge y \subseteq z \to x \cup_z y = y$
- $x \subseteq z \wedge y \subseteq z \wedge z \subseteq w \to x \cup_z y = x \cup_w y$
- $(x \cup_w y) \cup_w z = x \cup_w (y \cup_w z)$.

1.3 The Empty Category

P 3 $\exists y \forall x(y \subseteq x)$

Definition 3. *P_4 guarantees the existence of a category which is a subcategory of any category. This category is uniquely determined and is called the empty category. It is denoted by $\oslash_c$.*

Theorem 2. $x \subseteq \oslash_c \leftrightarrow x = \oslash_c$.

1.4 Objects of a Category

Definition 4. *unit(x)*$=_{def.} x \neq \oslash_c \wedge \forall u(u \subseteq x \to u = x \vee u = \oslash_c)$.

Definition 5. *x is an object of z* $=_{def.}$ *unit*$(x) \wedge x \subseteq z$.

1.5 Morphisms of a Category

Definition 6.
• *f is a strict morphism of z between x and y* $=_{def.}$ *(x is an object of $z \wedge y$ is an object of $z \wedge f \subseteq z \wedge \forall u(u \subset f \leftrightarrow u \subseteq x \cup_z y)$).*
• *x is a morphism object of z between x and x* $=_{def.}$ *x is an object of z.*

Definition 7.
• *$f\colon x \Leftrightarrow_z y$ is an abbreviation for f is a strict morphism between x and y in z, if $x \neq y$.*
• *$x\colon x \Leftrightarrow_z x$ is an abbreviation for x is an object morphism between x and x in z.*
• *f is a morphism of z is an abbreviation for $\exists x \exists y(f\colon x \Leftrightarrow_z y)$.*

comp(z, x, y) that we read x is composable with y in z, will be written $x\mathrm{comp}_z y$. Moreover, instead of $\mathrm{o}(z, x, y)$, we write $x\mathrm{o}_z y$.

Definition 8. *x and y are composable in z is an abbreviation for $x comp_z y \vee y comp_z x$.*

P 4 $(z \subseteq w \wedge x$ is morphism of $z \wedge y$ is morphism of $z \wedge x\mathrm{comp}_z y) \rightarrow (x\mathrm{comp}_w y \wedge x\mathrm{o}_z y = x\mathrm{o}_w y)$

P 5 $(x$ is morphism of $z \wedge y$ is morphism of $z \wedge x\mathrm{comp}_z y) \rightarrow (x\mathrm{o}_z y$ is morphism of $z)$

P 6 $(x$ is an object of $z \wedge y$ is morphism of $z) \rightarrow ((x\mathrm{comp}_z y \rightarrow x\mathrm{o}_z y = y) \wedge (y\mathrm{comp}_z x \rightarrow y\mathrm{o}_z x = y))$

P 7 $(f\colon x \Leftrightarrow_z x \wedge g\colon x \Leftrightarrow_z x \rightarrow f\mathrm{comp}_z g) \wedge (f\colon x \Leftrightarrow_z x \wedge g\colon x \Leftrightarrow_z y \rightarrow$ 'f and g are composable in z')

P 8 Let f, g, and h be morphisms of z. In this case, it follows that:
• $(f\mathrm{comp}_z g \wedge g\mathrm{comp}_z h \rightarrow (f\mathrm{o}_z g)\mathrm{comp}_z h)$
• $(f\mathrm{comp}_z g \wedge (f\mathrm{o}_z g)\mathrm{comp}_z h \rightarrow g\mathrm{comp}_z h \wedge f\mathrm{comp}_z(g\mathrm{o}_z h))$
• $(f\mathrm{comp}_z g \wedge (f\mathrm{o}_z g)\mathrm{comp}_z h \rightarrow (f\mathrm{o}_z g)\mathrm{o}_z h = f\mathrm{o}_z(g\mathrm{o}_z h))$

Theorem 3. If x and y are distinct objects of a category z, and $f\colon x \Leftrightarrow_z y$, then either $f\mathrm{comp}_z x$ or $f\mathrm{comp}_z y$, and only one of these two formulas is valid.

Definition 9. *f is a morphism of z from x to y (or $f\colon x \rightarrow_z y$)* $=_{def.}$ *$f\colon x \Leftrightarrow_z y \wedge f comp_z x$.*

Theorem 4. Let f and g be morphisms of a category z. Thus: $f\mathrm{comp}_z g \leftrightarrow \exists u \exists v \exists t(u, v, t$ are objects of $z \wedge g\colon u \rightarrow_z v \wedge f\colon v \rightarrow_z t)$ $f\mathrm{comp}_z g \rightarrow ((f\colon u \rightarrow_z v \wedge g\colon v \rightarrow_z t) \rightarrow (f\mathrm{o}_z g)\colon v \rightarrow_z t)$.

Theorem 5. $(f\colon x \rightarrow_z y \wedge f\colon x' \rightarrow_z y') \rightarrow x = x' \wedge y = y'$.

Definition 10. *If $f\colon x \rightarrow_z y$, then x is called the source of f, and is denoted by $\alpha(f)$, and y is called the sink of f, and is denoted by $\beta(f)$. ($\alpha(f)$ and $\beta(f)$ do not depend on the category z.)*

Definition 11. *$f\,comp\,g =_{def.} \exists z$ (f is a morphism of $z \wedge g$ is a morphism of $z \wedge f\,comp_z g$).*

P 9 $((f, g$ are morphisms of $z \wedge f$compg$) \rightarrow (f, g$ are morphisms of $w \wedge (f \circ_z g = f \circ_w g))) \rightarrow z \subseteq w$

1.6 The Category of Two Morphisms

P 10 unit$(x) \wedge$ unit$(y) \rightarrow \exists z(x \subseteq z \wedge y \subseteq z \wedge \forall f(f$ is morphism of $z \rightarrow f = x \vee f = y))$

Definition 12. *If unit(x) and unit(y), then $[x, y]$ is the category universally described by the axiom P14. When $x = y$, $[x, x]$ is denoted by $[x]$.*

Theorem 6. If unit(x), then $[x] = x$. If x, y are objects of z, then $x \cup_z y = [x, y]$.

Theorem 7. If z is a category and $f\colon x \rightarrow_z y$, then:
- x, y are the sole objects of z.
- f, x, and y are the sole morphisms of f.

1.7 The Postulates of Identity

We shall write I(x) as follows: I_x. Given the category x, I_x is a category associated with x, satisfying the conditions:

P 11 $(I_x = I_y \rightarrow x = y) \wedge (\text{unit}(I_x))$

1.8 The Postulate P_{17}

Let F(f) be a formula of OB_0 in which the variables x, y, and w do not appear, and let F_1 and F_2 be the following formulas:

F_1: $\forall x \forall y(x, y$ are morphisms of $z \wedge x\text{comp}_z y \wedge F(x) \wedge F(y) \rightarrow F(x \circ_z y))$

F_2: $\forall x(x$ is morphism of $z \wedge F(x) \rightarrow F(\alpha(x)) \wedge F(\beta(x)))$
Under these conditions, the following formula is an axiom of OB_0:

P 12 $\forall z(F_1 \wedge F_2 \rightarrow \exists w(w \subseteq z \wedge \forall f(f$ is morphism of $w \leftrightarrow f$ is a morphism of $z \wedge F(f))))$

The preceding postulate constitutes a kind of comprehension scheme: it guarantees that some categories can be formed, starting with a previous category.

Theorem 8. $\forall z \exists y \forall u(u$ is morphism of $y \leftrightarrow u$ is an object of $z)$.

Definition 13. *$O(z)$ denotes the category of the objects of the category z.*

1.9 Functorial Categories

We call functorial categories the categories of a particular kind. They are implicitly defined by special axioms, containing the unary predicate symbol *funct*. If x is a category, funct(x) means that x is a functorial category.

P 13 $(\text{funct}(x) \ \wedge \ y \subseteq x \rightarrow \text{funct}(y)) \ \wedge \ (\text{funct}(\text{I}x)) \wedge (\text{funct}(z) \ \wedge \ x$ is an object of $z \rightarrow \exists u(\text{I}_u = x))$

P 14 $\text{funct}(x) \wedge \text{funct}(y) \rightarrow \exists z(\text{funct}(z) \ \wedge \ x \subseteq z \ \wedge \ y \subseteq z \ \wedge \ \forall u(\text{funct}(u) \ \wedge \ x \subseteq u \ \wedge \ y \subseteq u \rightarrow z \subseteq u))$

Using axiom P_{19}, we easily define the functorial sum of two functorial categories x and y, which is denoted by $x\mathbf{V}y$.

Theorem 9. $x\mathbf{V}y = y\mathbf{V}x$; $(x\mathbf{V}y)\mathbf{V}z = x\mathbf{V}(y\mathbf{V}z)$; $x \subseteq y \rightarrow x\mathbf{V}y = y$, where x and y are functorial categories.

Definition 14. *If $x = I_u$, we denote u by $E(x)$.*

Theorem 10.
- $\text{E}(\text{I}_x) = x$.
- x is an object of a functorial category $\rightarrow \text{I}_{E(x)} = x$.
- x, y are objects of functorial categories $\rightarrow (\text{E}(x) = \text{E}(y)) \leftrightarrow x = y$.

1.10 Functors

The following two definitions are basic:

Definition 15.
- *f is a functor $=_{def.} \exists z(funct(z) \ \wedge \ f$ is a morphism of $z)$*
- *f is a functor of x into $y =_{def.} \exists z(funct(z) \ \wedge \ f\colon x \rightarrow_z y)$*
- *$f\colon x \rightarrow y =_{def.} f$ is a functor of x into y*

Definition 16.
If f is a functor, we pose:
- *$d(f) =_{def.} E(\alpha(f))$ (domain of f)*
- *$c(f) =_{def.} E(\beta(f))$ (codomain of f)*

Theorem 11. If f and g are morphisms of the functorial categories z and w, respectively, then $f\text{comp}g$ implies that $f\text{o}_z g = f\text{o}_w g$. Moreover, if z and w are functorial categories, then $z \subseteq w$ if and only if every morphism of z is also a morphism of w.

Definition 17. *If f and g are composable functors, we state: $f o g =_{def.} f o_z g$, where $z = f\mathbf{V}g$.*

1.11 The Category of Subcategories of a Given Category

The symbol P is unary and P(x) denotes the category of subcategories of x. The postulates governing P are as follows:

P 15 $\forall x(\text{funct}(\mathrm{P}(x)) \ \wedge \ (u \subseteq x \leftrightarrow \mathrm{I}_u$ is an object of $P(x)) \ \wedge \ (x \subseteq y \leftrightarrow P(x) \subseteq P(y))$

The expression '$\exists! x F(x)$' is an abbreviation for 'there is a unique x such that $F(x)$'.

P 16 $(x \subseteq y \ \wedge \ y \subseteq z \rightarrow \exists! f(f{:}\mathrm{I}_x \rightarrow_{P(z)} \mathrm{I}_y)) \ \wedge \ (\exists f(f{:}\mathrm{I}_x \rightarrow_{P(z)} \mathrm{I}_y) \rightarrow x \subseteq y)$

Definition 18. If $x \subseteq y$, then $\mathrm{I}_{xy} =_{def.}$ the f such that $f{:}\mathrm{I}_x \rightarrow_{P(y)} \mathrm{I}_y$. ($\mathrm{I}_{xy}$ is called functor inclusion of x into y.)

Theorem 12. $I_{xx} = I_x$.

P 17 $(x \subseteq z \ \wedge \ y \subseteq z \ \wedge \ g{:}\mathrm{I}_x \rightarrow \mathrm{I}_y \wedge \mathrm{I}_{xz} = \mathrm{I}_{yz} o g) \rightarrow (x \subseteq y \wedge \mathrm{g} = \mathrm{I}_{xy})$

Theorem 13. Given the category z, there exists a unique category M(z) $\subseteq P(z)$, satisfying the condition: u is a morphism of M(z) if and only if there exists a morphism $f \subseteq z$ such that $u = \mathrm{I}_f$. M(z) is called the category of morphisms of z.

1.12 Image by a Functor

Let A(g) and B(g) be the following formulas:

- A(g): g is a functor $\wedge \ d(\mathrm{f}) = d(\mathrm{g}) \wedge \mathrm{c}(g) \subseteq \mathrm{c}(f) \wedge \mathrm{I}_{c(g)c(f)} o g = f$
- B(g): $\forall h \ (\mathrm{A}(h) \rightarrow c(g) \subseteq c(h) \ \wedge \ I_{c(g)c(h)} o g = h)$.

Under these conditions, we have:

P 18 f is a functor $\rightarrow \exists g \ (\mathrm{A}(g) \ \wedge \ \mathrm{B}(g))$

Definition 19. *Let f be a functor and x its domain. The functor g of P 28 is called a functor restriction of f to its image. We call c(g) the image of f, which is denoted by $f(x)$.*

Definition 20. *Let f be a functor, x its domain and $u \subseteq x$. Thus, the image of u by f, denoted by $f(u)$, is the image of the functor $f o I_{ux}$, and the restriction of $f o I_{ux}$ to its image is called the restriction of f to u and $f(u)$.*

Theorem 14. If x, y, and u are categories such that $x \subseteq y$, and $u \subseteq x$, then the image of u by I_{xy} is u, and the restriction of I_{xy} to u and $\mathrm{I}_{xy}(u)$ is I_u.

1.13 Functors and Their Images

P 19 f is a functor $\wedge$ unit(u) $\wedge$ $u \subseteq$ d(f) $\rightarrow$unit$(f(u))$

P 20 f is a functor $\wedge$ b is morphism of d$(f) \rightarrow f(b)$ is morphism of c(f)

P 21 (f is a functor $\wedge$ g is morphism of d(f) $\wedge$ h is morphism of d(f) $\wedge$ $g$comp$h$) $\rightarrow$ ($f(g)$comp$f(h)$ $\wedge$ $f(g\text{o}_{d(f)}h) = f(g)\text{o}_{c(f)}f(h)$)

P 22 f, g are functors $\wedge$ $f$comp$g$ $\wedge$ $u \subseteq$ d$(g) \rightarrow f(g(u)) \subseteq fog(u)$

P 23 (f, g are functors $\wedge$ d(f) = d(g) $\wedge$ c(f) = c(g) $\wedge$ $\forall u$(u is morphism of d$(f) \rightarrow f(u) = g(u)$)) $\rightarrow f = g$

Theorem 15. (a) If f is a functor and x is an object of $d(f)$, then $f(x)$ is an object of $c(f)$. (b) If f is a functor and $h\colon x \rightarrow_{d(f)} y$, then $f(h)\colon f(x) \rightarrow_{c(f)} f(y)$.

Theorem 16. If f and g are functors such that $f$comp$g$ and $u \subseteq$ d(g), then $f(g(u)) = fog(u)$.

1.14 Sets

Definition 21. *The category x is called a set if x is unitary, functorial and $E(x)$ has only object morphisms. The functors between sets are called functions.*

Clearly, we have:

Theorem 17. If x is a category all of whose morphisms are object morphisms, then I_x is a set.

Definition 22. *$I_{\oslash c}$, which is a set, is called the empty set and denoted by $\oslash$. ($\oslash_c$ is the empty category, as we have already seen.)*

Definition 23. *$x \in y =_{def.} x$ is an object of $E(y)$.*

Definition 24. *$x \sqsubseteq y =_{def.} E(x) \subseteq E(y)$, where x and y are sets. ($x \sqsubseteq y$ is read x is a subset of y, or x is part of y.)*

Theorem 18. $x \sqsubseteq y \leftrightarrow \forall z(z \in x \rightarrow z \in y)$, where x and y are sets.

Theorem 19. If x and y are sets, then $x = y \leftrightarrow \forall z(z \in x \leftrightarrow z \in y)$.

Theorem 20. $\forall x(x \notin \oslash)$.

Definition 25. *x is a unitary set if unit $(E(x))$.*

Theorem 21. x is a unitary set $\leftrightarrow \exists! y(y \in x)$.

Theorem 22. If x is a unitary set, then its only element is $E(x)$. (x is an element of y means $x \in y$.)

Theorem 23. If unit(x), then since $\mathrm{E}(\mathrm{I}_x) = x$, I_x is a unitary set whose only element is x.

Definition 26. *If x is a unitary category, then I_x is denoted by $\{x\}$. Obviously, $x \in \{x\}$.*

Given the unitary categories x and y, there exists the category $[x, y]$ whose only morphisms are x and y. Therefore, $\mathrm{I}_{[x,y]}$ is a set; $z \in \mathrm{I}_{[x,y]} \leftrightarrow z = x \vee z = y$.

Definition 27. $\{x, y\} =_{def.} I_{[x,y]}$.

When x is a unitary category, $\{x, x\} = \mathrm{I}_{[x,x]} = \mathrm{I}x = \{x\}$.

Definition 28. $\langle x, y\rangle = \{\{x\}, \{x, y\}\}$ *(ordered pair).*

Theorem 24. If x and y are unitary categories, then $\langle x, y\rangle$ is a set. If, in addition, x' and y' are unitary categories, then $\langle x, y\rangle = \langle x', y'\rangle \leftrightarrow x = x' \wedge y = y'$.

Theorem 25. Let x be a category and let us denote by k the category $I_{O(P(E(x)))}$. Thus: k is a set $\wedge \forall u(u \in k \leftrightarrow \mathrm{unit}(u) \wedge \mathrm{funct}(u) \wedge \mathrm{E}(u) \subseteq \mathrm{E}(x))$.

Definition 29. $\wp(x) =_{def.} I_{O(P(E(x)))}$ *(set of the subsets of the set x).*

Theorem 26. Let x be a set. We have: $y \in \wp(x) \leftrightarrow y \sqsubseteq x$, for every set y.

1.15 Complete Categories of Sets

Definition 30. *u is called a complete category of sets (ccs) if the following conditions are met:*
(1) funct(u)
(2) x is an object of $u \rightarrow (y \in x \rightarrow y$ is an object of $u)$
(3) x is an object of $u \rightarrow x$ is a set
(4) x, y are objects of $u \rightarrow \{x, y\}$ is an object of u
(5) x is an object of $u \rightarrow \wp(x)$ is an object of u.

Since the empty category satisfies the preceding conditions, there exists a *ccs*. In the next definition, $\forall'$ and $\exists'$ are the quantifiers restricted to the objects of the category u.

Definition 31. *A ccs is called normal if we have, for any formula $R(x, y)$:*
$\forall y \exists' z \forall x(R(x, y) \rightarrow x \in z) \rightarrow \forall' w \exists' v \forall x(x \in v \leftrightarrow \exists y(y \in w \wedge R(x, y)))$.

Note. The above property corresponds to Bourbaki's scheme of selection and union (Bourbaki [1968], p. 69). In set theory, with the help of Bourbaki's scheme, we prove the separation scheme, the substitution scheme and the axiom of union.

1.16 ZF and OB_0

Let ZF' denote Zermelo-Fraenkel set theory without the axioms of infinity, regularity, and choice. It is not difficult to prove that the axioms of ZF' are verified in a normal *ccs*, when they are interpreted in an obvious way (see Oubiña [1966] and [1969], and Volkov [1997]).

Theorem 27. Any normal *ccs* constitutes a 'model' of ZF'.

If we impose the conditions that a normal *ccs* u satisfies infinity, regularity, and choice, we obtain a 'model' of ZF in OB_0. Conversely, if in ZF' we define the notions of category theory (see Oubiña [1966] and [1969]), then the axioms of OB_0 are satisfied in ZF', through a convenient interpretation.

Theorem 28. OB_0 can be interpreted in ZF'.

OB_0 is a powerful theory, and it can be taken as the basis for topos theory, the theory of sheaves, the functorial theory of structures, etc., including all extant, usual mathematics. Moreover, since it does not rely on sets, OB_0 provides us with a new ontology of mathematics.

Remark. Let us indicate an intuitive interpretation of OB_0 in terms of a first-order theory whose primitive concepts are those of domain, counter-domain, and composition (in the usual, intuitive categorial-theoretic sense;[1] see Hatcher [1982]); let us also adopt the usual senses of morphism and category.

The *categories* of OB_0 can be conceived of as "aggregates" of morphisms. It is worth noting that "aggregate" is not being used here in any set-theoretic sense. The notion of subcategory can be interpreted in a natural way as "aggregates" of morphisms, and for a given morphism $f\colon a \to b$, the subcategories are f, a, and b. The *union* is not the set-theoretic union, but is more general, involving morphisms. The *null category* has no morphism. The *unitary categories* are the identity morphisms, and the *objects* of a category are the identity morphisms of this category. Given a morphism f: $f\colon a \to b$, its objects are a, b, and $[a, b]$. The notion of being *composable* can be expressed in terms of domain and counter-domain; "f is composable with g in z" means that g's counter-domain is the same as f's domain, and f and g are subcategories of z.

The *identity* over a given category x (notation: I_x) can be interpreted as the identity functor over x. *Functorial categories* are "aggregates" whose morphisms are *functors*. In particular, functors are also functorial categories, since they correspond to "aggregates" with a single functor. Moreover, the objects of functorial categories are identity functors (of type I_x).

Given a particular x, let us denote by $E(x)$ the only u such that $I_u = x$. A unitary category X is a *set* if, and only if, X is a functorial category and

[1] Here we follow Lawvere's ideas. As McLarty points out: "Lawvere believes 'intuitive' categories and spaces and other structures are just as real (or more accurately, just ideal) as 'intuitive' sets" (McLarty [1990]).

$E(X)$ only has object morphisms. We can think of sets as the I_u with u only having objects.

The following question immediately arises: are the members of sets of OB_0 also sets of OB_0? Not necessarily; there may well be a set I_u such that u has an object I_x and x does not have only objects. Thus, sets of OB_0 do not correspond exactly to what we usually call sets. But we can introduce the notion of a *normal complete category of sets* (normal *ccs*). Every object of a normal *ccs* is a set and the elements of objects of a normal *ccs* are also objects of the normal *ccs*. Moreover, a normal *ccs* is closed by the pair set and the set of parts. A normal *ccs* is what most resembles, in OB_0, the universe of sets; in other words, it is easy to relate the notion of a universe of sets with the concept of normal *ccs*. Furthermore, given that the objects of a normal *ccs* are sets and so are the elements of objects, morphisms between objects are functions. Therefore, a normal *ccs* can be associated with the usual category *SET* (of sets and functions). Finally, if we add certain conditions to the definition of normal *ccs*, we can obtain a concept equivalent to the concept of *topos*.

2 The Paraconsistent Theory OB_1

In this section we outline a paraconsistent category theory OB_1. We stress the fact that the technical details are relevant, but the most important aspect of the present work is the method used to build OB_1. There are infinitely many possible paraconsistent systems, and each of them is developed taking into account the intended applications. However, our outline is enough to substantiate the philosophical conclusions of the last section of this paper.

The underlying logic of OB_1 is the first-order predicate calculus $\mathbf{C}_1^=$ (see da Costa [1974] and [1997]; on paraconsistent logic, see, for example, da Costa, Béziau and Bueno [1995]).

2.1 The Calculus $C_1^=$

$\mathbf{C}_1^=$ is a paraconsistent predicate calculus with identity. Its primitive symbols are the following: (1) individual variables, as usual; (2) predicate symbols; (3) function symbols; (4) connectives: $\rightarrow$, $\wedge \vee$, $\neg$, $\leftrightarrow$ (this last symbol is defined as usual); (5) the quantifiers: $\forall$, $\exists$ (6) individual constants (i.e., 0-place function symbols); (7) identity: $=$.

The syntactical notions, such as those of formula and term, are the usual ones. The postulates of $\mathbf{C}_1^=$ are classified into three groups: (1) the propositional postulates (which characterise the propositional calculus); (2) the postulates of quantification (which, with the postulates of $\mathbf{C}_1$, constitute the postulates of the predicate calculus $\mathbf{C}_1^*$); (3) the identity postulates. A^o is an abbreviation for $\neg(A \wedge \neg A)$.

Propositional postulates. They are those of the classical positive logic, plus the following:

$\neg_1) B^o \to ((A \to B) \to ((A \to \neg B) \to \neg A))$
$\neg_2) A^o \wedge B^o \to (A \to B)^o \wedge (A \wedge B)^o \wedge (A \vee B)^o$
$\neg_3) A \vee \neg A.$

We define $\neg^* A$ as $\neg A \wedge A^o$. ($\neg^*$ is a strong negation, and it has all the properties of classical negation.)

Quantificational postulates. They are composed of the quantificational postulates of classical positive predicate logic, plus:
$\forall^o)$ $\forall x(A(x))^o \to (\forall x(A(x))^o)$
$\exists^o)$ $\forall x(A(x))^o \to (\exists x(A(x))^o)$
K) $A \leftrightarrow B$, where A and B are congruent formulas in the sense of Kleene [1952], p. 153, or one of them is obtained from the other by the suppression of vacuous quantifiers.

Postulates of identity. The common ones.

The concepts of proof, deduction, (formal) theorem, etc., are clear adaptations of Kleene [1952].

A theory T is a set of sentences closed under deduction. T is trivial if it coincides with the set of all sentences; otherwise, T is said to be nontrivial. T is inconsistent if there exists a formula A such that A and $\neg A$ are both members of T; otherwise, T is said to be consistent. Roughly speaking, a logic is said to be paraconsistent if it can be the underlying logic of inconsistent but nontrivial theories.

Theorem 29. In $\mathbf{C}_1^=$ we have:
$\vdash A \vee \neg A$
$\vdash \neg\neg A \to A$
$\vdash A \vee \neg^* A$
$\vdash \neg^*\neg^* A \leftrightarrow A$
$\vdash A \to (\neg^* A \to B)$
$\vdash A \wedge \neg^* A \to B$
$\vdash (A \leftrightarrow \neg^* A) \to B$
$\vdash A^{oo}$
$\vdash A^o \to (\neg A)^o$
$B^o, A \to B \vdash \neg B \to \neg A$
$B^o, A \to \neg B \vdash B \to \neg A$
$\vdash (A \to \neg A) \to \neg A$
$\vdash \forall x A(x) \leftrightarrow \neg^* \exists x \neg^* A(x)$
$\vdash \exists x A(x) \leftrightarrow \neg^* \forall x \neg^* A(x).$

Proof. See da Costa [1974].

Theorem 30. The following schemes and rules are not provable in $\mathbf{C}_1^=$.
$\neg(A \wedge \neg A)$
$A \rightarrow (\neg A \rightarrow B)$
$(A \wedge \neg A) \rightarrow B$
$(A \wedge \neg A) \rightarrow \neg B$
$A \vee B, \neg A/B$
$A \rightarrow \neg\neg A$
$A \rightarrow B, A \rightarrow \neg B/\neg A$
$(\neg A \rightarrow B \wedge \neg B) \rightarrow A$
$\neg A \leftrightarrow \neg(A \wedge A)$
$\neg A \leftrightarrow \neg^* A$
$A \leftrightarrow \neg\neg A$
$\forall x A(x) \leftrightarrow \neg\exists x \neg A(x)$
$\exists x A(x) \leftrightarrow \neg\forall x \neg A(x)$.

Proof. See da Costa [1974].

$\mathbf{C}_0$, $\mathbf{C}_0^*$, and $\mathbf{C}_0^=$ denote, respectively, the classical propositional calculus, the classical first-order predicate calculus, and the classical first-order predicate calculus with equality.

$\mathbf{C}_1^=$ (and also $\mathbf{C}_1$ and $\mathbf{C}_1^*$) possess a semantics of valuations in relation to which $\mathbf{C}_1^=$(respectively, $\mathbf{C}_1$ and $\mathbf{C}_1^*$) is correct and complete. Informally, the semantics of valuations can be outlined as follows.

Let $\mathbf{A}$ be a set, and $\mathfrak{L}(A)$ the diagram language of $\mathbf{C}_1^=$ associated with $\mathbf{A}$. When $\mathbf{A} = \oslash, \mathfrak{L}(\mathbf{A}) = \mathfrak{L}$, the language of $\mathbf{C}_1^=$.

A valuation in $\mathfrak{L}(A)$ is the characteristic function of a certain kind of maximal nontrivial theory. A sentence F is true in a valuation v if $v(\mathrm{F}) = 1$; otherwise it is false, i.e., $v(\mathrm{F}) = 0$. The valuation v is a model of a set of sentences Γ if $v(\mathrm{S}) = 1$, for each S in Γ.

Let $\Delta \cup \{G\}$ be a set of sentences of $\mathfrak{L}$. We say that G is a semantic consequence of Δ, and write $\Delta \models G$, if every model v of Δ is such that $v(\mathrm{G}) = 1$, in all diagram languages $\mathfrak{L}(A)$. In $\mathbf{C}_0^=$, it is easy to verify that a valuation in $\mathfrak{L}(\mathbf{A})$ determines a first-order structure whose universe is A, and conversely. In the case of $\mathbf{C}_1^=$, a valuation in $\mathfrak{L}(\mathbf{A})$ similarly individualises a first-order structure, but given the structure, the valuation is not determined.

2.2 The Theory OB$_1$

Let F be a formula of $\mathbf{C}_1^=$. F* is the same formula F, in which the occurrences of $\neg$ are replaced by $\neg^*$.

The underlying logic of OB$_1$ is $\mathbf{C}_1^=$, and its primitive symbols are those of OB$_0$. The specific axioms of OB$_1$ are the following: if F is an axiom of OB$_0$, then A and A^* are axioms of OB$_1$. In general, every concept of OB$_0$ leads to two corresponding concepts in OB$_1$, one obtained with $\neg$ (weak negation), and another with $\neg^*$ (strong negation).

Theorem 31. Let F be a formula of the language of OB_0. Then, if F is a theorem of OB_0, F^* is a theorem of OB_1.

Proof. Consequence of the postulates of OB_0 and OB_1, and the fact that $\neg^*$ constitutes a classical negation (see da Costa [1986a]).

Theorem 32. OB_0 is contained in OB_1.

Proof. Consequence of the fact that $\mathbf{C}_1$ is a subcalculus of $\mathbf{C}_0^=$.

Theorem 33. OB_1 is a paraconsistent system.

Proof. If we add Russell's set to OB_1, i.e., the set $R = \{x : x \notin x\}$, where $x \notin x$ means $\neg(x \in x)$ (weak negation), the resulting system is not trivial (see da Costa [1986*a*]).

So, OB_1 is a very strong system, in which we can develop not only all classical mathematics, but also study the properties of "inconsistent" mathematical structures, such as Russell's set. In OB_1, we can construct other types of "inconsistent" structures, which are not set-theoretic in nature. For instance, we may introduce the definition

$$x \prec y =_{def.} x \text{ is not an object of } I_y,$$

and consider the category K such that

$$x \prec K \leftrightarrow \neg(x \prec x).$$

As a consequence, we have:

$$K \prec K \leftrightarrow \neg(K \prec K),$$

and

$$K \prec K \wedge \neg(K \prec K),$$

a certain form of Russell's paradox. This and other issues will be left to future works (including the full development of OB_1).[2]

3 Philosophical Remarks

After putting forward a paraconsistent formulation of category theory, we would like to discuss some philosophical issues raised by this construction as well as by paraconsistency in general.

---

[2] A paraconsistent category theory has been articulated by Mortensen (see Mortensen [1995]). However, OB_1 is enterely distinct from, and contains, Mortensen's construction.

The construction we have just presented make clear that, at least at the level of mathematics, there are inconsistent but nontrivial theories. In other words, there are theories T in which both a formula A and its negation $\neg A$ are theorems of T, and some formula B of T is *not* a theorem. In our view, this fact provides an important support for paraconsistency, since it shows that the attempt at accommodating inconsistencies by devising appropriate inconsistent but nontrivial theories is by no means empty or unrealisable. On the contrary, it provides a distinctive perspective to the issues under consideration. Instead of retaining classical logic, and avoiding the inconsistency by rejecting one or another of the premises which generate it – making more or less *ad hoc* moves – we retain the inconsistency, change the underlying logic to a paraconsistent one, and study the properties of the "inconsistent object" so "generated". The important feature, as the paraconsistent category theory discussed above shows, is that these "inconsistent objects" have certain determined properties and lack others: it is simply *not* the case that everything goes with regard to them. So, as opposed to what happens in the case of classical logic, there is a whole new domain of investigation determined by the formulation of paraconsistent logic: the domain of the inconsistent.

Now, the issue arises as to the status of the resulting theory: is it *true*? Can we say that there are true inconsistencies? The answer depends on several considerations. (1) What is the notion of truth used in this context? (2) What kinds of objects are we considering (mathematical objects or physical objects)? (3) What notion of existence is assumed? And how are ontological commitments to be spelled out? Of course, the examination of these issues involves particular philosophical accounts, and we cannot do better here than consider them in fairly general terms. But we hope to say enough in order to make clear the approach we favour with regard to them.

According to some authors, the answer to the question *'Are there true contradictions?'* is affirmative (see Priest [1987]). The examples given by Priest are the logical and semantic paradoxes, statements about moving objects (objects subject to change), and certain views in the foundations of mathematics. In order to articulate an "ontic" or "realist" view about true contradictions, Priest advocates (i) a strong notion of truth – truth *simpliciter* understood in the correspondence sense, (ii) a classical view about existence (as the range of bound variables), and (iii) an extended claim as to the domain of his theory – which incorporates both mathematical and physical objects. (Of course, in order to avoid triviality, Priest adopts a paraconsistent logic; see his system LP, see Priest [1987].) So Priest's approach countenances classical views about truth and existence, and applies them to a wide-ranging domain. In our view, Priest's commitment to several doctrines is by no means fortuitous: in order to be adequately accommodated, inconsistencies require a whole package of logical and philosophical doctrines (indeed, a research programme). Of course, there are stronger and weaker programmes, some are closer, some further from classical proposals.

It seems to us that, in retaining classical notions of truth and existence, Priest's proposal became committed to metaphysical views which are decidedly strong. Given the use of truth in the correspondence sense, and the claim that our claims about the world (be it the "empirical" or the "mathematical" world) are to be true, Priest's view is *ipso facto* committed to all objects which are postulated in these claims. In particular, his proposal is committed to "inconsistent objects" in the physical world: the objects to which our inconsistent but true physical theories refer. But how can their existence be established?

The argument to this effect assumes, of course, the classical account of ontological commitment: we are ontologically committed to whatever our bound variables range over. And, in the case of inconsistent theories, this criterion leads us to postulate objects which both have and lack a given property (for instance, the liar sentence is both true and false, Russell's set is both a member of itself and it is not etc.). The same goes for theories about the physical world.

In our view, this argument is not so conclusive. First, this criterion of ontological commitment is *not* independent of particular philosophical assumptions. It comes as part of a philosophical programme – Quine's view – and it has built into it, as it were, a given logic: *classical* first-order logic. It goes without saying that, as such, it is at odds with Priest's own dialetheic approach, in which a paraconsistent logic is advocated. Moreover, Quine's criterion is *not* independent of logic: if we change the underlying logic of a given theory, we change the entities we are quantifying over. This can be seen in several ways. If we move to second-order logic, we are allowed to quantify over predicates and relations. As a result of its strong expressive power, several mathematical theories can be better formulated (in particular, as is well known, arithmetic and analysis are categorical). Because of this, several nominalist proposals (such as those developed in Field [1980] and Hellman [1989]) have adopted second-order logic as part of their nominalisation strategies of science and mathematics. The idea is that, by increasing the strength of the logic, we can decrease our ontological commitments. Secondly, using paraconsistent logic, we are allowed to quantify over certain constructions (such as Russell's set) which are impossible in classical logic, given its identification of inconsistency with triviality.

Our point here is that Quine's slogan – to be is to be the value of a variable – can only have any force once a particular logic is admitted. Quine knows that, of course. The problem is that his view assumes a logic (classical first-order logic) which is not the most adequate to deal with inconsistencies in a heuristically fruitful way.

We suggest to address the inconsistency issue differently. We may well explore the rich representational devices allowed by the use of paraconsistency in inconsistent domains, but withholding any claim to the effect that there are "inconsistent objects" in reality. Whether the world is indeed "inconsistent"

– assuming that there is a sensible formulation of this claim – is something we would rather be *agnostic* about. Just as empiricists (such as van Fraassen [1980]) are agnostic about (the existence of) unobservable entities in science, we are agnostic about the existence of true contradictions in nature. And one of the reasons in support of this claim is an *underdetermination* argument. Given the hierarchy of paraconsistent systems presented in da Costa [1974], there are always infinitely many paraconsistent logics which can be used to accommodate a given "phenomenon" – whether it is an "inconsistent" reasoning or an "inconsistent" theory (see da Costa, Bueno and French [1998b]). Which of these paraconsistent logics reflects *the* logic of the world? There is no argument based on purely observational terms that could establish this in general. We can select, of course, one of these logics on *pragmatic* grounds, but these grounds are certainly not enough to establish a substantive claim about the world. For instance, if one of these logics make the modeling of the inconsistency in question easier, why should this be taken as a reason for this logic to be *true*? Simplicity may well be a sensible criterion to adopt on pragmatic grounds, but the claim that a logic selected on this basis is (likely to be) true, is to confuse pragmatic and epistemic issues. Why should the world conform to our cognitive limitations? Of course, it might well do. But to establish this demands an argument that goes beyond what we observe: it requires a metaphysical claim of the simplicity of reality. However, to a certain extent, this is as strong as the claim that there are true contradictions, in the sense that both make substantial assertions about the world that transcend empirical observation. So, both are metaphysical claims.

In our view, an alternative programme of interpretation of inconsistencies can be devised in which no commitment to this kind of metaphysics is required. The idea is first to avoid the claim that inconsistent theories are *true*; they are *quasi-true* at best (see da Costa and French [1990] and [1993], and da Costa, Bueno and French [1998*a*]). The notion of quasi-truth receives a straightforward formal treatment (see Mikenberg, da Costa and Chuaqui [1986], da Costa [1986*b*], and Bueno and de Souza [1996]). But for our needs it suffices to say that a sentence α is quasi-true if it models adequately only part of a given domain D, leaving open the "complete" description of the latter.[3]

[3] We add in a footnote some further details. The investigation of a certain domain Δ of knowledge involves the study of the relations among its objects. Nonetheless, the information about such objects is frequently "incomplete", since one does not know whether certain relations concerned hold for all the objects of the relevant domain (or for all the n-tuples thereof). In order to accommodate this situation, da Costa and French advanced the concept of a *partial relation*; that is, a relation which is not defined for every object, or n-tuple of objects, of its domain. More formally, we can characterise an n-place partial relation R over D as a triple $\langle R^1, R^2, R^3 \rangle$, where R^1, R^2, and R^3 are mutually disjoint sets, with $R^1 \cup R^2 \cup R^3 = D^n$, and such that R^1 is the set of n-tuples that belong to R; R^2 of those n-tuples that do not belong to R; and R^3 those n-tuples for which

it is not defined whether they belong or not to R. (Notice that if R^3 is empty, R is a standard n-place relation which can be identified with R^1; see da Costa and French [1990], p. 255, note 2.)

However, in order to accommodate our patterns of modeling information, something more than partial relations is required: an appropriate notion of *structure* should also be presented. This notion is similarly thought of as incorporating the "openness" typical of our epistemic situation, where we usually face "incomplete" information. Based on a partial relation, da Costa and French formulate the concept of a *partial* (or *pragmatic*) *structure*. It is a set-theoretical structure of the form $\langle D, R_i, P\rangle_{i\in I}$, where D is a nonempty set (representing the objects used in the systematisation of the relevant domain of knowledge Δ, whose study we are concerned with), $(R_i)_{i\in I}$ is a family of partial relations defined over D, and P is a set of sentences (representing what is known, or taken to be known, about D, including laws and observable statements).

In terms of partial structures, we can formulate a particular notion of truth, which extends Tarski's account, and leads to the characterisation of the concept of *quasi-truth*. Since the Tarskian semantics was constructed only for total structures, in order to use it, it is necessary that a total structure be obtained from a partial one, 'extending' its partial relations. Such 'extended' structures are called *normal structures*. More formally, given a partial structure $A = \langle D, R_i, P\rangle_{i\in I}$, we say that the structure $B = \langle D', R_i', P'\rangle_{i\in I}$ is an A-*normal* structure if: (1) $D = D'$; (2) every constant of the language in question is interpreted by the same object both in A and in B; (3) R_i' extends the corresponding relation R_i(in the sense that, as opposed to the latter, the former is defined for every n-tuples of objects of its domain); and (4) if $\alpha \in P'$, then α is true in B (in the Tarskian sense).

It should be noticed that, given a partial structure, it is not always the case that is possible to extend it into a normal one. Necessary and sufficient conditions for this result are as follows (see Mikenberg, da Costa and Chuaqui 1986). Given a partial structure $A = \langle D, R_i, P\rangle_{i\in I}$, for each partial relation R_i, we construct a set M_i of atomic sentences and negations of atomic sentences such that the former corresponds to the n-tuples that satisfy R_i, and the latter to those n-tuples that do not satisfy R_i. Let M be $\cup_{i\in I} M_i$. Therefore, a pragmatic structure A admits an A-normal structure if, and only if, the set $M \cup P$ is consistent. In other words, the extension of a pragmatic structure A to a total A-normal structure B is possible whenever the extension of the partial relations is made in such a way that the consistency between the new extended relations and the accepted basic propositions P is guaranteed.

Normal structures are introduced in order to present an interpretation of the language in question. This strategy was also employed by Tarski in his characterisation of the concept of truth: the latter was defined in a *structure*. And the same feature is employed in the formulation of the concept of quasi-truth. We say that a sentence a is *quasi-true* in a partial structure $A = \langle D, R_i, P\rangle_{i\in I}$, according to an A-normal structure $B = \langle D', R_i', P'\rangle_{i\in I}$, if α is true in B (in the Tarskian sense). If a is not quasi-true in S according to B, we say that a is quasi-false (in S according to B).

Notice that the concept of quasi-truth is an as *if* concept: if a sentence α is quasi-true, it describes the domain in question *as if* its description were true.

(Of course, in a precise sense, α is consistent with a true description of D.) In claiming that, with regard to inconsistent theories, all we need is to determine their quasi-truth, we are in position to provide a "formal underpinning" to our agnosticism with regard to true contradictions. We are suggesting a change in the notion of truth to the weaker notion of quasi-truth, withholding then the commitment to "inconsistent objects" (see also da Costa and French [1993]), given that there are several distinct (A-normal) structures which describe the domain under consideration, and such objects are not countenanced in all of them.[4] Moreover, we also suggest to revise Quine's slogan about ontological commitment, making explicit its dependence on the underlying logic (see also da Costa and French [1991]). In this way, it becomes clear that it is *not* the only criterion to adjudicate between alternative logics.

However, if we may not be committed to an ontology of actual inconsistent physical objects, are we committed to inconsistent *mathematical* entities? This depends, of course, on how we interpret the theory OB_1. Does it provide a *true* description of the inconsistent category-theoretical "world"? Again, if all we claim is to have described this "world" in at best quasi-true terms, we will not be committed to those entities. There is, of course, a whole story to be told here, but for the present purposes it suffices to note that it is possible to provide an entirely syntactic formulation of set theory, in which OB_1 can be embedded,[5] such that the only commitment is to a countable language. Thus, in a certain sense, we do not require a special commitment to mathematical objects either (see da Costa and Bueno [1998]).

So, the "package" we suggest to accommodate inconsistencies is characterised by (1) the claim that inconsistent theories are *quasi-true* at best; (2) an agnosticism with regard to the existence of true contradictions and a nominalism about inconsistent mathematical entities; and (3) a re-evaluation of Quine's view about ontological commitment, emphasising its dependence on the underlying logic.

In our view the striking feature of this "package" is its logical pluralism, on the one hand, and the fact that it allows us to make sense of the several uses and applications of paraconsistent logic with no commitment to actual "inconsistent objects". The logical pluralism derives from the claim (3) above.

Being consistent with the basic knowledge (represented by the set of accepted sentences P) of this domain, α grasps some of the main bits of information about it, without committing us to the acceptance of the further bits (formulated by the extended A-normal structure) as *true*.

[4] Descartes once remarked (in his *Principles*, iv, 204) "that, with regard to those things that our senses cannot perceive, it suffices to explicate how they can be". We would say the same with regard to inconsistent theories. The notion of quasi-truth allows us to formally accommodate this point, since it is strictly weaker than truth, and does not commit us to anything *beyond* the assertion that certain structures are *possible, vis à vis* a given paraconsistent logic.

[5] This construction has to be articulated in a paraconsistent set theory; that is, one in which the underlying logic is paraconsistent (see da Costa [1986*a*]).

Depending on the domain we are considering, different kinds of logic may be appropriate. For instance, if we want to model constructive features in mathematical reasoning, an intuitionistic logic is the best alternative; if we are concerned with inconsistent bits of information, the use of a paraconsistent logic is the natural option. In particular, we do *not* reject classical logic: it has its own domains and applications. To this extent, while dealing with distinct domains, paraconsistent logic and classical logic are complementary rather than rivals. (They become rivals only when applied to the *same* domain. The rivalry derives from the fact that they provide different accounts of the logical connectives.)

But in applying paraconsistent logic, as we have done when formulating OB_1, we do not have to be committed to the existence of "inconsistent entities" – this is the point of our claims (1) and (2) above. We can always use the resources provided by this logic only to help us in drawing consequences from inconsistent theories without triviality, but with no commitment to the *truth* of the theory in question; it can be at best quasi-true.

In this way, a noncommittal (agnostic) interpretation of paraconsistency can be presented, making sense of the application of paraconsistent logic articulated here. "Inconsistent objects", either mathematical or physical, can be accommodated without requiring an ontology which includes them. In particular, a paraconsistent category theory is suggested here, but it is not necessary to countenance the existence of the entities the theory is taken to be about.

References

1. J. Bosch: *Seminar on Differentiable Manifolds* [in Spanish] (Comissión de Investigación Cientifica de la Prov. de Buenos Aires, Buenos Aires 1964)
2. N. Bourbaki: *Theory of Sets* (Addison-Wesley, Boston, Mass. 1968)
3. O. Bueno, E. de Souza: 'The Concept of Quasi-Truth'. Logique et Analyse **153-154** (1996) pp. 183–199.
4. H. Burkhardt, B. Smith (eds): *The Handbook of Ontology and Metaphysics* (Philosophia Verlag, Munich 1991)
5. N.C.A. da Costa: 'On the Theory of Inconsistent Formal Systems'. Notre Dame Journal of Formal Logic **15** (1974) pp. 497–510
6. N.C.A. da Costa: 'On Paraconsistent Set Theory'. Logique et Analyse **115** (1986a) pp. 361–371
7. N.C.A. da Costa: 'Pragmatic Probability'. Erkenntnis **25** (1986b) pp. 141–162
8. N.C.A. da Costa: 'Paraconsistent Mathematics'. In: *Frontiers of Paraconsistent Logic*, ed. by D. Batens, C. Mortensen, G. Priest and J.-P. Van Bendegem, (Research Studies Press, Baldock, Hertfordshire, England; Philadelphia, PA 2000)
9. N.C.A. da Costa, J.-Y. Béziau, O. Bueno: 'Aspects of Paraconsistent Logic'. Bulletin of the Interest Group in Pure and Applied Logics **3** (1995) pp. 597–614

10. N.C.A. da Costa, O. Bueno: Nominalism and Syntactic Models of Set Theory. Unpublished manuscript, Department of Philosophy, University of São Paulo, and Department of Philosophy, University of Leeds (1998), forthcoming
11. N.C.A. da Costa, O. Bueno, S. French: 'The Logic of Pragmatic Truth'. Journal of Philosophical Logic **27** (1998) pp. 603–620
12. N.C.A. da Costa, O. Bueno, S. French: 'Is There a Zande Logic?'. History and Philosophy of Logic **19** (1998b) pp. 41–54
13. N.C.A. da Costa, S. French: 'The Model-Theoretic Approach in the Philosophy of Science'. Philosophy of Science **57** (1990) pp. 248–265
14. N.C.A. da Costa, S. French: 'Ontology and Paraconsistency'. In: Burkhardt and Smith (eds.) (1991) pp. 656–658
15. N.C.A. da Costa, S. French: 'Towards an Acceptable Theory of Acceptance: Partial Structures and the General Correspondence Principle'. In: French and Kamminga (eds.) (1993) pp. 137–158
16. H. Field: *Science Without Numbers: A Defense of Nominalism* (Princeton University Press, Princeton, N.J. 1980)
17. S. French, H. Kamminga (eds.): *Correspondence, Invariance and Heuristics: Essays in Honour of Heinz Post* (Reidel, Dordrecht 1993)
18. W.S. Hatcher: *The Logical Foundations of Mathematics* (Pergamon Press, New York 1982)
19. G. Hellman: *Mathematics Without Numbers: Towards a Modal-Structural Interpretation* (Clarendon Press, Oxford 1989)
20. S.C. Kleene: *Introduction to Metamathematics* (North-Holland, Amsterdam 1952)
21. C. McLarty: 'The Uses and Abuses of the History of Topos Theory'. British Journal for the Philosophy of Science **41** (1990) pp. 351–375
22. I. Mikenberg, N.C.A. da Costa, R. Chuaqui: 'Pragmatic Truth and Approximation to Truth'. Journal of Symbolic Logic **51** (1986) pp. 201–221
23. C. Mortensen: *Inconsistent Mathematics* (Kluwer, Dordrecht 1995)
24. L. Oubiña: Logico-Axiomatic Theory of Categories. [in Spanish] PhD thesis, National University of La Plata (1966)
25. L. Oubiña: 'Axiomatic Theory of Categories [in Spanish]' *Cahiers de Topologie et Géométrie Différentielle* X (3) (1969) pp. 375–394
26. G. Priest: *In Contradiction* (Nijhoff, Dordrecht 1987)
27. B.C. van Fraassen: *The Scientific Image* (Clarendon Press, Oxford 1980)
28. A. Volkov: Theory of Categories and Theory of Sets. [in Portuguese] MA Thesis, University of São Paulo (1997)

Combinatory Logic, Language, and Cognitive Representations

Jean-Pierre Desclés

The aim of logic is not only a study of correct reasoning or a mathematical analysis of formal systems:

> *"The aim of mathematical logic is (...) to analyse logic with a view to obtaining a greater command over its use and more profound understanding of its nature"* (Curry, 1929, Seldin, p. 9),

but it is also a criticism of concepts and thoughts:

> *"(...) The objective of logic is the explanation of thought; if there are thoughts excluded from the explanation, then it is deficient. Furthermore, it is exactly these meaningless concepts which lead to paradoxes; if one excludes them, one can avoid the paradoxes but not explain them. That something is a concept is the only requisite for its being subject to treatment in logic.(...) Among concepts as given here, some will be "meaningful" and others "meaningless"; the main aim of combinatory logic is to distinguish these two kinds."* (Curry, 1929, Seldin, p. 15)

Thus, it is necessary to take in account what is the "significance" of such a theory for the "analysis of thought". The construction of an abstract formal theory is based on a very simple prelogic (*Urlogik*). Through this prelogic the questions that one is accustomed to place in the usual intuitive prelogic can be answered by symbolic reasoning.

My claim is: *Curry's combinatory logic is a prelogic.*[1]

In accordance with this view, the essential purpose of logic[2] is the construction of an abstract theory (strictly formalised when it is possible), such that when its fundamental notions are properly interpreted, we obtain an analysis of those universal principles in accordance with which valid thinking

[1] It is true that other conceptual systems are candidates for being a prelogic. For instance, the mathematical theory of topoi is also a "good candidate": there are deep links between combinatory logic and topoi theory.

[2] Mathematical logic may be defined as an application of the formal methods of mathematics to the domain of logic.

goes on. The term analysis here means that a certain body of knowledge is exhibited and derived from a much simpler body assumed at the beginning. Two questions lead to this analysis. The first is the problem of formulating the foundations of logic as precisely as possible. The exhibition of the complicated rules of the prelogic as a synthesis made from rules of much simpler character is a step in that direction. Combinatory logic has the capacity to be a foundation for abstract theory of all logics and mathematics, including functions (predicates, relations) or arbitrarily many operators and free variables.

> *"Combinatory logic is a branch of mathematical logic which concerns itself with the ultime foundations. Its purpose is the analysis of certain notions of such basic character that they are ordinarily taken for granted. These include the process of substitution, usually indicated by the use of variables; and also the classification of the entities constructed by these processes into types or categories, which in many systems has to be done intuitively before the theory can be applied. It has been observed that these notions, although generally presupposed, are not simple; they constitute a prelogic, so to speak, whose analysis is by no means trivial."* (Curry, 1958 : 1)

The second question is the explanation of paradoxes:

> *"There is reason for believing that the true source of our difficulties with these puzzles lies in the prelogic itself, and therefore that an analysis of the prelogic will contribute to understand them."* (Curry, 1958 : 1)

I think that combinatory logic is a very useful formalism to represent the meanings of concepts and to study the categorisation process.[3]

The first characteristic feature of this formalism is the following: combinatory logic[4] does not use bound variables. Indeed, in studying the general notion of substitution, Curry considers that variable is unclear, ambiguous, and difficult to manipulate (with the confusion of bound variables). A deep analysis of "variable" is given by Curry and Feys (1958 : 52-57).

The second characteristic feature is: according to my deep understanding of combinatory logic, all entities are "objects of thought" or "general concepts" without *a priori* restrictions. These "objects of thought" are expressed by means of elementary and complex operators built with a basic operation. The basic operation, called *application*, is binary: the first argument is an operator, it applies to a second argument, an operand, to build up another operator, the result of the application.

[3] ...with typical and atypical instances, in using new quantifiers more adequate to quantification in natural languages than classic fregean quantifiers (Desclés, Guentcheva, 1997, 1998).

[4] Church's λ-calculus, which is close to combinatory logic, uses bound variables.

Third feature: combinatory logic is not exactly an alternative logic but rather a fundamental logical system, used to define, to analyse, and to compare, in the same formal framework, other logical systems, as well as classical systems (first-order systems) and as less classical systems (intuitionist logic, modal logic, paraconsistant logic, etc.).

Pure combinatory logic studies operative forms or abstract processes. The topic of *illative combinatory logic* is the inference process from a given set of "logical constants" (connectors between propositions, negations, modal operators, quantifiers, etc.). Combinatory logic as prelogic is the trunk of a tree. Different branches with leaves grow out from the top of the trunk: illative logics and also other logical systems for analysing conceptualisation processes, etc. Combinatory logic is able to express the building of concepts from more elementary concepts and to classify the concepts in different categories (or types). It supplies us with a general theory of formal systems which are applied to different domains as arithmetic and recursive functions; calculus of propositions with different negations (classical, intuitionist, minimal negations); calculus of predicates with quantifiers (fregean or no fregean quantifiers[5]); building up sentences in natural languages from linguistic units (words, morphemes,...) in order to represent the meaning;[6] building up programs from elementary operations in functional (or applicative) programming;[7] analysis of knowledge representations[8] and concepts in cognitive sciences, psychology, philosophy, semiotic, etc. Combinatory logic is a cornerstone or a paradigm in formalised systems used in logic, programming languages, and natural languages analysis:

> *"We first argue that the theories of functional languages and Applicative Universal Grammar (AUG) share a common underlying paradigm. By paradigm we mean a pattern of thought, a system of concepts and assumption underlying a group of theories, regardless of how they differ from one another in detail. Some of the fundamental ideas of this common paradigm can be traced to the works of Haskell B. Curry on combinatory logic and its philosophical and linguistic implications."* (Shaumyan and Hudak, 1998 : 70)

In a first part, we develop some aspects for combinatory logic with a distinction between concatenative and applicative structures from a semiotic viewpoint (§1). The expressions of combinatory logic are expressions of an applicative language (§2). Combinators are specific abstract operators of combinatory logic (§3). Relations between expressions are governed by β-reductions generated from combinators (§4). The entities denoted by expressions of combinatory logic are "objects of thought" viewed as operators (§5).

[5] Desclés, Guentcheva, 1997, 1998.

[6] Shaumyan, 1977, 1987; Desclés, 1990, 1997.

[7] It is a new style for programming languages.

[8] Desclés, 1990.

The combinators are used to build up complex operators from more elementary operators (§6). It follows that combinatory logic is a prelogic. In (§7) we show how combinatory logic can be used to analyse some "meaningless expressions" and then with the notion of functional type (§8). We define cognitive representations by means of a very simplified example (§9).

1 Concatenative and Applicative Structures

Haskell Curry distinguished two kinds of semiotic structures: the structure of *concatenative systems of strings* used for instance in standard presentations of classical logical systems (first-order languages) and the *applicative structure of operator–operand relations* used in combinatory logic and other applicative languages. In an illuminating study Curry wrote:

> *"Combinatory logic was formulated as a system in which the formal objects were rather differently conceived than was the case in standard formalisations of logic. The standard formalisation at that time was to demand that formal objects be the expressions of some "object language"; this means that they be strings formed from the symbols of that object language by concatenation. In combinatory logic these formal objects, called obs, were wholly unspecified; it was merely postulated that there was a binary operation of application among them, that the obs be constructed from the primitive objects, called atoms, by these operations, and that the construction of an ob is unique. This means that the obs were thought of, not as strings of atoms, but as structures like a genealogical tree."* (Curry, 1961 : 65)

The application of an operator X to an operand Y may be presented by a functor [X,Y] involving parentheses, or by a prefix *Apply XY* without parentheses or also by the simple juxtaposition 'XY'. We would then have three symbolic systems for representing an applicative language; these would have the same grammatical structure, but would differ in their phenogrammatics. These presentations can be shown to be monotectonic (well-formed expressions indicating unique constructions). For analysing natural languages, we introduce a clear distinction between the symbolic *presentations* of expressions by means of strings and the *underlying structure* of linguistic units (words, phrases, sentences, discourses, etc.).

> *"We may (...) conceive of the grammatical structure of the language as something independent of the way it is represented in terms of expressions; and this grammatical structure can be studied by means of* ***F****[the operation of application] (and possibly other similar notions). (...) This gives us two levels of grammar, the study of grammatical structure in itself, and a second level which has much the same relation to the first that morphophonemics does to morphology. In*

order to have terms for immediate use I shall call these two levels tectogrammatics and phenogrammatics respectively; no doubt someone will propose better terms later." (Curry, 1961 : 65)

A linguist, Sebastian Shaumyan (1977, 1987), has developed this semiotic distinction with a deep analysis for natural languages in connection with combinatory logic and functional programming languages:[9]

> *"We can think of the syntactic structure of a sentence as something independent of the way it is represented in terms of syntagmatic units. In this way we come up with two levels of grammar, which I call genotype grammar and phenotype grammar. Genotype grammar comprises functional units – predicates, terms, modifiers – and abstract operator–operand relations between these units. Phenotype grammar comprises syntagmatic units – morphemes and words – and connections between them in terms of linear order and their morphological properties. (...) The rules of genotype grammar are invariant with respect to various possibilities of their realization by phenotype grammar. (...) The distinction between the genotype and phenotype levels is a paramount importance for linguistic theory, because this distinction puts us on the road to the solution of the basic question of the linguistic theory formulated above: What factors contribute to the similarities and differences between natural languages?"* (Shaumyan, 1987: 97)

> *"Indeed, in studying any language – a programming language, a language of mathematics or logic, or a natural language such as English or Russian – we must distinguish two different kinds of structure: the physical structure and the functional structure of language. The physical structure of programming languages is a reflection of the physical structure of the computer. The physical structure of natural language is the so-called phrase-structure, which reflects the action of the human machine, i.e. the organs of speech. Because of the dependence on the human organs of speech, natural language has a linear phrase-structure. However, the essential structure of natural language, its functional structure, is independent of the linear phrase-structure."* (Shaumyan and Hudak, 1998 : 71)

Thus, in order to analyse some language – a natural language or a formal language which express mathematical relations or logical formula, a certain kind of programming languages (functional programming languages) – we can consider it as a semiotic system and we must distinguish two different levels: the *presentation level* in a specific symbolic system and *the level of the underlying objects* which are *represented*[10]. As for a language, and more generally,

[9] See also Desclés (1990).

[10] In rewriting systems studied by Markov algorithm, Thue's relations between strings, Post systems, the structure of presentation is identical to the structure of

a formal system, we are going to clarify different semiotic concepts: *presentation, representation, interpretation.* According to Curry, a formal system can be viewed as an abstraction from its presentation, representation, and interpretation.

Let us put some highlight on these three concepts.

Presentation:[11]

"A particular enunciation of the primitive frame of a formal system will naturally employ symbols to designate primitive ideas; and it must describe how the closures of the functives are to be symbolised. Such a particular enunciation with a special choice of symbolism, we call a presentation of the primitive frame (and the formal system). It is clear that a formal system can be communicated only through a presentation. (...) We can, therefore, regard a formal system as something independent of this choice, and say that two presentations differing only in the choice of symbolism are presentations of the same formal system. In this sense a formal system is abstract with respect to its presentation". (Curry and Feys, 1958 : 20)

Representation:[12]

"A presentation of a primitive frame speaks about the obs, using nouns to name them; but it does not specify what determinate thing

---

underlying objects. The objects are always strings (sequences of symbols in some alphabet, oriented from left to right); the rewriting rules operate on strings. The classical Chomsky phrase grammar theory used string–rewriting systems (with symbols belonging to a vocabulary); in this theory, the distinction between presentation and the underlying object is irrelevant: an object is always a string. It is true that in some rewriting systems ("context–free" grammars or grammars of the type 2 in Chomsky's hierarchy), the derivation of a terminal string (a sentence) may be represented by a tree (the derivation tree) but the objects handled by the grammar, viewed as a rewriting system, are strings but not trees. This is not the case in the class of Chomsky's transformational grammars because the objects are derivation trees and the rules operate on trees.

[11] "Un système formel étant donné, on peut parler de ses présentations, de ses représentations et de ses interprétations. Une *présentation* d'un système formel est une énonciation particulière du cadre constitutif de ce système, de l'ensemble des conventions qui le définissent. En d'autres termes, c'est un choix particulier des symboles qui sont utilisés pour désigner les "idées primitives" du système, à savoir ses objets élémentaires (ou atomes), ses opérations et ses prédicats" (Jean Ladrière, *L'articulation su sens*, Aubier-Montaigne, 1970, p. 66).

[12] Une *représentation* d'un système est une correspondence établie entre les *objets* du système et certaines entités (qui peuvent être des symboles, des nombres, des idées, des objets physiques, ou tout ce que l'on voudra), de manière telle que deux entités différentes sont mises en correspondance avec deux objets différents." (Ladrière, op. cit. p. 66)

each ob is. If we assign a unique determinate thing to each ob in such a way that distinct things are always assigned to distinct obs, we have a representation of the system. The things assigned may be any kind of objects about which statements can be made: symbols (or symbol combinations), qualities, numbers, ideas, manufactured things, natural beings. The assignment of such a representation (or a change of representation once made) does not affect in any way the criterion of truth for elementary statements, and thus the predicates remain definite by the primitive frame without reference to any external "meaning". (...) A system for which no presentation is specified will be called abstract. (...) It follows (...) that the names of the obs in any presentation constitute a representation. We shall call such a representation an autonomous representation". (Curry and Feys, 1958 : 20)

Interpretation:[13]

> *(...) By an interpretation of a formal system we mean a correspondence between its elementary statements and certain statements which are significant without reference to the system. Let us call the latter statements contensive statements. (...) It is understood that one formal system may have an interpretation in another, or it may have a completely intuitive interpretation. (...)* (Curry and Feys, 1958 : 21-22)

A *valid interpretation*[14] is one such that every contensive statement which corresponds to an elementary theorem is true; in that case the system will also be said to be valid for the interpretation.

2 Combinatory Logic as an Applicative Language

Combinatory Logic is usually *presented as a concatenative system* (an operand always follows an operator in a presentative string) but the underlying operators are not strings but *applicative expressions* (operator applies to operand for building up a new object, the result of the application).

[13] "Une *interprétation* d'un système est une correspondance établie entre les *propositions* du système et certains énoncés qui ont une signification indépendamment du système (soit qu'ils constituent des propositions d'un autre système, soit qu'ils aient un sens intuitif, par exemple en tant que théorèmes d'une théorie mathématique intuitive ou en tant qu'énoncés empiriques relatifs à tel ou tel domaine d'expérience). La correspondance ne doit pas être telle que toutes les propositions du système aient une traduction". (Ladrière, op. cit. p. 66). The interpretation of one formal system in another is important in modern logic. Under certain conditions, this gives rise to 'models'.

[14] "Une interprétation est dite *valide* si tout énoncé correspondant à un théorème du système est vrai" (Ladrière, o.c. p. 66).

Combinatory logic is a specific applicative language.[15] An applicative language is a formal language whose basic operation is the application. All expressions of any applicative language are built up by means of this basic operation, called *application.*

The application is a primitive binary operation: the first argument is an operator, the second is an operand. The application of 'X' to 'Y' produces a new object Z which can be an operator or an operand. This operation is presented by means of an *applicative tree* as in Fig. 1.

$$\frac{X \qquad Y}{Z}$$

Fig. 1. XY⊳Z

The operation of application can run again with Z as operator (Fig. 2) or as operand (Fig. 3), hence, respectively, the results are V or V'.

$$\frac{\dfrac{X \qquad Y}{Z} \qquad U}{V}$$

Fig. 2. (XY)U⊳V

$$\frac{U' \qquad \dfrac{X \qquad Y}{Z}}{V'}$$

Fig. 3. U'(XY)⊳V'

For illustrating the applicative tree of Fig. 2, let us take the natural numbers; the instances of X, Y, Z, U, V are, respectively: *'act of adding ... and ...'*, '2', *'adding 2 and ...'*, '3', '5'; we obtain an applicative tree given in Fig. 4. Thus, the number is obtained by two successive applications: first, application of the operator *'act of adding ... and ...'* to the operand '2' producing

[15] There are other applicative languages: Church's λ-expressions are applicative expressions, programs of a functional programming language are applicative expressions. In (Desclés, 1990, 1997), following H.B. Curry and S.K. Shaumyan and the categorial grammar approach for studying natural languages, we have argued that the underlying structure of expressions in natural languages is applicative: all linguistic units are built up as applicative expressions by means of linguistic operators which apply linguistic operands.

the result *'adding 2 and ...'*, which is an operator; second, this operator is applied to the operand '3' producing the result '5'.

For illustrating the applicative tree of Fig. 3, let us take propositions; the instances of X, Y, Z, U, V are, respectively: *act of negating ... Socrates is mortal*; *Socrates is mortal is false*, *act of negating ... Socrates is mortal is false is false*; we obtain an applicative tree given in the Fig. 4 bis. The proposition *Socrates is mortal is false is false* is obtained by means of two successive applications: first, application of the operator *act of negating ...* to the proposition *Socrates is mortal* hence the new proposition *Socrates is mortal is false*; this proposition is the operand of the operator *act of negating ...* hence the proposition: *Socrates is mortal is false.*

act of adding... and... 2

adding 2 and ... 3

5

act of negating Socrates is mortal

act of negating 'Socrates is mortal' is false

"'Socrates is mortal' is false" is false

Fig. 4. Examples of applications

The application can be *presented* not by a tree but by a concatenative string, where operator and operand are juxtaposed: first, in first position, operator, and in second position, operand. To every applicative expression we may associate 'in a one-to-one correspondence' an applicative tree which represents its building by means of the operation of application. Every building may be expressed in a symbolic system by a concatenative expression with the following conventions of presentation: the operator is always prefixed before the operand. Thus, to the above presentations of applicative trees (Fig. 4) correspond the two following concatenative expressions:

(((*act of adding ... and ...*) 2) 3)

(*act of negating* (*act of negating* (*Socrates is mortal*))).

More generally, the application of the operator X to the operand Y is presented by means of the concatenative expression 'XY'. The juxtaposition of operator and operand which *presents* the applicative expressions *is not associative*. Indeed, the meaning of the two concatenative expressions '(XY)Z' and 'X(YZ)' is not the same. The first concatenative expression '(XY)Z' means that the operator X applies to Y and the result applies to Z; the second concatenative expression 'X(YZ)' means that the operator X applies to the result of the application of Y to Z.

When the application of the operator X to Y is computed, we can distinguish, within the concatenative presentation: 1°) the *applicative program*

'XY' *before* execution; 2°) *the result* Z *after* execution. We introduce a relation (called an *applicative evaluation*) between the application program XY not yet executed and the result Z after execution: XY ▷ Z. Thus, for the above examples, we obtain:

$$((\textit{act of adding} \ldots \textit{and} \ldots)\ 2)\ 3 \triangleright 5$$

act of negating (*act of negating* (*Socrates is mortal*)) ▷
"'*Socrates is mortal*' is false" is false'.

All *applicative expressions* are presented as follows from a given set of atoms:

(i) atoms are applicative expressions;
(ii) if X and Y are applicative expressions then (XY) is an applicative expression.

In order to simplify the concatenative expressions of applicative expressions, we delete all parentheses which can be restored and we introduce the left associativity property in applicative expressions:

$$\mathrm{XY} =_{def} (\mathrm{XY})$$
$$\mathrm{XYZ} =_{def} (\mathrm{XY})\mathrm{Z}\ \mathrm{X}(\mathrm{YZ}).$$

In an applicative language,[16] the specific action of an operator on operand can be unspecified inside the language.[17] We must remark that in any specific applicative expression, one occurrence of an expression can be an operator but in other occurrences, the same expression can be an operand. For instance, let us take the following applicative expression 'XX';[18] the first occurrence of X is an operator but the second occurrence is an operand. In the applicative expression 'X(YX)' the first occurrence of YX, is an operator whose operand is X, the second occurrence of X is operand.

[16] In combinatory logic all expressions are applicative expressions.

[17] "Recall that (...) it is stated that X is a function then (XY) represents the result of substituting Y for the first variable of X, but if X is not a function then nothing is said about the interpretation of (XY), which may be left uninterpreted. (...) Now consider the following interpretation under which (XY) is always interpreted provided that atomic terms are interpreted: the domain is the set of signals that can be fed into a computer. This set will include the signals for a program and for any data that may be used in running it. The job control language will provide an operation representing application. (...) For this interpretation, (XY) has the meaning for every X and Y in the domain. Furthermore, if X is a properly written program and Y an appropriate set of data, then (XY) will cause the computer to produce the output for the program X applied to the data Y. (...) Thus, to say that (XY) exists only means that it makes sense as computing instruction, not that it has a value." (Seldin, Curry's program, p. 27)

[18] In combinatory logic, it is possible to apply an operator to itself. There are no *a priori* restrictions on applicative operation. In set theory, when an operator is a function, we cannot apply a function to itself.

3 Combinators Are Abstract Operators

Combinatory logic adjuncts specific operators, called *combinators.*, to other applicative expressions generated from atoms:

> *"A basic role is played by certain operators [combinators] which represent combinations as functions of the variables they contain (perhaps along with other variables). The combinations in question are those formed from the variables alone by means of the operation [application of a function to an argument]".* (Curry, 1958 : 5)

An applicative language all of whose expressions, called combinatory expressions, are generated by means of application, from basic combinators and free variables interpreted by operators is pure. Every combinator is an abstract operator which builds up complex operators from more elementary operators. The action of a combinator is defined *inside* the language; it is defined by specific rules (introduction and elimination rules in Gentzen's style). We can represent the action of many combinators – but not all combinators[19] – by means of a transformation between trees, where every tree represents the actions of operators.

In combinatory logic, *basic combinators* belong to the set of atoms. Some combinators defined in terms of basic combinators are called *elementary combinators* because, on the one hand, they are very useful to build up more complex combinators and, on the other hand, their intuitive meaning is obvious. The elementary combinators represent abstract acts for combining operators together.

We give below intuitive interpretations of elementary combinators viewed as abstract operators that combine functions – or operators – in a functional space of functions[20]:

- **I** is an identity operator: **I**(f)(x) = f(x)
- **W** is a diagonalisation operator; it duplicates an argument of a binary operator: **W**(f)(x) = f (x,x)
- **B** is a functional composition which combines two unary operators as a functional composition defined on set theory functions: **B**(f,g)(x) = f(g(x))
- **C** is a converse operator, it permutes the arguments of a binary operator f: **C**(f)(x,y) = f(y,x)
- **K** is an operator which forms a constant-function: **K**(a)(x) = a
- **S** is a stronger composition operator: **S**(f,g)(x) = f(x, g(x))
- Φ is an in–parallel composition operator: Φ(f,g,h)(x) = f(g(x), h(x))

---

[19] "Regular combinators" can be represented as operators which act on trees. There are no regular combinators.

[20] The interpretation is given here in set–theory: every combinator is interpreted as a function of an abstract functional space. However, the notions of operator and combinator are more general than the concept of set theory functions.

- Ψ is another parallel composition operator: Ψ(f,g)(x,y) = f(g(x),g(y))
- $\mathbf{C}_*$ is an operator which changes any argument of an operator into a new operator whose operator becomes an argument: ($\mathbf{C}_*$ x)(f) = f(x).

These operators (except $\mathbf{C}_*$) may be viewed as operators whose actions are defined on a set of trees, each tree representing an operator with variable arguments. The result of the action of the combinator is again a tree.

All combinators are applicative expressions built up with only occurrences of combinators. For instance, the following applicative expressions

$$\mathbf{B}(\mathbf{BW})\mathbf{B},\ \mathbf{BCC},\ \mathbf{S}(\mathbf{KS})\mathbf{K},\ \mathbf{WS}(\mathbf{BWB})$$

are combinators whose actions can be specified exactly from the elementary actions of their components **B**, **C**, **W**, **K**.

I claim, following Hindley and Seldin (1986), that combinators are abstract operators which formalise the function-as-operation-process concept (as opposite to the set-of-ordered-pairs concept)[21]

> *"Perhaps the most important difference between operators and functions [in set theory] is that an operator may be defined by describing its action without defining the set of inputs for which this action produces results, i.e. without defining its domain. (...) A second important difference is that some operators have no restriction on their domain; they accept any inputs, including themselves. (...) The operator concept has never been successfully formalised, at least not in a theory that is both consistent and reasonably comprehensive.(...) But if a concept has not yet been formalised fully, we cannot infer that it does not exist. (...) Further insight and clarification are needed, but the operator concept is far too fertile a way of thinking to be simply abandoned".* (Hindley and Seldin, 1986 : 45)

Let us take three basic combinators **I**, **S**, and **K**. The actions of these basic combinators on any applicative expressions X, Y, Z, are defined (in Gentzen's style[22]) by means of the following rules (elimination and introduction rules):

[21] In (Desclés, 1980), an algebraic approach of the intuitive operator concept had been proposed with a recursive set T[Σ] of derived multioperators generated from a given set Σ of elementary operators by means of two general operations – called "greffe", "intrication" and "coprojections". This study was developed in the framework of the theory of categories. An interpretation of these multioperators on a set is a T[Σ]-algebra defined and studied by Lawvere. The general operations defined on operators and multioperators can be represented by means of combinators of Combinatory Logic.

[22] We follow here the Fitch presentation of combinatory logic (Fitch, 1974).

elimination rules **introduction rules**

$$\frac{\mathbf{I}X}{X}\ [\text{e-}\mathbf{I}] \qquad \frac{X}{\mathbf{I}X}\ [\text{i-}\mathbf{I}]$$

$$\frac{\mathbf{S}XYZ}{XZ(YZ)}\ [\text{e-}\mathbf{S}] \qquad \frac{XZ(YZ)}{\mathbf{S}XYZ}\ [\text{i-}\mathbf{S}]$$

$$\frac{\mathbf{K}XY}{X}\ [\text{e-}\mathbf{K}] \qquad \frac{X}{\mathbf{K}XY}\ [\text{i-}\mathbf{K}]$$

More generally, the action of any combinator is defined from the action of basic and elementary combinators by means of derived introduction and elimination rules.[23] Thus the following elementary combinators **W**, **B**, **C**, $\mathbf{C}_*$, Φ are derived from **I**, **S**, and **K** and expressed as applicative expressions with occurrences of **I**, **S**, and **K**. The derived elimination and introduction rules are as follows:

elimination rules **introduction rules**

$$\frac{\mathbf{W}XY}{XYY}\ [\text{e-}\mathbf{W}] \qquad \frac{XYY}{\mathbf{W}XY}\ [\text{i-}\mathbf{W}]$$

$$\frac{\mathbf{B}XYZ}{X(YZ)}\ [\text{e-}\mathbf{B}] \qquad \frac{X(YZ)}{\mathbf{B}XYZ}\ [\text{i-}\mathbf{B}]$$

$$\frac{\mathbf{C}XYZ}{XZY}\ [\text{e-}\mathbf{C}] \qquad \frac{XZY}{\mathbf{C}XYZ}\ [\text{i-}\mathbf{C}]$$

$$\frac{\mathbf{C}^*XY}{YX}\ [\text{e-}\mathbf{C}_*] \qquad \frac{YX}{\mathbf{C}_*\ X}\ [\text{i-}\mathbf{C}_*]$$

$$\frac{\Phi XYZU}{X(YZ)(YU)}\ [\text{e-}\Phi] \qquad \frac{X(YZ)(YU)}{\Phi XYZU}\ [\text{i-}\Phi]$$

Elementary combinators are defined in terms of basic combinators. For instance, we can show that we have the following definitions:[24]

$$[\mathbf{B}=_{def}\mathbf{S}(\mathbf{KS})\mathbf{K}], \quad [\mathbf{C}=_{def}\mathbf{S}(\mathbf{BBS})(\mathbf{KK})], \quad [\mathbf{I}=_{def}\mathbf{SKK}].$$

---

[23] It is easy to show that every combinator can be represented by a close λ-expression (of the Church's λ-calculus). This close λ-expression indicates how the result is built up from a sequence of operands.

[24] As for the definition $\mathbf{B}=_{def}\mathbf{S}(\mathbf{KS})\mathbf{K}$, see the second example given in the next paragraph.

4 β-Reductions (β-Expansions) of Combinators

To every elimination rule may be associated an internal evaluation of the action of the corresponding combinator in using a relation between combinatory expressions. This relation is called an elementary β-*reduction*. The converse relation of the elementary β-reduction is called an elementary β-*expansion*; it is associated to an introduction rule. Thus to the above rules for basic combinators **I**, **S**, and **K**, are associated the following elementary β-reductions and β-expansions between combinatory expressions:

β-reductions	β-expansions
$\mathbf{I}X \rightarrow_\beta X$	$X \leftarrow_\beta \mathbf{I}X$
$\mathbf{S}XYZ \rightarrow_\beta XZ(YZ)$	$XZ(YZ) \leftarrow_\beta \mathbf{S}XYZ$
$\mathbf{K}XY \rightarrow_\beta X$	$X \leftarrow_\beta \mathbf{K}XY$

The weak β-reduction system of combinators is defined as follows:

(i) basic combinators **I**, **S**, **K** are constants; they belong to the set of atoms (free variables are the other atoms of the system);

(ii) all combinatory expressions are applicative expressions generated from atoms;

(iii) a β-reduction, denoted $\rightarrow_\beta$, is defined between between combinatory expressions; this relation is the closure of elementary β-reductions, for any combinatory expressions X, Y , Z, U, generated by:

- β-reductions schemes associated to the combinators **I**, **S**, **K**:
  - [**I**] $\mathbf{I}X \rightarrow_\beta X$
  - [**S**] $\mathbf{S}XYZ \rightarrow_\beta XZ(YZ)$
  - [**K**] $\mathbf{K}XY \rightarrow_\beta X$
- reflexive scheme:
  - [ρ] $X \rightarrow_\beta X$
- deduction–rules:
  - [τ] $X \rightarrow_\beta Y$ and $Y \rightarrow_\beta Z \Rightarrow X \rightarrow_\beta Z$
  - [μ] $X \rightarrow_\beta Y \Rightarrow ZX \rightarrow_\beta ZY$
  - [ν] $X \rightarrow_\beta Y \Rightarrow XU \rightarrow_\beta YU$.

The weak $_\beta$-expansion system of combinators (respectively weak $_\beta$-equality with symmetry) are defined in the same way.

From basic and elementary combinators, we express all combinators. When a combinator applies, as operator, to operands, it builds up a new operator that applies to another operand... to obtain a final result. This calculus uses deduction–rules (monotony of application at left and at right and transitivity). Let us show some examples.

- **First example**: Let us consider the combinator '**B**(**BW**)**B**'. Let us apply this combinator to successive variables x, y, and z. The calculus, by successive β-reduction, is presented as follows:

1.	**B**(**BW**)**B**xyz	
2.	(**BW**)(**B**x)yz	[e-**B**], [ν]
3.	**W**(**B**xy)z	[e-**B**], [ν]
4.	(**B**xy)zz	[e-**W**]
5.	x(yz)z	[e-**B**], [ν]

We deduce the β-reduction, by the transitivity rule [τ]:
B(**BW**)**B**xyz $\rightarrow_\beta$ x(yz)z.

- **Second example**: **S**(**KS**)**K**xyz $\rightarrow_\beta$ x(yz)

1.	**S**(**KS**)**K**xyz	
2.	(**KS**x)(**K**x)yz	[e-**S**], [ν]
3.	**S**(**K**x)yz	[e-**K**], [ν]
4.	(**K**xy)(yz)	[e-**S**]
5.	x(yz)	[e-**K**], [ν]

- **Third example**: Some applicative expressions with combinators cannot be reduced. Let us take the combinator '**WWW**'. We have the successive β-reductions:

$$\mathbf{WWW} \rightarrow_\beta \mathbf{WWW} \rightarrow_\beta \mathbf{WWW} \rightarrow_\beta \ldots$$

- **Fourth example**: Sometimes, combinators have interesting properties, as in this example. Let the combinator

$$\mathbf{Y} =_{def} \mathbf{WS}(\mathbf{BWB})$$

Using $_\beta$-reductions (or elimination rules) and $_\beta$-expansions (or introduction rules), let us apply **Y** to an arbitrary operand X.

1.	**WS**(**BWB**)X	
2.	**S**(**BWB**)(**BWB**)X	[e-**W**], [ν]
3.	**BWB**X(**BWB**X)	[e-**S**], [ν]
4.	**W**(**B**X)(**BWB**X)	[e-**B**]
5.	(**B**X)(**BWB**X)(**BWB**X)	[e-**W**]
6.	X(**BWB**X)(**BWB**X))	[e-**B**]
7.	X(**S**(**BWB**)(**BWB**)X)	[i-**S**], [μ]
8.	X(**WS**(**BWB**)X)	[i-**W**], [μ], [ν]
9.	[**Y** $=_{def}$ **WS**(**BWB**)]	def of **Y**
10.	X(**Y**X)	rempl. definiens

We have the following β-equality: **Y**X = X(**Y**X). We will return later to this example.

5 Operators in Combinatory Logic Are "Objects of Thought"

What is the general meaning of entities (called "obs" by Curry) in combinatory logic? As for combinators, the answer is clear. Indeed, a combinator represents an abstract process for combining operators. To understand the discussion, let us recall how Frege has characterised the intuitive notion of concept. A concept is "unsaturated", it is "incomplete". For instance, every predicate is an incomplete concept. "Concept" (in Frege's sense) is opposite to saturated and complete "Object".[25] Every concept is an unsaturated operator. We generalise this notion: all operators will be viewed as denoting unsaturated entities. Thus, predicates, connectors, and quantifiers are operators.[26]

In pure combinatory logic, since the application operation is used without restriction, an operator can be applied to an operand, for instance to itself, to build up a new entity belonging to the system. I claim that a combinator (basic, elementary combinator or derived combinator) must always be interpreted as an operator[27] or as a function-as-operation-process.[28] Every combinator combines different entities to build up a new entity of the system. The meaning of an elementary combinator is intrinsic, that is, it is given

[25] It is the classic opposition between syncategorematic (unsaturated) and categorematic (saturated) expressions.

[26] The linguistic expressions of these operators are syncategorematic.

[27] "La logique combinatoire est un effort en vue de mettre sur pied une technique opératoire plus abstraite encore que celle qui se donnait cours dans les formalismes étudiés. (...). On y suppose que l'esprit est en présence d'objets absolument quelconques et indéterminés, et que penser consiste à leur appliquer mentalement quelque chose, n'importe quoi d'ailleurs, il n'a pas même pas lieu en principe de distinguer les symboles affectés aux objets et à ce que la pensée entend leur appliquer. (...) A proportion cependant où elle est déjà en présence de ces premières séries d'applications représentées symboliquement, la pensée est amenée à envisager des opérations plus déterminées, qui porteront mentalement, sur les choses ainsi appliquées les unes aux autres dans un certain ordre (...) Mais ces opérations sont elles-mêmes des applications de "quelques chose" (déterminément spécifié cette fois) à autre chose. On représentera donc ces opérations par des symboles d'opérateurs spéciaux traités exactement comme les symboles de choses appliqués à autre chose (...) Le premier objectif de la logique combinatoire est de constituer une théorie systématique des opérateurs composés de ces opérateurs simples [les combinateurs élémentaires]"(D. Dubarle, *Initiation à la logique*, 1957, p. 67, 68).

[28] Every combinator is represented in Church's λ-calculus by a closed λ-expression; the "meaning" of such a closed λ-expression is a "function as rule" (rather than as graph). A "function as rule" refers to the process of going from argument to value, a process coded by a definition. "The λ-calculus regards functions again as rules in order to stress their *computational* aspects" (H.P. Barendregt, *The lambda calculus, its syntax and semantics*, 1981, p. 3).

inside the language of combinatory expressions by specific rules for acting.[29] The other combinators act on operands by means of a β-reduction process executed from left to right. If the set of atoms contains only basic combinators, then the system of all combinators is *a pure combinatory logic* whose all of entities are clearly operators.

In the pure combinatory logic with variables no assumption is made of the different kinds of variables; free variables $x, y, z, \ldots$ are adjoined to basic combinators without giving an *a priori* interpretation to variables into domains. It follows that an adjoined variable can be combined without restriction with combinators and with all other variables. For instance, we can apply a variable x to itself to build up the combinatory expression 'xx'. The meaning of 'xx' is "x applies to itself". In this case, the first occurrence of 'x' is an operator and the second occurrence is the operand of the first occurrence. In this case, the variable must be interpreted as denoting an operator in an unspecified domain. The denotation of 'x' cannot be an "absolute operand", that is an object which would never be an operator. In the pure combinatory logic with adjoined variables, combinators are operators and every adjoined variable must be interpreted as an unspecified operator.

Now, in combinatory logic *with different kinds of variables and constants*, we can adjoin specific constants and variables related to a specific purpose. In this case, some constants and some variables may be *absolute operands*. For instance, a string of symbols, a proposition, a sentence, a complex object, a physical object, a state of a physical system, etc. can be viewed as objects; in this case, combinators operate on these constants and variables to build up new objects (for instance another string of symbols, a new proposition, a new sentence, a new complex object, a new physical object, a new state of a physical system, etc.). At the system of combinators, we can also adjoin definite operators with a specific meaning inside a domain. For instance, differential operator , operator of integration, "nabla operator", etc., grammatical operators can be adjoined to combinators to constitute an applied combinatory logic system with a specific interpretative domain. In these systems, the combinators combine specific operators of the domain.

Let us return to the pure combinatory logic whose adjoined variables have no restricted interpretation. What is the possible interpretation of these variables exactly? I claim that the variables must always be interpreted as "unsaturated operators" and never as "saturated objects".[30] All combinatory expressions are *(complex) operators* built up from atoms (basic combinators and free variables) by means of combinators which are also abstract operators. This interpretation is practically imposed upon it by general applicative

[29] In Gentzen's "natural deduction", the meaning of connectors, functors, and quantifiers is defined by introduction and elimination rules. For instance, according to the rules associated to the negation we obtain different logical systems (minimal negation, intuitionist logic, and classical logic).

[30] I do not agree completely on this point with J.-P. Ginisti (1997).

systems without different kinds of variables. Thus, the obs [31] in Curry's terminology, are only "objects of thought" and not at all "objects"[32] viewed, in the Frege sense[33], as "saturated" or "complete" entities:

> *"As is the case with most formal systems, combinatory logic was motivated by a certain interpretation, and consideration of this interpretation, although theoretically superfluous, helps to make the system more intelligible. The basic idea of the original interpretation is that with the obs of the system we can associate certain contensive notions called values. (...) These values can then be any sort of logical notion – propositions, functions, predicates, classes, relations, numbers etc. – Then application has the following interpretation. If X means a function X^* [or an operator in our approach], and Y means a Y^* suitable for the first argument of X^*, then XY means the result of specifying the first arguments of X^* to be precisely Y^*; if X^* is an n-place function then $(XY)^*$ is a (n−1)-place function (or a constant if n=1). This sort of interpretation is practically forced on us by the reduction to an applicative system (...)"* (Curry, Hindley, 72 : 9)

Questions arise immediately: is any combinatory expression built up with combinators and variables significant? Are certain combinatory expressions senseless? We quote again Curry:

> *"It is not necessary that there be a value for every ob, any more than it is necessary for the acceptability of a physical theory that every statement in it be capable of experimental test. Thus it seems*

[31] "Every ob is significant in the broadest sense of being an object of thought; significance in a narrower sense means belonging to some semantical category" (Curry, 1958 : 261).

[32] "When we have thus admitted objects without restriction as arguments and values of functions, the question arises as to what it is that we are here calling an object. I regard a regular definition as impossible, since we have here something too simple to admit of logical analysis. It is only possible to indicate what is meant. Here I can only say briefly: An object is anything that is not a function, so that an expression for it does not contain any empty place" (Frege "function and concept", in Geatch's translation, p. 32).

[33] "The expression of a function needs completion, 'is unsaturated'. (...) The argument is not to be reckoned in with the function, but serves to complete the function, which is 'unsaturated' by itself. (...) Objects stand opposed to functions. I therefore count as an object everything that is not a function: thus, examples of objects are numbers, truth-values (...). The names of objects – or proper names – are not therefore accompanied by argument-places, but are 'saturated', like the objects themselves" (Frege, *Grundgesetze*, vol. I, 1893, translation by Geatch, p. 152–154). "We thus see how closely that which is called a concept in logic is connected with what we call a function. Indeed, we may say: a concept is a function whose value is always a truth-value". (Frege "function and concept", in Geatch's translation, p. 30).

> *best to formulate pure combinatory logic for unrestricted combinations, and to leave to the illative theory all questions as to whether an ob has a value, and if so what sort of value (...) The notions ruled out as meaningless by certain systems of philosophy, as well as those corresponding to obs which have no value for the quasi-grammatical reasons (...) are not actually senseless to our intuition. We can understand as concepts. Even those which lead to contradiction do so only when one ascribes certain properties to them."* (Curry, Hindley, 1972 : 10)

From the above text, it follows that, in a first step, every combinatory expression built up without *a priori* restriction is "understood as a concept", in other words it is an (abstract) operator. In a second step, we can research at which operand this operator can be applied to construct a significant result.

6 Combinators Used to Express Complex Concepts

The combinators are used to define a new complex concept from more elementary concepts already defined. Let us take several examples.

- We can define a "negation of a concept" or a "negation a property" from the propositional negation N_0 : $[N_1 = \mathbf{B}N_0]$; we define also a "negation of a relation" from the "negation of a concept": $[N_2 = \mathbf{B}N_1]$ ..., as follows:

1.	$(N_1 f)x$	
2.	$[N_1 = \mathbf{B}N_0]$	def. N_1
3.	$\mathbf{B}N_0$ fx	replacement of N_1
4.	$N_0(fx)$	[e-**B**]

1.	$(N_2 r)yx$	
2.	$[N_2 = \mathbf{B}N_1]$	def. N_2
3.	$\mathbf{B}N_1$ ryx	replacement of N_2
4.	$N_1(ry)x$	[e-**B**]
5.	$[N_1 = \mathbf{B}N_0]$	def. N_1
6.	$\mathbf{B}N_0(ry)x$	replacement of N_1
7.	$N_0(ryx)$	[e-**B**]

- Let us take another example. The unary concept "being square root of 2 and being rational", is represented by the applicative expression:

$$(\Phi\ \&\ (\text{being square root of 2})(\text{being rational})).$$

This complex concept is built up by means of the combinator Φ that combines "into a whole" the two elementary unary concepts 'being square root of 2' and 'being rational' with the binary connector '&'. By a β-reduction we deduce a "normal form" (an applicative expression without combinator) from the applicative expression built up with the above complex operator applied to an operand x:

$$(\Phi \text{ (being square root of 2)(being rational)})(x)$$
$$\rightarrow_\beta \&((\text{to be squarre root of 2})(x))\ ((\text{being rational})(x))$$

In analysing the applicative structures of this complex operator and its associated β-reduction, we can see that two unary concepts act together "in parallel" onto the same operand 'x'.

– Combinators can be used to give a logical and deep analysis of some problematic concepts and in this way to contibute to a philosophical debate. For instance, the concept for characterizing God in the Anselm's *Proslogion* can be formalized in the framework of combinatory logic.[34] Indeed, in *Proslogion*, the famous *"unum argumentum"* starts with a name characterizing God and accepted by the "insipiens": *id quo nihil majus cogitari possit.* This linguistic name expresses a concept *quo nihil majus cogitari possit.* The argument runs by analysing not the denotation of the name but the meaning of the associated concept. The consequence of this logico-linguistic analysis is: "*id quo nihil majus cogit ri possit*" exists no only in *"in-solo-intellectu"* (as a mental representation) but also *"in re"*. Anselm produces another argument. From this name and concept it deduces another proposition about God: "*Quod non possit cogitare non esse*". This proposition is expressed by:

$$[\mathbf{Q}(X) =_{def} (N \mathbf{\ o\ } P \mathbf{\ o\ } N \mathbf{\ o\ } E)(X) = (\mathbf{B}N(\mathbf{B}P(\mathbf{B}NE)))(X)],$$

where 'P' is the concept "is thinkable (or possible)", 'E' the concept "exists", 'N' the negation of a concept. The formalization of the argument was possible because the combinatory logic takes into account abstract concepts whose meaning can be studied by means of a combination of more elementary concepts. In the framework of predicate calculus with an interpretation in extensional classes in set theory, the logical analysis of the Anselm's argument fails.

Let us return to the combinator **Y** with the definition of the combinator given above:

$$\mathbf{Y} =_{def} \mathbf{WS}(\mathbf{BWB}).$$

This combinator applies to an unspecified operand X; it builds up an expression whose mathematical interpretation is following: '**Y**X' *is a fix point of the operator* X. In other words, the combinator **Y** was been used to build up a certain "object of thought" '**Y**X' which is invariant by the action of the

[34] See J.-P. Desclés, "La double négation dans l'Unum Argumentum analysé à l'aide de la logique combinatoire" *Travaux du Centre de Recherches Semiologiques*, 59, pp. 33–74, Université de Neuchtel, September, 1991.

operator X. What is the meaning of this "object of thought" **Y**? Its meaning is clear; it is given by the mathematical point fix situation ($\mathbf{Y}\mathrm{X} = \mathrm{X}(\mathbf{Y}\mathrm{X})$, where X is an operator whatever). The meaning of **Y** is: "to build a fix point of any operator".

Now, let us suppose that X is an *operator of negation*, denoted N. In this case, the combinator '**Y**' builds up a new entity '**Y**N' from the negation N. The expression '**Y**N' cannot be interpreted as a proposition! Indeed, with this interpretation, a contradiction arises since a proposition cannot be related to its own negation and we cannot accept the following property $\mathbf{Y}\mathrm{N} = \mathrm{N}(\mathbf{Y}\mathrm{N})$ (property of fix point for N): '**Y**N' *must not be interpreted as a proposition.*

Let a property or arbitrary concept be f. We define the new property "to be non-self-applicative" or the concept of "non-self-applicability of a property whatever". We obtain on applying the concept F to the property f: $\mathrm{Ff} =_{def} \mathrm{N(ff)}$. By introducing the combinators **B** and **W**, we define, by abstraction of f, F in terms of a combinator which applies to the operator N:

1. | $\mathrm{N(ff)}$
2. | $\mathbf{B}\mathrm{Nff}$
3. | $\mathbf{W}(\mathbf{B}\mathrm{N})\mathrm{f}$
4. | $\mathbf{BWB}\mathrm{Nf}$
5. | $[\mathrm{F} =_{def} \mathbf{BWB}\mathrm{N}]$.

Thus, the concept F ("non-self-applicability of a property whatever") is defined from N by means of the combinator **BWB**. Let us examine the consequences of the "self-aplicability of the concept F":

1. | FF
2. | $\mathbf{BWB}\mathrm{N}(\mathbf{BWB}\mathrm{N})$
3. | $\mathbf{S}(\mathbf{BWB})\mathbf{BWB}\mathrm{N}$
4. | $\mathbf{WS}(\mathbf{BWB})\mathrm{N}$
5. | $[\mathbf{Y} =_{def} \mathbf{WS}(\mathbf{BWB})]$
6. | $\mathbf{Y}\mathrm{N}$.

It follows from the above that the concept "self-applicability of the non-self-applicability of a property whatever" is identical to the applicative expression '**Y**N'. But, since the combinator **Y** (to see above) is a constructor of fix point, we obtain:

$$[\mathrm{FF} = \mathbf{Y}\mathrm{N} = \mathrm{N}(\mathbf{Y}\mathrm{N}) = \mathrm{N(FF)}].$$

The above analysis is an applicative formulation of Russell's paradox. We deduce again that the applicative expression 'FF' cannot be interpreted as a proposition. However, it does not follow that the expression 'FF' is "meaningless" and would be eliminated as an "object of thought" by running away from it. In the next paragraph, we will return to this topic.

7 Meaningless and Types

There are concepts which are "contradictory in themeselves". Such are the above concept "self-applicativity of non-self-applicativity of a property" or "the greatest cardinal number", "the least undefined ordinal number", "round square", etc. These concepts are considered meaningless because of their inherent contradictions. But the contradictions do not lie in the concepts themselves, but in properties associated with them. Thus, the above mentioned concept "self-applicativity of non-self-applicativity of a property" is a contradictory *only if one maintains that it is a proposition*; "the greatest cardinal number" only if it really is a cardinal number, etc.

> *"Now let us consider meaningless objects in general. Do not same thoughts apply to them also? Yes, the meaningless of these concepts consists only in the existence of properties which do not hold of them. And certainly I may say more exactly that they do not belong to the usual categories. These categories are in fact presupposed as somewhat contensive (Inhaltlich) primitive concepts, and nothing is considered which does not belong to them".* (Curry, 1929, Seldin: 14)

The concept "self-applicability of the non-self-applicability of a property whatever" seems to be paradoxical since it is identical to its negation. However, it may be analyzed in logicians[35] by the same fashion the mathematicians recognize that "the square root of the number 2" is an irrational number or that there are complex numbers which are not "real solutions" of certain algebraic equations. As Curry says: "*We must stand and look it in the eye*".[36] Whitehead and Rusell maintained that "self-applicability of a property" is meaningless for every property. To say that this claim is about only the symbol 'f(f)' – where f is a property – is t o get away the question. According to combinatory logic, self-applicability is *a priori* licit and the fix-point situation [FF = N(FF)] does not mean that we must exclude the applicative expression 'FF' from a logical investigation because this "object of thought" would not exist. The consequence of the above analysis is only: the "objects of thought" 'FF' and '**Y**N' *must be excluded from the category of propositions.* This brings us to the necessity of formulating the notion of "proposition" and we must also classify the "objects of thought" in different categories or types.

The meaningless seems to have an objective content, but what exactly is not clear. Above all there is a sense in which anything thinkable has a mean-

[35] "We can no longer "explain" a paradox by running away from it; we must stand and look it in the eye. (...) The paradoxes are forced, so to speak, into the open, where we can subject them to analysis" (Curry, 58, p. 5).

[36] (Curry, 1958 : 258–262); (Ladrière, 1973, p. 44–47); (Desclés, 1990).

ing, namely a concept. Here a concept is to be understood as anything that can be identified or differentiated from other things.[37]

> *"Then it is an absolute nonsense to say that something does not exist as a concept; for before one can understand such a sentence one must have already understandood the thing as a concept. Therefore even the "meaningless" objects are concepts and as such have a meaning. (...) The objective of logic is the explanation of thought; if there are thoughts excluded from the explanation, then it is deficient. (...). That something is a concept is the only requisite for its being subject to treatment in logic. (...)."* (Curry, 1929, Curry and Hindley 1972, Seldin : 14, 15)

The pure combinatory logic does not use different categories of entities. In introducing types, we take into account a classification of the entities into different kinds – syntactical and semantical categories – and the rules which determine the ways in which the results of a combination process can be so classified, once the classification of the constituents is known. We turn to the development of a machinery for the classification of objects into categories. We need four sorts of primitive notions:

- certain primitive categories;
- means for constructing composite categories from primitives;
- assertive propositions assigning atoms to certain categories;
- rules for inferring the category of an applicative expression when those of its components are known.

We define all *functional types* which are categories by means of a new specific atom **F**, as follows:

(i) primitive types (primitive categories) are functional types;
(ii) if x and y are functional types then **F**xy is a functional type.

This specific atom **F** is interpreted as an operator which builds up a functional type from two given types. The assignment of a type x to an applicative expression X is designated by 'xX'. The rule for inferring the type of an applicative expression is: "If the type of the opertator X is **F**xy and the type of the operand is x, then the type of the result 'XY' is y".

The applicative scheme[38] becomes with assigned types:

$$\frac{\mathbf{F}\mathrm{xyX} \qquad \mathrm{xY}}{\mathrm{y(XY)}}$$

[37] Curry's footnote: "It is well worth remarking that a concept by this definition is an object and not a process of thought. Some examples of concepts are Bismarck, Göttingen, animal, umbrella, red, temperature, matter, substance, causality, entity (Etwas), function, the King of France, the greatest cardinal number, etc."

[38] We note the analogy with *'modus ponens'*:

$$\frac{\mathrm{p} \Rightarrow \mathrm{q} \quad \mathrm{p}}{\mathrm{q}}$$

From this rule, we derive other rules[39]:

$$\frac{\mathbf{F}\mathrm{xyX} \qquad \mathrm{y(XY)}}{\mathrm{xY}} \qquad\qquad \frac{\mathrm{xY} \qquad \mathrm{y(XY)}}{\mathbf{F}\mathrm{xyX}}$$

Evidently combinators, which are essentially means for building up combinations other than those built up by direct application, will have special intuitive properties in connection with such an **F**.[40]

For a logical approach of predication with predicates defined on individual entities, we take two kinds of primitive categories or primitive types $\underline{J}$ and $\underline{H}$ representing, respectively, the categories of individuals and propositions. In a simplified theory of types without relations or decriptive functions, the types would be simply:

- $\underline{J}$ category of individual entities;
- $\mathbf{F}\underline{J}\underline{H}$ category of predicates or classes of individuals, such as "being a square";
- $\mathbf{F}(\mathbf{F}\underline{J}\underline{H})\underline{H}$ category of properties of properties, such as finiteness, cardinal number, quantifiers;
- $\mathbf{F}(\mathbf{F}(\mathbf{F}\underline{J}\underline{H})\underline{H})\underline{H}$ category of properties of families of sets, such as being ordered with respect to inclusion.

We consider other systems of functional types, for instance:

- $\mathbf{F}\underline{H}\underline{H}$: category of one-place propositional connectives, such as negation;
- $\mathbf{F}\underline{H}(\underline{H}\underline{H})$: category of two-place propositional connectives, such as implication or conjunction;
- $\mathbf{F}\underline{J}\underline{J}$: category of functions defined on individuals, such as "the square of", "the factorial of";
- $\mathbf{F}(\mathbf{F}\underline{J}\underline{J})\underline{H}$: category of properties of functions on individuals, such as monotony;
- $\mathbf{F}\underline{J}(\mathbf{F}\underline{J}\underline{H})$: category of relations between individuals, such as equality, order;
- $\mathbf{F}\underline{J}(\mathbf{F}(\mathbf{F}\underline{J}\underline{H})\underline{H})$: category of relations between an individual and a property (or a class), such as possession of a property or membership of the class.
- $\mathbf{F}(\mathbf{F}\underline{J}\underline{H})\underline{H}$: category of simple (unrestricted) quantifiers;
- $\mathbf{F}(\mathbf{F}\underline{J}\underline{J})\underline{H}$: category of typicality constructor operator which associates to every concept an object, the typical object;

---

[39] Note that here we have not a direct analogy with derived rules from *modus ponens*:

$$\frac{\mathrm{p} \Rightarrow \mathrm{q} \qquad \mathrm{p}}{\mathrm{q}} \qquad\qquad \frac{\mathrm{p} \qquad \mathrm{q}}{\mathrm{p} \Rightarrow \mathrm{q}}$$

These rules are not correct. We must introduce the deductive relation (denoted '$\vdash$') to obtain correct logical rule (deduction theorem):

$$\frac{\mathrm{p} \quad \vdash \quad \mathrm{q}}{\vdash \quad (\mathrm{p} \Rightarrow \mathrm{q})}$$

[40] The type of schemes of combinators are:
Fxx**I**; **F**x(**F**yx)**K**; **F**(**F**yz)(**F**(**F**xy)(**F**xz))**B**; **F**(**F**x(**F**yz))(**F**y(**F**xz))**C**; **F**(**F**x(**F**xy))(**F**xy)**W**

- $\mathbf{F}(\mathbf{F}\underline{\mathrm{J}}\underline{\mathrm{H}})(\mathbf{F}\underline{\mathrm{J}}\underline{\mathrm{J}})$: category of determination constructor operator which associates to every concept a determination operator, that is a function defined on individual entities.

The aim of illative logic is to characterize the correct inferences in using propositions, connectors, predicates, terms and quantifiers. It is an applicative system without bound variables in combinatory logic. With this applicative formalism "without bound variables", illative logic introduces restricted quantifiers Π (restricted universal quantifier)[41] and Σ (restricted existential quantifier)[42] as operators whose operands are unary predicates. The type of these two quantifiers is $\mathbf{F}(\mathbf{F}\underline{\mathrm{J}}\underline{\mathrm{H}})(\mathbf{F}(\mathbf{F}\underline{\mathrm{J}}\underline{\mathrm{H}})\underline{\mathrm{H}})$. Each quantifier builds up a proposition ('ΠXY' or 'ΣXY') with the type $\underline{\mathrm{H}}$, from a predicate Y (an operator with the type $\mathbf{F}\underline{\mathrm{J}}\underline{\mathrm{H}}$) with a restriction given by another predicate X (an operator with the type $\mathbf{F}\underline{\mathrm{J}}\underline{\mathrm{H}}$). The action of theses quantifiers is given by inference elimination and introduction rules without using bound variables. For the restricted quantifiers, let us give the corresponding rules (with one or two premises):

$$\frac{\vdash \Pi\mathrm{XY} \qquad \vdash \mathrm{XU}}{\vdash \mathrm{YU}}\ [\text{e-}\Pi] \qquad \frac{\vdash \mathrm{XU} \qquad \vdash \mathrm{YU}}{\vdash \Pi\mathrm{XY}}\ [\text{i-}\Pi]$$

$$\frac{\vdash \Sigma\mathrm{XY}}{\vdash \&(\mathrm{XU})(\mathrm{YU})}\ [\text{e-}\Sigma] \qquad \frac{\vdash \mathrm{XU} \qquad \vdash \mathrm{YU}}{\vdash \Sigma\mathrm{XY}}\ [\text{i-}\Sigma]$$

In the syntax of natural languages, the syntactical categories are represented by syntactical types such as nominal phrase, sentence, intransitive verbs, transitive verbs, prepositions, adverbs, adjectives, ... The syntactical types and syntactical process are studied in the theory of Categorial Grammars.

8 Cognitive Representations

We analyse the meaning of a lexical predicate by breaking it up into semantic relations and operations, called semantico-cognitive primitives. Thus, the meaning of a lexical predicate is represented by a cognitive scheme, that is, by a combination of semantico-cognitive primitives by means of combinators. Let us take an example. The meaning of *'enters'* in the sentence:

(1) *The ball enters the room*

is a "combination" of the following primitives (relators and operators):

---

[41] This quantifier is the formalisation of the usual quantifier defined as follows: $(\forall \mathrm{U})\ ((\mathrm{XU}) \Rightarrow \mathrm{YU}))$.

[42] This quantifier is the formalisation of the usual quantifier defined as follows: $(\exists \mathrm{U})\ ((\mathrm{XU})\ \&\ (\mathrm{YU}))$.

- ϵ_0 means "localisation of object X in relation to the place LOC(Y)":

$$\mathrm{X}\ \epsilon_0\ \mathrm{LOC(Y)}$$

- INT means "topological operator which is applied to the place LOC(Y); it determines its the interior of this place": INT(LOC(Y))
- EXT means "topological operator which is applied to the place LOC(Y); it determines its the exterior of this place": EXT(LOC(Y))
- MOVT means "relator expressing a spatial movement for an object X, from a static location SIT_1 to another static location SIT_2:

$$\mathrm{MOVT}\ (\mathrm{SIT}_1)\ (\mathrm{SIT}_2)$$

The lexical predicate *enters'* is considered as a "complex predicate". Its meaning is expressed by the "combination" of preceding cognitive primitives MOVT, ϵ_0, EXT, INT by means of a particular combinator **X**. More precisely, we have:

(2) [*enters'* $=_{def}$ **X** MOVT ϵ_0 EXT INT]

The above relation between a lexical predicate and its meaning is an example of a lexical law: the right side (*definiens*) express the meaning of the left side (*definiendum*). The combinator **X** is given in terms of more elementary combinators **B**, **C**, Φ and Ψ as follows:

(3) [$\mathbf{X} = \mathbf{C}(\mathbf{BB}(\mathbf{B}\Psi(\mathbf{B}\Phi\Phi)))\mathbf{B}$]

This combinator **X** expresses an applicative program for a "combination" of semantic primitives MOVT, ϵ_0, EXT, INT. We do not give, here, how we obtain the exact form of the combinator **X** (see the annex).

Now, let us consider the applicative representation of the sentence *The ball enters the room*:

(4) *enters' the-room' the-ball'*

We replace in the applicative expression (4) the predicate *enters'* by its meaning '**X** MOVT ϵ_0 EXT INT'. In executing, by means of successive β-reductions, the applicative program by eliminating all elementary combinators which are components of the combinator **X**, we obtain as a final result the applicative expression (5):

(5) MOVT (ϵ_0 (EXT (*the-room'*)) (*the-ball'*)) (ϵ_0 (INT(*the-room'*)) (*the-ball'*))

The last applicative expression (5) is a "normal form" which cannot be reduced. This means that the entity "the-ball" is the topic of a spatial movement from an initial spatial location, in the situation 1 ("the-ball" is located outside of the place "the-room") to the new spatial location, in the situation 2 ("the-ball" is located inside of the place "the-room").

By abstracting from the above cognitive representation (5), we define a *cognitive scheme* (6) expressed in English as follows:

(6) "MOVEMENT FROM OUTSIDE A PLACE TO INSIDE THE SAME PLACE"

The scheme (6) can be expressed by a λ-expression (7) (from Church's λ-calculus):

(7) $\lambda z.\lambda x.\{\text{MOVT } (\epsilon_0 \text{ EXT}(\text{LOC}(z))\ x)\ (\epsilon_0 \text{ INT}(\text{LOC}(z))\ x)\}$

where 'LOC(z)' is a variable whose semantic type is: "place" and 'x' a variable whose semantic type is: "individual entity". The scheme (7) is an abstraction (built up by means of a *lambda*-abstractor) from the cognitive representation (5). Its intuitive meaning is given by (6). It expresses an *invariant meaning* for a class of "equivalent" lexical predicates. For example, the lexical predicate *enters'* is an instance of the cognitive scheme (6).

Thus, this cognitive representation associated with the meaning of entering *The ball enters in the room* is expressed by the following λ-expression (8) where an operator (a scheme with λ-abstractors) applies to two successive terms, the operands of the operator:

(8) $(\lambda z.\lambda x.\{\text{MOVT}(\epsilon_0 \text{ EXT}(\text{LOC}(z))\ x)\ (\epsilon_0 \text{ INT } (\text{LOC}(z))\ x)\})$
(the-room') *(the-ball')*

After two substitutions (β-reductions in λ-calculus):

1. $(\lambda z.\lambda x.\{\text{MOVT } (\epsilon_0 \text{ EXT } (\text{LOC}(z))\ x)\ (\epsilon_0 \text{ INT}(\text{LOC}(z))\ x)\})$
   (the-room') *(the-ball')*
2. $(\lambda x.\{\text{MOVT } (\epsilon_0 \text{ EXT}($*the-room'*$)\ x)\ (\epsilon_0 \text{ int } ($*the-room'*$)\ x)\})$ *(the-ball)*
3. MOVT $(\epsilon_0$ EXT*(the-room')* *(the-ball)* $(\epsilon_0$ int *(the-room'))* *(the-ball'))*

we obtain the applicative representation (5).

Now, in the sentence *The ball enters the room*, the denotation of the argument, "the-ball", *has no intentional control* whatever over the action entering "the-room", but in the sentence (9):

(9) *John enters the room*

the denotation of the subject, "John", as a human being, *controls his own spatial movement.* The cognitive scheme corresponding to *enter'* in *John enters the room* is not the same scheme but it is built up on the basis of the preceding scheme. We must adjoin the relational primitive CONTR of "intentional ability" which links an agent to a change process. The meaning of this primitive CONTR is given in (10):

(10) "the Agent has control over the process, that is, he has complete control over whether to initiate or stop it, or maintain it"

The semantic description of the sentence *John enters the room* is represented by the applicative expression (11):

(11) CONTR (MOVT $(\epsilon_0($EXT*(the-room'))(John'))* $(\epsilon_0($INT *the-room'))* *(John'))* *(John')*

This expression means that "John is the topic of a spatial movement from an initial spatial situation ($\prec$ "John" is located outside of the place "the-room" $\succ$) to the new spatial situation ("John" is located inside of the place "the-room") and John controls this movement".

Let us remark that the applicative expression (12):

(12) (MOVT(ϵ_0 (EXT(*the-room'*))(*John'*)) (ϵ_0 (INT *the-room'*))(*John'*)))

is an instance of the scheme (6). This expression (12) is embedded in the scheme of control relation (13):

(13) CONTR (....) John'

The cognitive scheme associated to enters' in *John enters the room* is obtained by abstracting from (11), hence the following λ-expression (14):

(14) λz.λx.{CONTR (MOVT (ϵ_0 EXT(LOC(z)) x) (ϵ_0 INT(LOC(z)) x)) (x)}

The meaning of this scheme is given in (15):

(15) "AN AGENT CONTROLS ITS MOVEMENT FROM OUTSIDE A PLACE TO INSIDE THE SAME PLACE"

When we apply this λ-expression (14) to the sequence of arguments (*the-room'*) and (*John'*) by means of two successive β-reductions of λ-calculus:

1. (λz.λx.{CONTR (MOVT (ϵ_0 EXT (LOC(z)) x) (ϵ_0 INT(LOC(z)) x))x}) (*the-room'*) (*John'*)
2. CONTR (MOVT (ϵ_0 EXT (*the-room'*) (*John'*)) (ϵ_0 INT(*the-room'*) (*John'*))) (*John'*)

we obtain the cognitive representation (12).

As above, we can define a combinator **Z** which combines together the primitives ϵ_0, EXT, INT, MOVT, and CONTR in "a whole" which becomes the result of an integration process. Now, the meaning of the lexical predicate *enters'* in the sentence *John enters the room* is given as follows:

(16) [*enters'* $=_{def}$ **Z** CONTR MOVT ϵ_0 EXT INT]

Here, we do not give the explicit form of the combinator **Z**.

These two examples of occurrences of the verb *enters'* in two different sentences *the ball enters the room* and *John enters the room* show that the same lexical item *enters'* is polysemic, it belongs to different related cognitive schemes. In this example, a scheme is embedded in another.

9 Conclusion

In conclusion, I would like to say that combinatory logic is not exactly an alternative but rather a logical framework for studying paradoxes, for developing different logical systems (systems of negation, of quantifiers, and so on), for analysing combinations of concepts into a complex concept and, also, for relating cognitive representations (or knowledge representations) to lexical predicates in a natural language. Furthermore, there is a close connection between combinatory logic, functional programming languages, applicative analysis of natural languages with an architecture where a specific level is a knowledge–representation level.

ANNEX

1. We give the reduction of the applicative representation of the sentence *the ball enters the room* to its normal form:

(1) MOVT (ϵ_0 (EXT(*the-room'*)) (*the-ball'*)) (ϵ_0 (INT (*the-room'*)) (*the-ball'*))

by using the lexical law:

(2) [enters' = def **X** MOVT ϵ_0 EXT INT]

that relates the predicate to its meaning expressed as an applicative expression. The exact form of the combinator **X** in the above lexical law is:

(3) [**X** $=_{def}$ **C**(**BB** (**BY BFF**))**B**]

All applicative expressions are written as prefixed expressions where the operator is always before the operand. Let us start with the applicative expression:

(4) ((enters' (the-room')) (the-ball'))

We have the following reduction, as in "Gentzen's natural deduction":

1.	enters' (the-room') (the-ball').	hyp.
2.	[enters' $=_{def}$ **X** MOVT ϵ_0 EXT INT]	lexical law
3.	[**X** $=_{def}$ **C** (**BB** (**B**Ψ(**B**$\Phi\Phi$)))**B**]	def. of **X**
4.	**X** MOVT ϵ_0 EXT INT (the-room") (the-ball')	remp. 2, 1
5.	**C**(**BB**(**B**Ψ(**B**$\Phi\Phi$)))**B**	
	MOVT ϵ_0 EXT INT (the-room') (the-ball')	remp. 3, 4
6.	**BB**(**B**Ψ(**B**$\Phi\Phi$))	
	MOVT **B** ϵ_0 EXT INT (the-room') (the-ball')	[**C**-e], 5
7.	**B**(**B**Ψ(**B**$\Phi\Phi$) MOVT)	
	B ϵ_0 EXTINT(the-room') (the-ball')	[**B**-e], 6
8.	**B**Ψ (**B** $\Phi\Phi$)	
	MOVT (**B** ϵ_0) EXT INT (the-room') (the-ball')	[**B**-e], 7
9.	Ψ(**B** $\Phi\Phi$ MOVT)	
	(**B** ϵ_0) EXT INT (the-room') (the-ball')	[**B**-e], 8
10.	1. **B** $\Phi\Phi$ MOVT	hyp.
	2. Φ(Φ MOVT)	[**B**-e], 10

11.	Ψ (Φ (Φ MOVT)) (**B** ϵ_0) EXT INT (the-room') (the-ball')	remp. 10, 9
12.	Φ (Φ MOVT) (**B** ϵ_0 EXT) (**B** ϵ_0 INT) (the-room') (the-ball')	[Ψ-e], 11
13.	Φ MOVT (**B** ϵ_0 EXT(the-room')) (**B** ϵ_0 INT (the-room') (the-ball'))	[Φ-e], 12
14.	MOVT (**B** ϵ_0 EXT(the-room' (the-ball'))) (**B** ϵ_0 INT (the-room' (the-ball')))	[Φ-e], 13
15.	1. **B** ϵ_0 EXT(the-room' (the-ball')) — hyp. 2. ϵ_0 (EXT(the-room') (the-ball')) — [**B**-e]	
16.	1. **B** ϵ_0 INT(the-room' (the-ball')) — hyp. 2. ϵ_0 (INT(the-room') (the-ball')) — [**B**-e]	
17.	MOVT (ϵ_0 (EXT(the-room') (the-ball'))) (ϵ_0 (INT(the-room') (the-ball')))	remp.,5,6,4

We obtain the reduction relation between the applicative representation of the sentence *the ball enters the room* and its normal form, that is an irreducible expression:

(6) enters'(the room (the-ball) $\rightarrow_\beta$ MOVT (ϵ_0 (EXT(the-room') (the-ball'))) (ϵ_0 (INT(the-room') (the-ball')))

This relation represents a compiling process (in an abstract functional programming language) of the sentence into its normal form, that is an "intermediate representation" for understanding this sentence (Desclés, 1990).

2. Now, we are going show how we research a combinator from a cognitive scheme which is applied to a sequence of arguments. Let us consider the cognitive scheme associated to the lexical predicate *enters'* in the sentence *John enters the room.* This scheme is given by the λ-expression:

(7) λz.λx.{ CONTR (MOVT (ϵ_0 EXT(LOC(z)) x) (ϵ_0 INT(LOC(z)) x)) (x) }

When the scheme applies to the sequence of arguments *the-room'* and *John'* we obtain the applicative cognitive representation (8):

(8) CONTR (MOVT (ϵ_0 EXT (*the-room'*) (*John'*)) (ϵ_0 INT (*the-room'*) (*John'*))) (*John'*)

by the two β-reductions:

1. (λz.λx.{CONTR (MOVT (ϵ_0 EXT(LOC(z)) x) (ϵ_0 INT(LOC(z)) x)) x}) (*the-room'*) (*John'*)

2. CONTR (MOVT (ϵ_0 EXT(the-room' (John'))) (ϵ_0 INT(the-room') (John'))) (John')

This applicative expression (8) is a normal form. By successive β-expansions (introduction of combinators), a complex combinator **Z** emerges from the normal form. When this combinator applies to a sequence of primitives, we obtain the normal form. The process which defines progressively this combinator **Z** takes as input a normal form (a cognitive representation instanced from a cognitive scheme;) as output, we obtain another applicative expression which expresses that the lexical predicate applies to its different operands. The lexical predicate is defined by means of a lexical law that relates this lexical predicate (the *definiendum*) and an applicative expression (a *definiens*) where a combinator combines different semantic and cognitive primitives or lexical predicates which had already been defined. This lexical law gives the intrinsic meaning of the lexical predicate.

1.	CONTR MOVT(ϵ_0 EXT(the-room') (John')) (ϵ_0 INT(the-room') (John')))(John')
2.	1. MOVT (ϵ_0 EXT(the-room') (John')) (ϵ_0 INT(the-room') (John'))
3.	2. [$\mathbf{X} =_{def}$ **C** (**BB** (**B** Ψ (**B** $\Phi\Phi$))) **B**]
4.	3. [enters$_1$' = **X** MOVT ϵ_0 EXT INT]
5.	4. enters$_1$' (the-room') (John')
6.	CONTR (enters$_1$' (the-room') (John')) (John')
7.	**B** CONTR (enters$_1$' (the-room')) (John') (John')
8.	**B** (**B** CONTR) (enters$_1$') (the-room') (John') (John')
9.	**BBB** CONTR (enters$_1$') (the-room') (John') (John')
10.	[enters$_{11}$' $=_{def}$ **BBB** CONTR (enters$_1$')]
11.	enters$_{11}$' (the-room') (John') (John')
12.	**W**(enter'$_{11}$ (the-room')) (John')
13.	**B W** (enters'$_{11}$) (the-room') (John')
14.	[enters'$_2$ = **B W** (enters'$_{11}$)]
15.	enters'$_2$ (the-room') (John')

We deduce from above the combinator **Z** such that:

$$[\ \text{enters'} \equiv \mathbf{Z}\ \text{CONTR MOVT}\ \epsilon_0\ \text{EXT INT}]$$

The exact formulation of **Z** is given as follows:

$$[\ \mathbf{Z} \equiv \mathbf{C}(\mathbf{B}\ \mathbf{B}^4(\mathbf{B}(\mathbf{B}(\mathbf{BW}))(\mathbf{BBB}))(\mathbf{C}(\mathbf{BB}(\mathbf{B}\Psi(\mathbf{B}\Phi\Phi)))\mathbf{B})]$$

This combinator integrates the primitives MOVT, ϵ_0, EXT, and INT into a whole. This combinator is an applicative program for combining the primitives. It is an indicator of the complexity of lexical synthesis from a description of a meaning by a scheme.

References

1. Ajdukiewicz, K.: Die syntaktische Konnexität (Studia Philosophica 1935) **1**, pp. 1–27
2. Asperti, A. and Longo, G.: *Categories, Types and Structures* (The MIT Press 1991)
3. Bach, E.: *Categorial Grammars as Theories of Language.* (Oehrle et al. 1988) pp. 17–34
4. Bar-Hillel, Y.: A quasi-arithmetical notation for syntactic description In *Language* 29 (1953) pp. 47–58
5. Barendregt, H.: *The Lambda-calculus: Its Syntax and Semantics* (North-Holland 1980)
6. Biskri, I. and Descles, J.-P.: Applicative and Combinatory Catégorial Grammar (from syntax to functional semantics) In *Acte du Colloque RANLP* (Bulgarie 1995)
7. Curry, H. B.: Grundlagen der kombinatorischen Logik (Inauguraldissertation) In *American Journal of Mathematics* pp. 509–536, 789–834 (1930)
8. Curry, H. B.: Some logical aspects of grammatical structures, Structure of language and its mathematical aspects In *Proceedings of the symposia in applied mathematics*, Vol. XII (American Mathematical Society 1961) pp. 56–68
9. Curry, B. H. and Feys, R.: *Combinatory Logic*(North-Holland 1958), Vol. I
10. Curry, B. H. and Hindley, J. R. and Seldin, J. P.: *Combinatory Logic* (North-Holland 1972), Vol. II
11. Descles, J.-P.: De la notion d'opération à celle d'opérateur ou à la recherche de formalismes intrinsèques In *Mathématiques et sciences humaines* **76** pp. 1–32 (1981)
12. Descles, J.-P.: Opérateurs/opération: Méthodes intrinsèques en informatique fondamentale. Applications aux bases de données relationneles et à la linguistique, Thèse d'Etat, Université René Descartes Paris (1980)
13. Descles, J.-P.: Théorème de Church-Rosser et structuration des langues naturelles In *Mathématiques et sciences humaines* **103** pp. 67–92 (1988)
14. Descles, J.-P.: *Langages applicatifs, langues naturelles et cognition* (Hermès, Paris 1990)
15. Descles, J.-P.: Raisonnements aspecto-temporels In *Travaux du Centre de Recherches Semiologiques* (Université de Neuchâtel septembre 1995) **63** pp. 50–91
16. Descles, J.-P.: La double négation dans l'Unum Argumentum analysé à l'aide de la logique combinatoire In *Travaux du Centre de Recherches Semiologiques, Université de Neuchâtel* (septembre 1991) **59** pp. 33–74
17. Descles, J.-P.: Dialogue sur la typicalité In *Modèles et concepts pour la science cognitive, hommage à J. F. Le Ny* (Presses Universitaires de Grenoble 1993) pp. 139–163
18. Descles, J.-P.: Auto-applicativité, paradoxes et logique combinatoire In *Les cahiers du CREA, épistémologie et anthropologie, autoréférence, identité et altérité, sous la direction de Lucien Scubla* (Crea, Ecole Polytechnique, Paris 1993) pp. 63–119
19. Descles, J.-P. and Biskri, I.: Logique combinatoire et linguistique: grammaire catégorielle combinatoire applicative In *Mathématiques, informatique et Sciences Humaines* **32** pp. 39–68 (1995)

20. Descles, J.-P. and Guentcheva, Z. and Shaumyan, S. K.: *Theoretical aspects of passivization in the framework of Applicative Grammar* (John Benjamins 1985)
21. Descles, J.-P. and Guentcheva, Z. and Shaumyan, S. K.: A theoretical analysis of reflexivization in the framework of Applicative Grammar In *Linguisticae Investigationes* **2** (1986)
22. Fitch, F.: *Elements of Combinatory Logic* (Yale University Press 1974)
23. Frey, L.: Langages logiques et processus intellectuels In *Modèles et formalisation du comportement*(Editions du CNRS, Paris 1967) pp. 327–345
24. Ginisti, J.-P.: *La logique combinatoire* (Presses Universitaires de France, Que sais-je? 1997)
25. Guentcheva, Z.: Présentation critique du modèle applicationnel de S.K. Shaumyan In *Documents de linguistique quantitative* **30** (Dunod 1976)
26. Hindley, J. R. and Seldin, J. P.: *Introduction to Combinators and Lambda-Calculus* (Cambridge Univ. Press 1986)
27. Ladriere, J.: Le symbolisme comme domaine de l'opératoire In *Articulation du sens, discours scientifique et parole de foi* (Aubier-Montaigne 1970) pp. 51–70
28. Lambek, J.: The Mathematics of Sentence Structure In *American Mathematical Monthly* **65** (1958) pp. 154–165
29. Lambek, J.: On the calculus syntactic types In *Proceeding of Symposia in Applied Mathematics* (American Mathematical Society, Providence, Rhode Island 1961) Vol. XII, pp.166–178
30. Lambek, J.: Deductive systems and categories. Cartesian closed categories, intuitionist propositional calculus and combinatory logic In *Category Theory and lambda-calculus. Lectures Notes in Mathematics, 274* (Springer-Verlag 1972) pp. 57–82
31. Lambek, J. and Scott, P. J.: *An introduction to higher order categorical logic* (Cambridge University Press 1986)
32. Lecomte, A. and Lamarche, F. and Perrier, G.: Logical Aspects of Computational Linguistics In *Second International Conference, LACL 7, Selected Papers, Nancy, France, Lecture Notes In Artificial Intelligence* (Springer Verlag, September 1998)
33. Martin-Löf, P.: *Intuitionistic Type Theory* (Bibliopolos, Naples 1984)
34. Montague, R.: *Formal philosophy: Selected papers of Richard Montague* ed. by R. Thomason, (Yale University Press, New Haven 1974)
35. Oehrle, R. T. and Bach, E. and Wheeler, D.: *Categorial Grammars and Natural Languages Structures* (D. Reidel 1988)
36. Schönfinkel, M.: *Über die Bausteine der Mathematischen Logik* **92** (Math. Annalen 1924)
37. Scott, D.: Some philosophical issues concerning theories of combinators In *Lecture Notes in Mathematics* 274 (Springer Verlag 1975) pp. 57–82
38. Seldin, J.-P. and Hindley, J. R.: *To H. B. Curry : Essays on Combinatory Logic, Lambda Calculus and Formalism* (Academic Press, London 1980)
39. Shaumyan, S. K.: *Applicational Grammar as a Semantic Theory* (Edimburgh Univ. Press 1977)
40. Shaumyan, S. K.: *A Semiotic Theory of Natural Language* (Indiana Univ. Press Bloomington 1987)
41. Shaumyan, S. and Hudak, P.: Linguistic, Philosophical, and Pragmatic Aspects of Type-Directed Natural Language Parsing In *Lecomte et al.* (1998) pp. 70–91
42. Steedman, M.: Combinators and Grammars In *Oehrle et al.* (1988) pp. 207–263

43. Stoy, J. E.: *Denotational Semantics. The Scott-Strachey Approach to Programming Language* (The MIT Press, Cambridge 1997)
44. van Benthem, J.: *Language in Action: Categories, Lambda and Dynamic Logic* (North-Holland, Amsterdam 1991)

Extending the Realm of Logic: The Adaptive-Logic Programme*

Diderik Batens

Summary. This paper reports on a development that involves a drastic broadening of the scope of symbolic logic. The central idea concerns dynamic proofs that explicate forms of reasoning for which no positive test is available. Two other forms of the dynamics of reasoning are briefly spelled out.

1 The Problem

It is well-known that much *interesting* actual reasoning cannot be understood in terms of logics of the usual type. The reason is two-fold. First, much actual reasoning is nonmonotonic.[1] This means that a conclusion may be withdrawn in view of *new information*. Moreover, much actual reasoning is dynamic: a conclusion may be withdrawn in view of the *increased understanding of the premises*.

To put it differently: much actual reasoning is affected by an external dynamics as well as by an internal dynamics. In both cases, a formerly drawn conclusion may be revoked. In the case of the *external* dynamics, the effect is caused by the fact that new information is gained while the reasoning is going on. In the case of the *internal* dynamics, it is caused by the reasoning process itself: as it proceeds, our insight into the premises increases. The external dynamics (nonmonotonicity) is a property of the inference relation. The internal dynamics only affects the actual reasoning, not the inference relation.

While the external dynamics is well-documented in the literature, this does not hold for the internal dynamics. Nevertheless, it is a most familiar phenomenon. All forms of ampliative reasoning (abduction, applications of analogy, etc.) rely on a given system of knowledge in order to extend it. Later, further reasoning may reveal that the extension is incompatible with the original system of knowledge – note that this involves a better understanding of the original system. Except for some special cases, the extension will be

* Research for this paper was supported by the Fund for Scientific Research – Flanders, and indirectly by the INTAS-RFBR contract 95–365. I am indebted to audiences in Adelaide, Brisbane, and Salzburg for helpful comments.

[1] A logic is monotonic if and only if $\Gamma \vdash A$ warrants $\Gamma \cup \Delta \vdash A$. Inductive reasoning is a simple and well-known example of nonmonotonic reasoning.

revoked. There is not much we can do about such 'mistakes'. Compatibility is not a decidable relation: there is no algorithm for "A is compatible with Γ". Even worse, there is no positive test for it.[2]

Even in the absence of a positive test, the derivability relation may, of course, be well-defined. Abstract definitions may be present with respect to some monotonic logic, with respect to a semantics, and with respect to dynamic proofs. Such definitions are nice and important, but computationally useless (even the ones in terms of dynamic proofs – see below). Does this mean that, from a computational point of view, we are completely at a loss? By no means, there may still be *criteria*. For example, proofs that have specific properties may warrant that A is compatible with Γ.

Let me recapitulate. Some derivability relations are decidable. Some are undecidable, but there is a positive test for them. For the derivability relations that interest us here – those characterized by an external dynamics and by an internal dynamics – there is no positive test, but there are criteria.

Often, no criterion applies. Even then, we have to decide and to act. Such decisions and actions need not be irrational. Quite to the contrary, present-day insights on rationality reject the possibility of absolute warrants and require us to take decisions in view of our *best available insights* – see [2] and [5].

Incidentally, this clarifies why the internal dynamics is not a property of the inference relation.[3] For example, in order to extend our knowledge, we need to make a judgement on the compatibility of some statement with our present knowledge. As (in general) no final judgement can be reached, the only alternative is to rely on a provisional judgement. The latter is provided by the dynamic proofs. As they proceed, they 'approach' the final judgement.

Logics of the usual type are monotonic and their proofs lack dynamics (in the above sense).[4] It follows that they are unable to make sense of such 'best insights'. So, the challenge is to find logics for the forms of reasoning that display an external as well as an internal dynamics.

---

[2] An algorithm for a property is a systematic procedure that, after a finite (but possibly very large) number of steps, leads to a "yes" if the property does apply, and to a "no" if it does not. A positive test is a similar procedure that leads, after finitely many steps, to a "yes" if the property applies, but may go on forever if it does not. The statement that there *is* or *is not* an algorithm (or a positive test) refers to existence in the mathematical sense, not to our knowledge.

[3] It is, however, connected to it. It is provable that the proof theory that characterizes a non-monotonic inference relation, necessarily displays an internal dynamics.

[4] But see Sect. 2 for a weaker dynamics that is common to all proofs.

2 Adaptive Logics: A Step Within Symbolic Logic Towards Actual Reasoning

When Aristotle discovered (or invented) deductive formal logic, his aim was clearly to get a hold on actual reasoning. Much has changed since Aristotle. The game is now played in formal languages and language schemes. Proof theory, formal semantics, and model theory have been developed, and the first is clearly separated from the latter two. The definition of abstract properties and relations is clearly distinguished from their (effective) decidability (or computability). A host of metatheoretic methods and results is available. And so on. All this has turned logic into a powerful and well-controlled instrument. In one respect, however, we did not make any progress. The actual reasoning forms explicated by the usual logical systems of our era are still of the same type as those Aristotle tried to explicate: they are monotonic and their proof theory is non-dynamic (in the sense of Sect. 1).

Adaptive logics overcome this limitation. They constitute a step, *within symbolic logic*, towards actual reasoning. They are formulated in formal languages of the usual sort – for most of them, just the language of **CL**. Their metatheory is phrased in the classical metalanguage. And many adaptive logics are just extensions of **CL**. I do not claim that it makes no sense to start from the opposite direction. To do so, however, would come to a start from scratch (as is illustrated by the multiplicity of approaches to 'argumentation'). Vagueness would threaten, and metatheoretic methods would be unavailable. By moving towards actual reasoning within symbolic logic, the enterprise profits from the available metatheoretic insights and techniques (and questions).

An adaptive logic is characterized[5] by an upper limit logic, a lower limit logic, and an adaptive strategy. The two logics are monotonic systems of the familiar sort. The upper limit logic defines a set of logical presuppositions. The lower limit logic, which is a subsystem of the former, drops some of these presuppositions. The intuitive idea is that a set of premises is interpreted 'as much as possible' in accordance with the presuppositions of the upper limit logic. As the phrase 'as much as possible' has several sensible interpretations, we need an adaptive strategy to settle on a precise meaning.

As all this sounds rather abstract, let us consider an intuitive example. Let **CL** be the upper limit logic, and let some paraconsistent[6] logic – let us call it **PL** – be the lower limit logic. **CL** presupposes that $A \wedge B$ is true

[5] As the adaptive logic programme is under development, and its domain constantly broadened, this characterization should be seen as provisional. Insights reached as late as 1999 forced the Ghent group to modify (once more) the characterization to the one in the text.

[6] For present purposes, take a logic to be paraconsistent if and only if it does not generally validate $A, \sim A \vdash B$; in words: the consequence set of inconsistent premises need not be trivial (contain all formulas). We shall meet a paraconsistent logic in Sect. 4.

iff both A and B are true, it requires that precisely one of A and $\sim A$ is true, etc.[7] **PL** gives up some of the presuppositions of **CL**, viz. that at most one of A and $\sim A$ is true (consistency). By choosing an adaptive strategy, we obtain an inconsistency-adaptive logic that interprets the premises 'as consistently as possible'. If they are consistent, the adaptive logic delivers all their **CL**-consequences; if they are inconsistent, it still delivers more than the **PL**-consequences.

The difference between the lower limit logic and the adaptive logic is best illustrated by an example. Suppose that **PL**, as most paraconsistent logics, does not validate Disjunctive Syllogism (from $A \vee B$ and $\sim A$ to derive B). Then neither r nor s is a **PL**-consequence of $\{p, \sim q, \sim p, p \vee r, q \vee s\}$. However, where any interpretation of these premises requires that both p and $\sim p$ are true, their maximally consistent interpretations make $\sim q$ true and q false. So, as $q \vee s$ is bound to be true, those maximally consistent interpretations all make s true. In other words, according to adaptive logics based on **PL**, s is and r is not a consequence of the premises – see Sect. 4 for details. This nicely illustrates that the lower limit logic invalidates certain *rules* that are valid on the upper limit logic – in the previous example Disjunctive Syllogism. The adaptive logic invalidates certain *applications* of those rules, depending on the specific premises. In the previous example, the adaptive logic validates the application of Disjunctive Syllogism to $\sim q$ and $q \vee s$, but invalidates its application to $\sim p$ and $p \vee r$.

The previous paragraph concerns the (abstract) derivability relation, which is easily seen to be nonmonotonic – if $\sim q$ is added to the premises, s is no longer derivable. The more complex the premises, the less people will be able to see whether the premises require a formula to behave inconsistently. And it is, in general, undecidable whether premises (from the first–order language) require some formula to behave inconsistently. This is where, and why, the dynamic proof theory comes in.

The dynamic proof theory is implemented as follows. If A follows from previous steps by the upper limit logic, but not by the lower limit logic, one may derive A at (what I shall call) a conditional line. At any stage of the proof, such lines are either 'marked' (as OUT) or nonmarked in view of the formulas that, at that stage, occur at the unconditional lines. In general, a marked line may be nonmarked at a later stage, and *vice versa.*

This proof format requires one to distinguish between *derivability at a stage* and *final derivability.* The first denotes that a formula is derived on a nonmarked line in some proof. A formula is *finally derived* at a line in a proof if and only if any extension of the proof in which the line is marked, may be further extended in such a way that the line is nonmarked.[8] A for-

[7] I phrase this in semantic terms to simplify the discussion.

[8] As shown in [9], this expression may be taken at face value for the Reliability Strategy, but involves a reference to infinite extensions for the Minimal Abnormality Strategy.

mula is *finally derivable* from a set of premises if it is finally derived in a proof from these premises.

All adaptive logics have a semantics that is *characteristic* with respect to final derivability (A is finally derivable from Γ if and only if A is a semantic consequence of Γ). It is instructive to graphically represent the relation between the models of the lower limit logic – **LLL** for short – the upper limit logic – **ULL** for short – and the adaptive logic. In Fig. 1, the 'ovals' represent the **LLL**-models and the **ULL**-models (as indicated), the circles represent the **LLL**-models of Γ, and the smallest fields (the small oval on the left and the intersection of the oval and the circle on the right) represent the adaptive models of Γ.

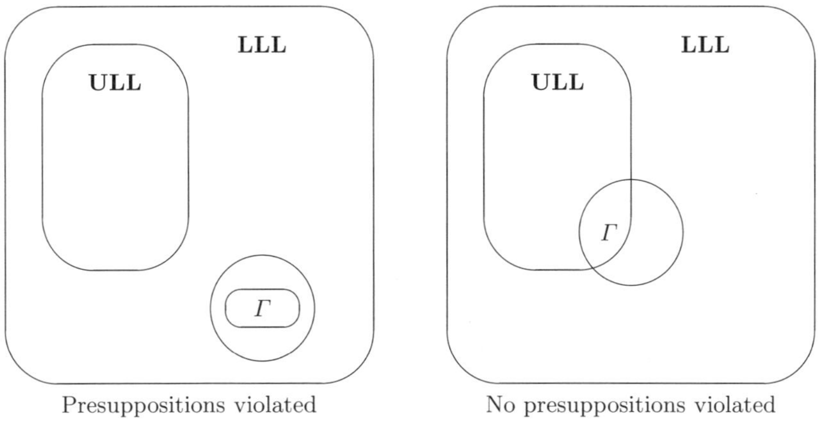

Fig. 1. Relations between the models

Keep in mind that a smaller set of models in general warrants more semantic consequences. Note also that, as Γ has no **ULL**-models on the left-hand side, its **ULL**-consequence set is trivial. Where **AL** refers to the adaptive logic and $Cn_{\mathbf{L}}(\Gamma)$ denotes the set of **L**-consequences of Γ, the figure illustrates (the provable statement) that, $Cn_{\mathbf{AL}}(\Gamma) = Cn_{\mathbf{ULL}}(\Gamma)$ if Γ has **ULL**-models, and that (in general) $Cn_{\mathbf{LLL}}(\Gamma) \subset Cn_{\mathbf{AL}}(\Gamma) \subset Cn_{\mathbf{ULL}}(\Gamma)$ otherwise.

Given that the semantics is characteristic for final derivability, one may wonder whether the dynamics of the proofs is real – that is, whether derivability at a stage is not an artificial notion. The question is central, as derivability at a stage is supposed to explicate the dynamics of the actual reasoning. The answer is provided in [8]. The result may be summarized as follows. Each proof at a stage determines a 'block analysis' of the formulas that occur in the proof: the minimal discriminations in and identifications of formulas that have to be made in order for the proof to be correct.[9] On the block-analysis, each proof corresponds to a list of block-formulas – formulas composed of 'blocks' that are internally unanalysed. It is then proved that derivability at a stage (of a block proof) is sound and complete with respect to the block semantics. This shows that derivability at a stage in a (usual) proof corresponds to *final* derivability in the corresponding block proof. Thus, the block analysis offers an explication for the insight in the premises that is provided by a stage of a proof. It follows that the dynamics of the proofs is real: only by continuing the proof does the insight in the premises increase. The block analysis, moreover, shows that derivability at a stage indeed offers the best insights in the premises available at that time – see the last paragraphs of Sect. 1.

Applying the block approach to adaptive logics also resulted in criteria for final derivability (other such criteria were found by studying tableau methods – see [14]).

Allow me a short digression. The block approach applies to any logic. The main gain is an exact measure of the information provided by a proof at a stage. This eliminates at once such riddles as omniscience. And the block approach offers much more. For example, a means to explicate meaning change, and the possibility to devise languages that are not build from nonlogical constants with a fixed meaning, but (as is the case for natural languages) from nonlogical constants the meaning of which may be modified later. I refer especially to [8] for further details. Let me stress that all this does not entail any loss in precision at the symbolic level – some reflection on Hilbert's programme reveals at once that symbolic logic requires exactness at the level of the symbols, not at the level of their meanings.

3 Possible Objections

Some will object that forms of reasoning as the ones discussed in Sect. 1 should not be studied in terms of logics. A superficial reply is that there is no reason to stick to the accidental choice made by Aristotle. But there is more. Only in the twentieth century, we did discover that first-order logic is undecidable. After gaining this unexpected insight, logicians did not decide to restrict their attention to decidable fragments of **CL**. Similarly, it was discovered in this century that there is no warrant for the consistency of arithmetic, that any first-order theory axiomatization of arithmetic is incomplete, that no consistent system of arithmetic contains its own truth theory,

[9] In order, for example, to derive $p \wedge q$ from q and $q \supset (p \wedge q)$, one has to see that the latter formula is composed of the two blocks connected by an implication, and one has to identify the first block with the other premise, and the second block with the conclusion. In terms of the block analysis, $[p \wedge q]^5$ is derived from $[q]^2$ and $[q]^2 \supset [(p \wedge q)]^5$, where the only function of the numbers is to indicate which formulas have been identified.

etc. All these limitative theorems did not cause logicians and mathematicians to turn away from the study of arithmetic, or to restrict their attention to specific fragments. So, why should we stop applying all those beautiful logical and metalogical insights when we find that some forms of reasoning are nonmonotonic and that their proof theory is dynamic?

There are further reasons to consider adaptive logics as belonging to the field of logic. These logics may be, and actually are, studied by absolutely standard means. The metatheory is always classical and relies heavily on standard techniques from the study of **CL**. Moreover, most adaptive logics are extensions of **CL**, just as modal logics. The only present exceptions are $\mathbf{LP}^m$ from [28] and **ANA** from [22], and even these may easily be transformed to extensions of **CL**. Also, the forms of reasoning that adaptive logics intend to explicate, are by no means restricted to so-called everyday reasoning. There is overwhelming evidence from the history of the sciences that, in mathematical as well as in 'empirical' sciences, those types of reasoning play a central role: reasoning about compatibility, abductive inferences, reasoning from inconsistent premises, etc. – see, for example, [17, 21, 24, 25, 26, 27, 30] on inconsistencies.[10]

With respect to deductive inference, the traditional view stresses the distinction between logic and formalization. Statements from natural languages have to be formalized before logic can be applied to them. How this formalization should proceed is not determined by logic. This is a good but not unproblematic point. First, it easily leads to the existence of a 'Logical Heaven' that is fully disconnected from the logical forms at the level of natural languages. As I argued in [10], such a realm is unreachable for humans and irrelevant to their reasoning. Moreover, no formalization will ever lead to an adequate treatment, within a logic of the usual sort, of the dynamic forms of reasoning mentioned in Sect. 1. For example, no sensible way to formalize an ampliative argument will make it correct according to a logic of the usual sort.

Ever since Aristotle, formal logic was seen as an attempt to get a hold on actual reasoning. The study of formal logic has forced us to develop a metatheory about abstract properties and relations. Although this move directs the attention away from actual reasoning, it was necessary and important; it led to a mature view on logic. (Precisely for this reason, it is applied to adaptive logics.) However, the sense of the whole enterprise, including its abstract aspects, is to get a hold on reasoning, including its limitations (such as the absence of a positive test). Some logicians tend to forget this. They are responsible for the fact that 'logic' becomes gradually more irrelevant to natural languages, to the philosophy of science, and even to the sciences.

[10] If empirical scientists would realize what **CL** really comes to, and how poor an instrument it is, they would cry out for something better.

4 Inconsistency-Adaptive Logics

These are the oldest adaptive logics, discovered in 1979 by considering the following problem. Let $T = \langle \Gamma, \mathbf{CL} \rangle$, in which Γ is a set of nonlogical axioms, be a theory that was meant to be consistent – this is why **CL** is the underlying logic – but at some point turns out to be inconsistent. There are well-known examples of such theories from the history of mathematics – Frege's set theory is one of them – as well as from the history of the 'empirical' sciences. In such a case, we want to replace T by some consistent improvement of it. In order to do so, however, we cannot start from scratch. We have *to reason* from T. We have to figure out at which points T is inconsistent, and at which points it is unproblematic.

Obviously, **CL** is useless to reason from T (as it turns inconsistent theories into trivial ones). Unlike what one might expect, paraconsistent logics are also inadequate to reason from T. The point is – see [3] – that they are too weak: they invalidate a large set of rules of inference that are correct according to **CL**. As a starting point for devising a consistent improvement of T, we want T 'in its full richness, except for the pernicious consequences of its inconsistency'. Where T is inconsistent, we want to restrict its consequences in order to avoid triviality. But where it is consistent, we want *all* classical consequences. In other words, we want an interpretation of T that is as consistent as possible.

Consider again Disjunctive Syllogism. According to most paraconsistent logics, it is invalid. As some models verify both A and $\sim A$, B is not true in all models of $A \vee B$ and $\sim A$. Now, remember what our goal was: an interpretation of T that is as consistent as possible. If both A and $\sim A$ are derivable from T, we do not want to apply Disjunctive Syllogism because this would lead us into triviality. But if *only* $\sim A$ is derivable from T, we *do* want Disjunctive Syllogism to apply.

This suggests an inference relation with the property: if both $A \vee B$ and $\sim A$ are derivable from Γ, then B is derivable from Γ, unless A is derivable from Γ. This property, however, is flatly circular. At this point, my training by Leo Apostel paid dividends. The idea is all right, but it should not be applied to the abstract inference relation, but to the concrete proofs. Moving to the level of the proofs at once delivered a noncircular approach: if both $A \vee B$ and $\sim A$ have been derived from the theory, and A has not been derived (at that stage of the proof), derive B. At a later stage, A may be derived from the theory. So, we better make the derivation of B a conditional one: we note that the line is correct provided A has not been derived from the theory. If, at a later stage, A is derived, we *mark* the line at which B is (thus) derived: the condition turns out not to be fulfilled at that later stage, and hence B turns out, at that stage, not to be derivable from the theory.

There is, alas, a slight complication. From some theories, $A \wedge \sim A$ is not derivable, but $(A \wedge \sim A) \vee (C \wedge \sim C)$ is. In some cases (see [3]) the approach suggested above leads to indeterministic proofs (and that is the last thing logicians want). So, we have to slightly change our policy with respect to marking lines.

There are several 'strategies' for interpreting a theory as consistently as possible. The formerly suggested strategy is the Simple Strategy. It is all right in some cases (see Sect. 6) but useless with respect to handling inconsistencies (except for some very special lower limit logics). The two most obvious strategies for the latter are Reliability and Minimal Abnormality. Suppose that the proof contains $(A_1 \wedge {\sim}A_1) \vee \ldots \vee (A_n \wedge {\sim}A_n)$,[11] but does not contain any formula obtained by dropping one or more disjuncts. The Reliability Strategy then considers each of $A_1, \ldots, A_n$ as unreliable, and requires all lines that have one of these A_i in their condition, to be marked. The Minimal Abnormality is slightly more complicated from a proof theoretic point of view, but much simpler from a semantic point of view. Where $Ab(M)$ is a (precise) measure for the abnormalities verified by the model M, M is a minimally abnormal model of Γ iff Γ has no model M' such that $Ab(M') \subset Ab(M)$.

Let us consider a specific lower limit logic, viz. **CLuN** – about the weakest sensible paraconsistent logic. Syntactically, it is characterized by the (full) positive fragment of **CL** – all rules that do not contain a negation – together with the axiom $A \vee {\sim}A$.[12] The $\sim$-negation in **CLuN** is extremely poor, and $\sim$-inconsistencies are isolated from each other. This is nicely illustrated by the semantics, for which there is no room here.

A first central idea that led to the discovery of inconsistency-adaptive logic was the following. Consider again Disjunctive Syllogism. While some **CLuN**-models do not verify $((A \vee B) \wedge {\sim}A) \supset B$, they *all* verify:

$$(A \wedge {\sim}A) \vee (((A \vee B) \wedge {\sim}A) \supset B).$$

This idea may be generalized. Whenever D is **CL**-valid, there are $C_1, \ldots, C_n$ $(n \geq 0)$ such that (see note 11 on the "$\exists$"):

$$\exists(C_1 \wedge {\sim}C_1) \vee \ldots \vee \exists(C_n \wedge {\sim}C_n) \vee D$$

is **CLuN**-valid.

The first displayed formula suggests the following interpretation: whenever both $A \vee B$ and ${\sim}A$ are derivable from the premises, then B is derivable from them, *unless* $A \wedge {\sim}A$ is derivable from the premises.

As announced, the circularity is removed by the second central idea: the move from the level of the abstract derivability relation to the concrete level of proofs (lists of formulas). This results in the rule: if $A \vee B$ and ${\sim}A$ have

[11] At the predicative level, we should consider $\exists(A_1 \wedge {\sim}A_1) \vee \ldots \vee \exists(A_n \wedge {\sim}A_n)$, where $\exists(A_i \wedge {\sim}A_i)$ abbreviates $(A_i \wedge {\sim}A_i)$, preceded (in some preferred order) by an existential quantifier over each individual variable free in A_i.

[12] The language is taken to include bottom ($\bot$), which may be seen as the conjunction of all formulas. It is characterized by the axiom $\bot \supset A$ (everything is derivable from $\bot$). As classical negation may be defined by $\neg A =_{df} A \supset \bot$, **CLuN** is actually an *extension* of **CL**.

been derived in the proof, then one may add B to the proof with the condition that A 'behaves consistently'. The latter phrase is interpreted by the adaptive strategy[13] that governs the marking of conditional lines.

Several inconsistency-adaptive logics have been studied. There is a variety of strategies as well as a variety of lower limit logics. While the poor paraconsistent logic **CLuN** maximally isolates inconsistencies, the maximal paraconsistent logic **CLuNs** reduces all inconsistencies to the level of primitive formulas.[14] In general, poorer paraconsistent logics isolate inconsistencies, and hence deliver the richer adaptive consequence set – remember that inconsistencies *prevent* the application of classical rules. So, this suggests that adaptive logics based on poorer lower limit logics are more efficient to interpret inconsistent premises as consistently as possible.

I should add that the most efficient logics are not always to be preferred. This is nicely illustrated in work by Joke Meheus – for example [22]. The adaptive logic **ANA** is based on the extremely rich paraconsistent logic **AN**. The latter not only validates all negation-reducing rules (see the last footnote for some examples), but validates all 'analysing' rules of **CL**. This even includes Disjunctive Syllogism. The price to be paid is that **AN** does not validate rules such as Addition (from A to derive $A \vee B$) and Irrelevance (from A to derive $B \supset A$). (Note that **AN** is rather close to the filtration systems of Paul Weingartner – see his paper in this volume.) Meheus has argued that, while adaptive logics based on **CLuN** may be more efficient, **ANA** offers a more realistic approach for reconstructing concrete cases from the history of the sciences.

A very popular inconsistency-handling mechanism is provided by the so-called Rescher–Manor approach from [29]: divide inconsistent sets of premises into maximal consistent subsets, and define derivability relations in terms of them. Later, further derivability relations were defined – see [16]. The somewhat peculiar approach[15] followed by Guido Vanackere in [32] led to the insight that all those consequence relations are defined by adaptive logics that have **CLuN** as their lower limit logic – see [12]. This not only shows that adaptive logics offer a unified framework for studying inconsistency-handling mechanisms, but also provides a (dynamic) proof theory for all the inference relations defined within the Rescher–Manor approach.[16] In different work, [7] and [18], it has been shown that several 'mixed' nonmonotonic logics (the popular systems concerning Tweety, the Nixon diamond, etc.) may be reconstructed as combinations of adaptive logics with nonlogical preferences.

[13] In general, more cautious strategies (such as Reliability) offer less consequences (than, for example, Minimal Abnormality), but have advantages from a computational point of view.

[14] In **CLuNs**, $\sim\sim A$ is equivalent to A, $\sim(A \vee B)$ is equivalent to $\sim A \wedge \sim B$, etc.

[15] The distinctive feature here is not the strategy, but the formalization of the premises: classical negation is used in the premises, and each premise is preceded by "$\sim\neg$".

[16] This includes inference relations in which nonlogical preferences play a role.

The metatheory of all these inconsistency-adaptive logics has been thoroughly studied. This involves not only Soundness and Completeness, Compactness, etc., but also properties that are specific for the adaptive logics. Let me just mention a few: Cautious Monotonicity (if $\Gamma \vdash A$ and $\Gamma \vdash B$, then $\Gamma \cup \{A\} \vdash B$), Proof Invariance (if $\Gamma \vdash A$, then any (dynamic) proof from Γ may be extended in such a way that A is finally derived in it), and Strong Reassurance (any paraconsistent model M that is not an adaptive model, is eliminated *by* an adaptive model M' – all inconsistencies verified by M' are verified by M).[17]

5 Further Corrective Adaptive Logics

Consider a set of premises that contains $p \supset q$, $\sim p \supset q$, and $\sim q$, and no further formulas in which p or q occur. An inconsistency-adaptive logic will consider q as inconsistent, and will handle the set in a suitable way. There is, however, another alternative. It is possible that q is consistent, but that p is negation-*incomplete* in that both p and $\sim p$ are false.

Similarly, a set of premises containing p, q, and $\sim(p \wedge q)$ may, but need not, be considered as inconsistent. The alternative is to consider conjunction as abnormal, viz. to apply a logic that requires consistency but allows for gaps with respect to conjunction. This means that exceptions on Adjunction (from A and B to derive $A \wedge B$) are tolerated. (Remember Kyburgh's 'conjunctivitis' – see [20] and elsewhere.)

What is going on here? The peculiar properties of **CL** cause all abnormalities to surface as inconsistencies. This, however, does not entail that the source of the abnormality is inconsistency. In different terms: in **CL** the symptom is invariably inconsistency, but the causes may be diverse.

The best approach to this diversity seems to be semantic. We may distinguish between gluts with respect to negation – these are inconsistencies: for some A, both A and $\sim A$ are true – and gaps with respect to negation – these are cases of incompleteness: for some A, neither A nor $\sim A$ is true. Similarly, we may distinguish gluts with respect to conjunction – $A \wedge B$ true while A or B is false – and gaps with respect to conjunction – $A \wedge B$ false while both A and B are true. The same applies for the other connectives, for identity, and for the quantifiers.

There is an important philosophical point to all this. The properties of **CL** caused classical logicians to see consistency as the sole criterion of normality. Only during the last twenty years has it became clear that the real (general) problem with logical abnormalities is not consistency but triviality. Unfortunately, logicians who developed alternative logics either share the abhorrence for inconsistency (like the intuitionists) or, still influenced by **CL**,

[17] It was shown in [11] that this property fails for Graham Priest's $\mathbf{LP}^m$ from [28]. Recently, I became convinced that Priest's variant to the Minimal Abnormality Strategy is misguided – see [13].

see inconsistency as the only form of logical abnormality. In other words, even the latter fail to see the possible role of other logical abnormalities.[18]

The above variety of abnormalities leads to a rather obvious variety of corrective adaptive logics. The abnormalities may be combined, and thus give rise to adaptive logics that tolerate several kinds of abnormalities.

An important breakthrough was realized by Guido Vanackere in [31] (and in forthcoming work). He applies the adaptive approach to ambiguity. Ambiguities pertain to the *non*logical constants, but nevertheless cause logical abnormalities – the **CL**-closure of a set of demonstrably ambiguous premises is trivial. Vanackere approached the phenomenon by indexing the *occurrences* of nonlogical constants. Thus, the set of premises $\{p, \sim q, \sim p, p \vee r, q \vee s\}$ becomes $\{p^1, \sim q^1, \sim p^2, p^3 \vee r^1, q^2 \vee s^1\}$. Quite obviously, triviality does not arise if **CL** is applied to the latter. The next idea is that logical normality is interpreted, quite naturally, as the absence of ambiguity. Thus, for example, $p^1 \equiv p^2$ expresses normality, and its negation expresses abnormality. The resulting adaptive logics *localize* the possible abnormalities in a set of premises. With respect to the example, the conclusion is that the first two occurrences of p are bound to be ambiguous, and hence that r cannot be derived (as $p^3 \equiv p^1$ leads to just as normal an interpretation as $p^3 \equiv p^2$).

Obviously, ambiguity-adaptiveness may be combined with adaptivity with respect to logical constants. An interesting as well as astonishing result is obtained by devising a logic that is adaptive with respect to *all* logical abnormalities. This is interesting because we do not know beforehand which abnormalities may occur in a text (or set of premises). So, this gives us the nonmonotonicity and dynamics that occur in actual reasoning in general.[19] The adaptive logic is astonishing because of its metatheoretic properties. Indeed, the lower limit logic **CL∅** validates not a single inferential rule. It is the absolutely minimal logic, viz. the empty one. So, the adaptive logic **A**CL∅ 2, obtained from **CL∅** by the Minimal Abnormality Strategy, tolerates all sorts of logical abnormalities. However, when it is applied to a normal set of premises (that is, one that has **CL**-models), it still delivers *all* classical consequences of the premises – that is: if Γ has **CL**-models, $Cn_{\mathbf{ACL\emptyset 2}}(\Gamma) = Cn_{\mathbf{CL}}(\Gamma)$. Thus ends, in beauty, the survey of corrective adaptive logics.

6 Some Other Adaptive Logics

Much actual reasoning is nonmonotonic as well as dynamic, but is not in any sense corrective. All *ampliative* reasoning belongs to this category. The

[18] Even relevant logicians (see especially [1, §80.2]) identify all logical abnormalities as inconsistencies (even negation-incompleteness, except when they preach in the language of the heathen).

[19] As everywhere in the present text, I disregard the role of nonlogical preferences.

idea to extend the adaptive approach to such logics, as well as most of the outlines are due to Joke Meheus. The study is completed for logics that concern compatibility (it 'follows from' Γ that A is compatible with it; $\Gamma \cup \Delta$ and $\Gamma \cup \Theta$ are compatible extensions of Γ) and abductive logics. Studies are forthcoming on aspects of analogy (see [23]), vagueness, induction, etc.

Moving to such logics requires that one reinterprets normality. The point is that (in practice) all corrective adaptive logics have **CL** as their upper limit logic (which defines normality). Obviously, **CL** is a natural standard for normality in view of its popularity, its Post-completeness, its functional completeness, etc. Ampliative adaptive logics require that we devise upper limit logics that extend **CL**. This leads to the situation sketched on the left–hand side in Fig. 2. In order to recognize both sorts of logics as adaptive, the situation has to be seen as on the right hand side of that figure.

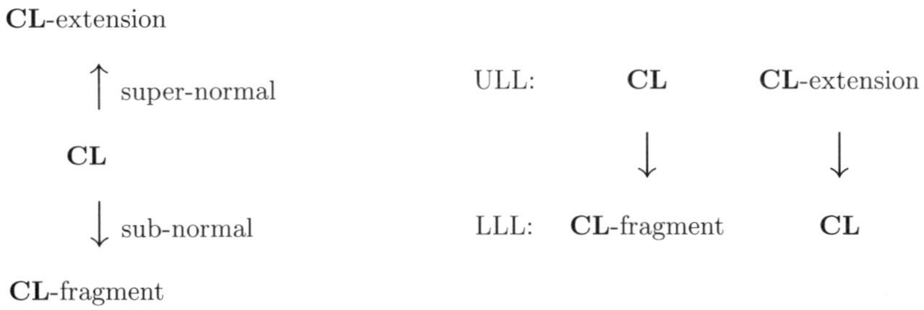

Fig. 2. Corrective *versus* ampliative adaptive logics: two views

Let me briefly spell out the way in which compatibility is handled in [15]. (This choice is inspired by the extreme simplicity of the case.) Let $\Gamma^{\Box} = \{\Box A \mid A \in \Gamma\}$. The lower limit logic is **S5** (handled in [15] in terms of an adequate subset of the usual models). The upper limit logic, $\mathbf{S5^P}$, introduces the normality requirement that $\Diamond A$ (A is possible) is a theorem whenever A is **S5**-contingent – the logic is obtained by adding to **S5** the rule: "If $\nvdash_{\mathbf{S5}} \neg A$, then $\vdash_{\mathbf{S5^P}} \Diamond A$".[20] As **S5** is undecidable, the abnormality of an **S5**-model of $\Gamma^{\Box}$ is actually measured by: $Ab(M) = \{A \mid M \text{ verifies } \Box A\}$.[21] It is provable that A is compatible with Γ iff $\Diamond A$ is verified by all minimally abnormal models of $\Gamma^{\Box}$.

[20] $\mathbf{S5^P}$ is a remarkable system in many respects. Its theorems are not closed under Uniform Substitution: $\vdash_{\mathbf{S5^P}} \Diamond p$ but $\nvdash_{\mathbf{S5^P}} \Diamond(p \& \neg p)$. Also, $\mathbf{S5^P}$ verifies the Principle of Predication from [19, p. 185].

[21] For all **S5**-models M, A belongs to $Ab(M)$ whenever $\Box A$ is an **S5**-theorem. Hence these formulas do not interfere with the definition of minimally abnormal **S5**-models of $\Gamma^{\Box}$.

The dynamic proof theory is readily obtained from this. In view of the **S5**-theorem $\Diamond A \vee \Box{\sim}A$, one may introduce into the proof from $\Gamma^{\Box}$, for any A, $\Diamond A$ on the condition that $\Box{\sim}A$ is not derivable from $\Gamma^{\Box}$. Incidentally, this allows me to illustrate the Simple Strategy. Let B be derived at line i on the condition that $\Box{\sim}A$ is not derivable from $\Gamma^{\Box}$. If $\Box{\sim}A$ is **S5**-derivable from $\Gamma^{\Box}$, it may later be derived and, as a result, line i will be marked, and will be marked *forever*. Other ampliative logics, for example abduction logics, are just a trifle more complicated.

The typical dynamics of all the aforementioned logics is a *revising* one: if the condition is no longer fulfilled, the line is marked (either provisionally or finally). While this paper was being written (long after the Salzburg meeting that led to this book), a further, anticipatory dynamics was discovered: lines are introduced to indicate that a formula may be derived on the condition that some other formula is derivable. This dynamics is common to all logics, just as the dynamics that was described in terms of blocks. A very simple application concerns proof heuristics. (Human proof heuristics is notoriously undocumented – see [4] and [6] for exceptions.)

Suppose that ${\sim}A \supset B$ and ${\sim}B \vee C$ have been derived and that the aim is to derive $A \vee D$. The latter is derivable from A by Addition; A may be obtained from ${\sim}A \supset B$ and ${\sim}B$ by Modus Tollens; and ${\sim}B$ may in turn be obtained from ${\sim}B{\vee}C$ and ${\sim}C$ by Disjunctive Syllogism. Of course, $A{\vee}D$ may also be derived from D, which possibly may be derived from other formulas in the proof; or $A \vee D$ may be derived in yet another way.

Such considerations sometimes lead to a rather complex reasoning that fully occurs outside of the proofs themselves. The anticipatory dynamics allows one to carry out this reasoning in the proofs themselves, and hence to control it in a systematic way. If our aim is to derive A from Γ, and ${\sim}A \supset B$ has already been derived, one may derive A conditionally, viz. on the condition that ${\sim}B$ is derivable. By writing down the line, one remembers that deriving ${\sim}B$ is useful in view of the aim. Thus, the proof heuristics may be pushed down to the level of the proof itself.[22]

7 In Conclusion

Adaptive logics are specifically interesting with respect to actual reasoning, including its provisional aspects. They are required for obtaining a formal grasp on the dynamics that characterize reasoning forms for which no positive test is available. The preceding survey is far from complete. Moreover, new types of adaptive logics, new formal properties, and new application contexts have been discovered regularly during the past ten years. So, the present survey might be quite provisional.

[22] By this move, the proper proof heuristics is not emptied but simplified: try to conditionally or, where possible, unconditionally derive the conditions that occur in the proof.

I have introduced three forms of dynamics. Two of them are not specific to adaptive logics, but were discovered while solving problems of adaptive logics. All three are important for understanding and explicating human reasoning, both in everyday contexts and in scientific contexts. They are also important for artificial intelligence, and, from a more theoretical perspective, for the foundations of logic, language, and mathematics. Together with the internal dynamics of the adaptive logic programme, this provides a strong motivation for deepening and extending the amme. It also supports the position that logics are central, not only for the philosophy of science, but also for the sciences themselves.

References

1. A.R. Anderson, N.D. Belnap, Jr., J.M. Dunn: *Entailment. The Logic of Relevance and Necessity* Vol. 2. (Princeton University Press, Princeton 1992)
2. D. Batens: 'Meaning, Acceptance, and Dialectics'. In: *Change and Progress in Modern Science* ed. by J.C. Pitt (Reidel, Dordrecht 1985) pp. 333–360
3. D. Batens: 'Dynamic Dialectical Logics'. In: *Paraconsistent Logic. Essays on the Inconsistent* ed. by G. Priest, R. Routley, J. Norman (Philosophia Verlag, München 1989) pp. 187–217
4. D. Batens: 'Natural heuristics for proof construction. Part I: Classical propositional logic'. In: Logique et Analyse **127–128** (1989) pp. 337–363 appeared (1992)
5. D. Batens: 'Do we Need a Hierarchical Model of Science?' In: *Inference, Explanation, and Other Frustrations. Essays in the Philosophy of Science* ed. by J. Earman (University of California Press 1992) pp. 199–215
6. D. Batens: *Logicaboek. Praktijk en theorie van het redeneren* (Garant, Leuven/Apeldoorn 1992)
7. D. Batens: 'Inconsistency-adaptive Logics and the Foundation of Nonmonotonic Logics. Logique et Analyse **145** (1994); pp. 57–94 appeared (1996)
8. D. Batens: 'Blocks. The Clue to Dynamic Aspects of Logic'. Logique et Analyse **150–152** (1995) pp. 285–328, appeared (1997)
9. D. Batens: 'Inconsistency-adaptive Logics'. In: *Logic at Work. Essays Dedicated to the Memory of Helena Rasiowa* ed. by E. Orłowska Physica Verlag (Springer, Heidelberg, New York 1999) pp. 445–472
10. D. Batens: 'Zero Logic Adding up to Classical Logic'. Logical Studies **2** (1999) p. 15. (Electronic Journal:`http://www.logic.ru/LogStud/02/LS2.html`)
11. D. Batens: 'Minimally abnormal models in some adaptive logics'. Synthese **125** (2000) pp. 5–18
12. D. Batens: 'Towards the unification of inconsistency handling mechanisms'. *Logic and Logical Philosophy* **8** (2001) pp. 5–31
13. D. Batens: 'Linguistic and ontological measures for comparing the inconsistent parts of models'. Logique et Analyse, **165–166** (1999) pp. 5–23
14. D. Batens, J. Meheus: 'A Tableau Method for Inconsistency-adaptive Logics'. In: *Automated Reasoning with Analytic Tableaux and Related Methods.* Lecture Notes in Artificial Intelligence Vol. **1847** ed. by R. Dyckhoff (Springer, Berlin 2000) pp. 127–142

15. D. Batens, J. Meheus: 'The adaptive logic of compatibility'. Studia Logica **66** (2000) pp. 327–348
16. S. Benferhat, D. Dubois, and H. Prade: 'Some Syntactic Approaches to the Handling of Inconsistent Knowledge Bases: A comparative Study. Part 1: The Flat Case'. Studia Logica **58** (1997) pp. 17–45
17. B. Brown: 'How to be Realistic About Inconsistency in Science'. In: Studies in History and Philosophy of Science **21** (1990) pp. 281–294
18. K. De Clercq: 'Two new Strategies for Inconsistency-adaptive Logics'. Logic and Logical Philosophy, in print
19. G. Hughes, M. Cresswell: *An Introduction to Modal Logic* (Methuen, London, New York 1972, first published 1968)
20. H.E. Kyburg: 'Conjunctivitis'. In: *Induction, Acceptance, and Rational Belief.* ed. by M. Swain (Reidel, Dordrecht 1970) pp. 55–82
21. J. Meheus: Adaptive Logic in Scientific Discovery: The Case of Clausius. Logique et Analyse **143–144** (1993) pp. 359–389 appeared (1996)
22. J. Meheus: 'An Extremely Rich Paraconsistent Logic and the Adaptive Logic Based on it'. In: *Frontiers of Paraconsistent Logic* ed. by D. Batens, C. Mortensen, G. Priest, and J.P. Van Bendegem (Research Studies Press, Baldock 2000) pp. 189–201
23. J. Meheus: 'Analogical reasoning in creative problem solving processes: Logico-philosophical perspectives'. In: *Metaphor and Analogy in the Sciences* ed. by F. Hallyn (Kluwer, Dordrecht 2000) pp. 17–34.
24. J. Meheus: 'Inconsistencies in scientific discovery. Clausius's remarkable derivation of Carnot's theorem'. In: *History of Modern Physics* ed. by H. Krach, G. Vanpaend, P. Marage (Tournhout, Brenols 2002) pp. 143–154
25. N. Nersessian: 'Inconsistency, generic modeling, and conceptual change in science'. In: *Inconsistency in Science.* ed. by J. Meheus (Kluwer, Dordrecht, 2002) pp. 197–212
26. J. Norton: *The Logical Inconsistency of the Old Quantum Theory of Black Body Radiation* Philosophy of Science **54** (1987) pp. 327–350
27. J. Norton: 'A Paradox in Newtonian Gravitation Theory'. PSA **2** (1992) pp. 421–420
28. G. Priest: 'Minimally Inconsistent **LP**'. Studia Logica **50** (1991) pp. 321–331
29. N. Rescher, R. Manor: 'On Inference From Inconsistent Premises'. Theory and Decision **1** (1970) pp. 179–217
30. J. Smith: 'Inconsistency and Scientific Reasoning'. Studies in History and Philosophy of Science **19** (1988) pp. 429–445
31. G. Vanackere: 'Ambiguity-adaptive Logic'. Logique et Analyse **159** (1997); appeared pp. 261–280 (1999)
32. G. Vanackere: **HL2**. 'An inconsistency-adaptive and inconsistency-resolving logic for general statements that might have exceptions'. Journal of Applied Non-Classical Logics **10** (2000) pp. 317–338

Comments on Jaakko Hintikka's Post-Tarskian Truth

Gerhard Heinzmann

This paper has two parts: first, I discuss some questions concerning the game theoretical framework of Hintikka's IF-Logic; secondly, I consider IF first-order logic with regard to Poincaré's criticism of modern logic.

Part I

In philosophical circles, the most prominent argument against the spirit of Game-Theoretical Semantics (GTS) has been popularised not only by Quine but also by certain constructivists. It runs along the following lines: formal semantics for first-order languages, especially model-theoretic ones, uses in its metalanguage set-theoretical means which are essentially stronger than the formal system of the object-language.

In a first step, before his creation of IF-Logic, Hintikka refutes this objection with the following fact:

GTS gives the general frame work for natural languages – and not only for first-order languages. In particular, it gives the frame for the widespread feature of informational independence of the semantics of current expressions reproducible by partially ordered "branching quantifiers", but not reducible to linear prefixes in ordinary first-order languages.

Example 1.

Consider the following formula with partially ordered quantifiers:

$$\begin{matrix} \forall x \exists y & \\ & S(x,y,z,u) \\ \forall z \exists u & \end{matrix}$$

If you try to linearize, $\exists y$ should depend on $\forall x$ but not on $\forall z$:

a) $\forall x \exists y \forall z$

and $\exists u$ should depend on $\forall z$ but not on $\forall x$:

b) $\forall z \exists u \forall x$

But of course a) and b) are incompatible [cf. Hintikka 1998, 5, 6].

As is well known, truthfulness in GTS is defined as the existence of a winning strategy for the initial verifier [Hintikka 1996, 26]. The notion of strategy is itself amenable to a formulation in logical terms [Hintikka 1996, 30]:

Example 2:

$$\forall x \exists y S(x, y) \tag{1}$$

is true iff one can find a function f (called a *Skolem function*) with

$$\exists f \forall x S(x, f(x)) \tag{2}$$

The equivalence of (1) and (2) is a variant of the axiom of choice [Hintikka 1996, 40]. In this view, the second-order translation from winning strategies via Skolem functions in an explicit formulation of a truth definition for natural languages seems to give a strong argument for considering:

i) the validity of the axiom of choice as purely logical;
ii) a fragment of second-order logic as logic and not, as Quine thinks, mathematics.

In a second step, Hintikka created a logical system in order to express in a first-order logic linearly partial ordered quantifiers.

Example 3:

The formula

$$\begin{matrix} \forall x \exists y & \\ & S(x, y, z, u) \\ \forall z \exists u & \end{matrix}$$

is translated as

$$(\forall x // \exists u)(\forall z // \exists y) \exists y \exists u S(x, y, z, u)$$

where $(\forall x // \exists u)$ indicate the independence of $\exists u$ from $\forall x$.

He calls the resulting logic *independence friendly* (IF-Logic) and proved for it the surprising and revolutionary result that the game-theoretical truth predicate of IF-first-order language containing elementary arithmetic is definable in that language itself in the sense '"that there is a complex predicate (of the Gödel numbers of the sentences of the given first-order language) which applies to a number if and only if it is the Gödel number of a true sentence" [Hintikka 1996, 118sq.]. We just have to be careful with respect to the informational independence of quantifiers in their role of lying numbers as numbers and numbers as Gödel codifications. Then, although according to Tarski's impossibility theorem, the $\sum_1^1$ -sentence corresponding to the truth of a first-order sentence *cannot* be retranslated in a normal-form-sentence of first-order logic whose quantifiers are linearly ordered by inclusion, it *can* nevertheless be formulated in IF first-order language. Now, Hintikka argues that truth is definable in the object-language. Therefore the metalanguage is no longer stronger than the object language and, according to Hintikka's sketch, model theory, which uses essentially a truth definition, can be done in logic and no longer presupposes set theory or mathematics. Nevertheless, apart from the fact that IF logic requires noncompositional semantics if one preserves the normal meaning of the quantifiers [cf. Hintikka 1998, 11], one might be tempted to think [cf. Rivenc in [Rigal 1998], 166sqq.] that the thus-suggested game theoretical definition of quantifiers presupposes already normal interpreted quantifiers in the metalanguage. Indeed, Hintikka seems to be able to reply with ease to accusations of circularity [ibid., 319]: the metalanguage quantifiers are by no means explicitly defined but *learned* in the game by means of examples. And this is in fact what I always learned with respect to the game-theoretical approach to truth invented by Lorenzen and Lorenz.

Most importantly, the law of bivalence asserts that there always exists a winning strategy for one of the two players of the game. This fails in IF first-order logic because the only IF formulas whose contradictory negation is expressible in the IF-language are precisely the ordinary first-order formulas. Indeed, the failure has nothing to do with some particular subject matter (future), with infinity or with the limitations of human knowledge [Hintikka 1996, 132 sqq.] and it has an important consequence: it entails that IF-logic fails the T-schema. In fact:

(T) Π is true $\leftrightarrow p$

where Π is a description of "p". (T) is equivalent to

(T*) (Π is true $\wedge\, p) \vee$ (Π is not true $\wedge\ \neg p)$,

which entails $p \vee \neg p$, in other words, the T-schema satisfies the law of the excluded middle [Hintikka 1996, 138/139].

Hence, the defect for which the Tarski-type truth definition is blamed by Hintikka concerns by no means the constructivist's point that it leaves unexplained the activities by means of which we actually verify and falsify [Hintikka 1996, 22]: games are exclusively on the level of strategies.

In the dialogical approach of Lorenzen and Lorenz, the bivalence principle is constructively reduced to the decidability of win or lose for individual plays of a game. The fact that, in general, the concept of a true proposition within a theory is neither decidable nor recursively enumerable even if we start with value definite, i.e. decidably true or false elementary propositions, has led Lorenzen and Lorenz "to drop the classical concept of proposition — according to the classical conception propositions are entities which can be either true or false — and to introduce a wider concept of proposition: [...] For entities to be propositions there must be a stipulation of rules which define possible argumentations [..] ending with win or loss for either player who both argue for or against that entity. This set of rules shall be called a dialogue-game" [Lorenz, 1980]. Truth of a proposition gets defined as existence of a winning strategy in a dialogue game about this proposition. So the mastery of meaning of a sentence learned on the level of single plays, won or lost, produces an effective learning process concerning winning-strategies, specifying which strategies are better.

One has plainly to agree with Hintikka's criticism of Davidson who is trying to connect truth conditions and language learning by recursive truth clauses and Frege's compositionality principle instead of the use of our language [cf. Saarinen, 280]. This criticism is evident because by means of a winning-strategy the concept of truth will only in special cases become recursively enumerable or even decidable. Indeed, in Hintikka's semantics the strategy is defined as a concept of the *mathematical theory* of games [cf. e.g. Saarinen 36]. GTS is a realistic truth-conditional theory where truth is defined independently of any epistemic category. The existence of a winning-strategy is an objective fact concerning a given model of the language [cf. e.g. Hintikka 1994, 172]. This means the existence of a set of functions which says nothing of the knowledge of them not even if such a knowledge is in principle possible [Hintikka 1994, 13/14, note 6]. There is consequently only an assumed representation *of* and an assumed learning process *between* the abstract players *Myself* and *Nature*. Therefore, I suppose that in Hintikka's game the concept of strategy is only available by logical analysis and not by learning, because he did not go back to single plays. I'm not yet convinced that Hintikka's link between truth definition and the sense of verification and falsification in his games can be related to the teaching and learning of truth so defined. To put the same point in other terms: in order that game-theoretical truth really does what Tarski-type definitions were accused by the Wittgensteinian of not doing [cf. Hintikka 1996, 120sq.], one should reformulate it in dialogical-game language. My Saarbruecken colleague Shahid Rahman has expressed the semantics of many non-classical logics in dialogical form. Now, the IF-particle-rule for quantifier-attacks, for example, should be defined along the following lines:

R1

i) The attack is as usual if the quantifier is not restricted
ii) If the quantifier is restricted and if there are only independent quantifiers, one has first to attack the first nonrestricted quantifier
iii) If there are independent and dependent quantifiers [for example: $(\forall x//\exists u)(\forall z//\exists y)\exists y \exists u S(x,y,z,u)$], then one should open a special dialogue context in order to imitate partial information [attack-description: *acbd* for $R(abcd)$].

Example: $(\forall x//\exists y)\exists y S(x,y) \rightarrow \exists y \forall x S(x,y)$ is a logical implication:

	O	P	
		$(\forall x//\exists y)\exists y S(x,y) \rightarrow \exists y \forall x S(x,y)$	
1	$(\forall x//\exists y)\exists y S(x,y)$	$\exists y \forall x S(x,y)$	2
3	?	$\forall x S(x,a)$	6
5	$(\forall x//\exists y) S(x,a)$	? [1]	4
7	[6] b?	$S(b,a)$	10
9	$S(b,a)$	b? [5]	8

The numbers without brackets keep track of the moves. The numbers between brackets [ ] indicate which move of the partner is thereby attacked.

How does the proof appear in GTS? The formula should be translated in its negation normal form (all its negation-signs immediately precede atomic formulas): Sandu proved the theorem that

$$\neg(\forall x//\exists y)\exists y S(x,y) \vee \exists y \forall x S(x,y)$$

is equivalent (as in classical first-order logic) to

$$(\exists x//\forall y)\forall y \neg S(x,y) \vee \exists y \forall x S(x,y).$$

The game associated with $((\exists x//\forall y)\forall y \neg S(x,y))$ is

Falsifier b
– – – – – *Rab.*
Verifier a

So the strategy for the verifier will amount to a strategy for $\forall y \exists x \neg S(x,y) \vee \exists y \forall x S(x,y)$, which is true.

Part II

The following considerations are suggestive rather than demonstrative in character. My aim is to suggest a plausibility argument for IF-logic in connection with Poincaré's foundational remarks.

It is well known that, according to Poincaré's polemic with the logicists, mathematical reasoning does not proceed by logical means. Indeed, once the distinction between context of discovery and context of justification is admitted, it seems that Poincaré is condemning mathematical logic for not providing psychological conviction for the former, whereas the logicists are concerned with the latter. So Poincaré and the logicists are trying to achieve different objectives. This is Warren Goldfarb's claim in his article *Poincaré against the logicists* [cf. Goldfarb 1985].

In reality, Poincaré's use of intuition does not reduce to psychology in the modern sense of this term: he uses the term "psychological" in all the situations where the dimension of understanding is involved. In this sense, psychology concerns the epistemological question of developing standards of clarity for conceptual presuppositions.

Formulated in modern terms, Poincaré holds that the varieties of formal logical theories don't express the structure essential for understanding mathematics [cf. SM, 149 (159)] in the context of justification. Genuine mathematical knowledge is, in general, not the result proved by means of a series of analytical deductions, called by him *verifications*. Mathematical reasoning is to be distinguished from traditional Logic [SH, 3 (32)], because it has itself a kind of creative virtue, exemplified by the principle of complete induction. So the question arises: is it possible to deduce the principle of complete induction from the new set-theoretical logic? If it is not possible, how have we to conceive the creative virtue of mathematical reasoning and, if it is possible, why should we call this new theory "logic"? I argued in my book on Poincaré [cf. Heinzmann 1995] that all proofs of complete induction violate certain forms of Poincaré's predicativity requirement and Charles Chihara argued in his article *Poincaré and Logicism* that it is by no means evident that axioms expressed in purely logical terms are laws of logic [cf. Chihara 1996].

I shall now reconsider these problems with respect to IF-first-order logic.

Concerning the positive answer about the creative virtue of mathematical reasoning, it is widely known that first-order logic without individual variables including higher-order entities cannot deal with the most characteristic modes of concept and inference in mathematics such as complete induction, well-ordering or power-set formation. So, mathematical induction is not generated but only represented by indefinite repetition of different levels [cf. Heinzmann 1987, 72], which means that the separation between object and symbol is not yet accomplished: it postulates a survey of a potentially reiterated stroke-concatenation or something analogous and a survey of a potentially reiterated *modus ponens*. An analogous principle consists of the fact that mathematical symbols can be read with respect to different contents without the possibility of checking this ambiguity on the level of the given notation. Here we have a typical case of partial information. Poincaré suggests that this difficult situation including semantic ambiguity should be overcome by the introduction of aesthetic feeling in mathematics. The mastering of simultaneous reasoning about different contents, provoked by the lack of perfect information in one field, requires the acquisition of a practice and can be made explicit only by model-theoretical means.

Now, if there is no longer any need to resort to set theory for the purpose of doing model theory of first-order logic or of expressing complete induction, and if, on the contrary, these tasks can be expressed in IF first-order language, equivalent to a $\sum_1^1$ fragment of second order language, respectively, insofar as mathematical induction is concerned, to a $\sum_1^1 \cup \prod_1^1$ fragment, we have found an argument for the thesis that Poincaré's Philosophy of Mathematics is, with respect to its inner structure, more technically than philosophically disconnected from his positive work in mathematics. Why?

As pointed out by Hintikka, "the main strategy of set theorists is to capture sets as fully as possible as extensions of formulas of the language of set theory" [Hintikka 1996, 178]. This has been expressed by the axiom of comprehension through the following equivalence

$$\forall x(x \in a \longleftrightarrow F[x]). \tag{3}$$

If $F[x]$ is now an independence-friendly formula containing informational independent quantifiers it fails to obey the law of bivalence and consequently (3) cannot be true. So runs Hintikka's argument for the incompatibility of classical set theory and IF first-order logic [cf. Hintikka 1996, 179]. Poincaré's refusal to accept the basic axiom of set theory is based on a diametrically opposed argumentation but yields the same result: the failure of the law of bivalence for $F[x]$. According to Poincaré, the reductionist program of Logicism seems to waver already from the antinomies occurring in the new logic containing set theory. He argues that the constitution of sets is based on a methodological inversion which concerns the philosophical presuppositions. The cause of the antinomies resides in an implicit recourse to a "false" intuition regarding abstract entities. The resulting solution is to limit mathematical definitions and proof-procedure to predicative ones, so that (3) has to be predicative. He gives two definitions of this concept: a first well-known one says that no bound variable occurring in F admits a as a value or admits one of the constant individuals with the reference to which x is defined. This idea was translated in 1956 by Hintikka in a symbolic language form, containing certain limitations for F, symbolised by $F+$.

Nevertheless, there is a second definition in Poincaré saying "that a coordination is predicative if the classification upon which it is based is predicative. The classification in turn is predicative if it is not changed by the introduction of new elements". This is the case for the first definition: The classification of x's into such as to make a predicate $F+$ true in contradiction to those which make it false is accomplished by the introduction of new elements, since $F+$ may contain bound variables which range over all the elements (with the exception of x) [cf. Hintikka 1956, 244/245 and Heinzmann 1985]. So, according to Poincaré, the failure of the bivalence principle with respect to $F+$, is caused by a phenomenon of partial information. It implies the existence of too many sets, whereas the same effect implies, according to Hintikka's IF-Logic, the existence of too few sets [cf. Hintikka 1996, 176, 177].

To preserve mathematical activity in its classical form, Poincaré is obliged to postulate the principle of complete induction as an *a priori* principle. Then it can be shown — which was especially done by Kreisel and Feferman — that predicatively definable sets according to the two impredicativity definitions should be included somewhere in an initial segment of the class of second order predicates characterised by

$$\sum_1^1 \cap \prod_1^1$$

formulas, i.e. the first level of the so called analytical hierarchy. On the other hand, IF first-order logic is equivalent to a proper extension of it, that is the $\sum_1^1$ fragment or, so to speak, the dual form of predicative analysis, the $\sum_1^1 \cup \prod_1^1$ fragment [Hintikka 1996, 191]. These facts now give one more argument in favour of IF first-order logic. The predicativity requirement is in its very philosophically convincing argumentation too strong for mathematical praxis: the induction principle is not predicatively provable. It is only the first level of the analytical hierarchy which gives the means for mathematical creativity in terms of the possibility to express model-theoretical notions and mathematical induction.

Notwithstanding predicativity, Hintikka shares with Poincaré the important point that the classical, set-theoretical abstraction-principle fails because its defining predicate does not satisfy the bivalence principle. To be practicable in mathematics, Poincaré's predicativity-restrictions, half constructive, half classic, have to be extended by *a priori* principles, incompatible with systematic means. On the contrary, Hintikka's solution is from the outset systematically neither constructive nor classic. It shows that Poincaré's suggestion to take creativity as a criterion for separating logic from mathematics is no longer possible. So, one part of his criticism of modern logic is dependent on its special technical form. Indeed, I think the core of Poincaré's criticism of logic was related to a criticism of the axiomatic method. And on this point his position is very timely: not only is in a full-formalism truthfulness quite different from deducibility, but IF first-order logic does not even admit of a complete axiomatization [cf. Hintikka 1996, 65].

References

1. Ch.S. Chihara: 'Poincaré and Logicism'. In: Greffe et al. (1996) pp. 435–446
2. W. Goldfarb: 'Poincaré against the Logicists'. In: *History and Philosophy of Modern Mathematics.* ed. by W. Aspray and P. Kitcher (Minnesota Press, Minneapolis 1985) pp. 61–81
3. J.L. Greffe, G. Heinzmann, K. Lorenz: *Henri Poincaré. Science and Philosophy.* (Akademie Verlag/Blanchard, Berlin, Paris 1996)
4. G. Heinzmann: *Entre intuition et analyse. Poincaré et le concept de prédicativité.* (Blanchard, Paris 1985)

5. G. Heinzmann: 'Philosophical Pragmatism in Poincaré'. In: *Initiatives in Logic, Reason and Argument.* ed. by J. Srzednicki, vol. 2 (Nijhoff, Dordrecht, Boston, Lancaster 1987) pp. 70-80
6. G. Heinzmann: *Zwischen Objektkonstruktion und Strukturanalyse* (Vandenhœck & Ruprecht, Göttingen 1995)
7. J. Hintikka: 'Vicious Circle Principle and the Paradoxes'. The Journal of Symbolic Logic. **22** (1956) pp. 245–249
8. J. Hintikka: *Fondements d'une théorie du langage* (PUF, Paris 1994)
9. J. Hintikka: *The Principle of Mathematics Revisited* (Cambridge University Press, Cambridge 1996)
10. J. Hintikka: *Post-Tarskian Truth* Manuscript (1998)
11. J. Hintikka, J. Kulas: *The Game of Language. Studies in Game-Theoretical Semantics and Its Applications* (Reidel, Dordrecht, Boston 1983)
12. K. Lorenz: 'Main Ideas of Dialogic Logic'. *Studies in Logic, Grammar and Rhetoric I* (Warsaw University, Bialystok Branch, Warsaw 1980) pp. 17–34
13. H. Poincaré: SM *Science et méthode* (Flammarion, Paris 1908) Translation: Science and Method. (Thomas Neldon & Sons, London, New York s. d.)
14. E. Rigal (ed.): *Jaakko Hintikka. Questions de logique et de phénoménologie* (Vrin, Paris 1998)
15. E. Saarinen: *Game-Theoretical Semantics* (Reidel, Dordrecht, Boston 1979)

Part III

Alternative Logics Motivated by Problems of Application to Science

Applied Logics for Computer Science

Paul Gochet and Pascal Gribomont

Logic is intensively applied in many areas of computer science. Successful applications often lead to the question whether alternative logics could be more adequate than classical propositional and first-order logics. Not surprisingly, answering this question usually amounts to a trade-off between the expressive power of the new logic (usually stronger than the expressive power of classical logic) and its decidability properties (usually weaker). From this point of view, formal methods for program design and verification are an interesting field of applied logic and computer science since one can take advantage of the high expressive power of specific logics and also of the algorithmic tools available in classical logic; two examples will be given. We also illustrate how logic can be used directly as a programming language.

1 Hoare Logic for Sequential Programs

Hoare logic is probably the most well-known alternative logic used in computer science. This axiomatic system allows us to express many interesting program properties and their proofs. An especially useful feature of Hoare's system is the fact that a typically operational piece of reasoning about computations generated by some program can easily be translated into a mathematical proof of a property of this program. However, the correctness of a program relies on mathematical and logical truths, which are best expressed and proved in a suitable classical first-order theory. In fact, Hoare's system allows us to take the best parts of two worlds: the strong expressive power of Hoare logic itself, and the effective proof strategies available in classical first-order logic.

1.1 A Sorting Algorithm and Its Specification

We illustrate Hoare logic with a classical example, the *insert-sort algorithm.* (Additional examples and more details are given in [9]. Classical textbooks on this topic include [5] and [15]; see also [6].) The underlying algorithmic idea is elementary and best explained with an example (Fig. 1).

The left part of the figure is a "macro-description" of a typical computation. It displays the effect of the (macro-)step number i: the first unsorted

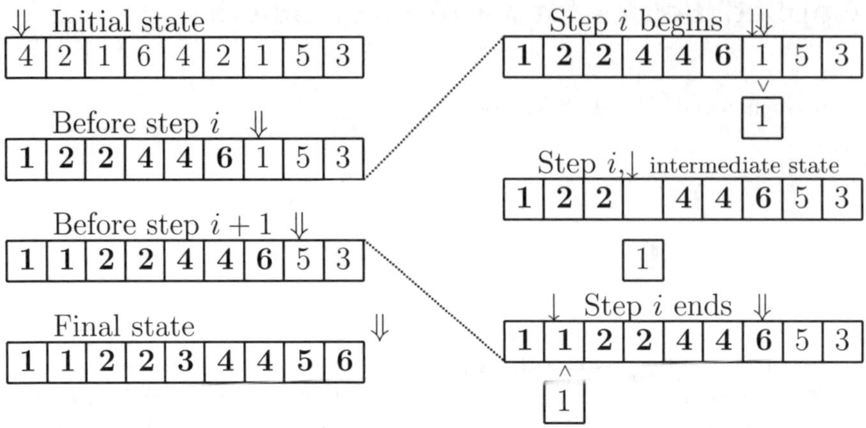

Fig. 1. Insert-sort: the principle

value, i.e., 1, is sorted, that is, inserted at the correct place within the left, already sorted part of the array. The net-effect of step i is that the size of the sorted part of the array was $i-1$ before the step, and i afterwards. If the size of the whole array is n, the array will be sorted in n steps.

The right part of the figure is a "micro-description" of the step number i: the new element to be sorted is extracted from the array; it is then compared with the sorted elements (bigger, rightmost ones first), and as long as the new element is smaller, bigger elements are pushed one cell to the right. The process ends when the new element reaches its final place.

It looks now rather easy for a competent program writer to translate this algorithmic idea into an algorithm ... but as every computer user knows, programs are likely to contain mistakes! It is therefore appropriate to document a program in a proper way (the connection between the data and the result should be specified precisely), and sometimes to validate the program, that is, to prove in a formal way that all computations satisfy the specification. In this area as in many areas of pure and applied mathematics, the specification is usually formalized into a sentence written in some appropriate logic.

Let us introduce some terminology and notation. Arrays and their mathematical models are denoted by uppercase letters. An array X of bounds u and v is a collection of $v-u+1$ memory cells denoted $X[u], \ldots, X[v]$; the mathematical model is a function with domain $\{u, \ldots, v\}$. There is one such function for each computation state; in particular, X_0 and X_f correspond to the initial and the final state, respectively.[1] Some array predicates are now defined; here A and B denote arbitrary arrays (in fact, their models).

[1] The array is empty if $v < u$. Models of the array X are also denoted X, possibly with subscripts or superscripts; $X[k]$ denotes a memory cell and $X(k)$ denotes its contents.

The notation $equal(A, i, j, B, k, \ell)$, also written $A[i : j] = B[k : \ell]$, stands for the formula

$$(j - i = \ell - k \wedge \forall r\,[0 \leq r \leq j - i \Rightarrow A(i + r) = B(k + r)])\,.$$

Array A_f is the sorted version of array A_0 if, first, A_f is a permutation of A_0 and, second, A_f is ordered. We introduce two pieces of notation to deal with the first condition; $card(A, i, j, z)$ stands for

$$|\{r : [i \leq r \leq j \wedge A(r) = z]\}|\,,$$

where $|S|$ denotes the *cardinal* of set S, i.e., if S is finite, the number of elements in S. Furthermore, $perm(A, i, j, B, k, \ell)$ stands for

$$\forall u\,[u \in \mathbb{Z} \Rightarrow card(A, i, j, u) = card(B, k, \ell, u)]\,.$$

The second condition will make use of the following definition:

$$ord(A, i, j) =_{def} \forall r, s\,[i \leq r \leq s \leq j \Rightarrow A(r) \leq A(s)]\,.$$

Hoare notation can now be used to specify the program as the *triple*

$$\{equal(A, 1, n, A_0, 1, n)\}\ SORT\ \{perm(A, 1, n, A_0, 1, n) \wedge ord(A, 1, n)\} \quad (1)$$

which expresses exactly what we would expect from a sorting algorithm: "If the value of the array A to be sorted is A_0 before running the algorithm, then this value will be the ordered permutation of A_0 after running the algorithm".

1.2 Hoare's Axiomatic System

More generally, if P and Q are formulas and if π is a program (or a program fragment), then the triple

$$\{P\}\,\pi\,\{Q\}$$

expresses that, for all finite computations of π, if the initial state satisfies P, then the final state satisfies Q; formulas P and Q are the *precondition* and the *postcondition*, respectively. (Nothing is said about the possible existence of infinite computations.) Formulas like P and Q model relations between the values of the program variables; we call them *assertions*.

When Q and π are fixed, the set $\{P : \{P\}\pi\{Q\}$ is valid$\}$ has a weakest member, denoted $wlp[\pi; Q]$ (weakest liberal precondition). Similarly, when P and π are fixed, the set $\{Q : \{P\}\pi\{Q\}$ is valid$\}$ has a strongest member, denoted $slp[P; \pi]$ (strongest liberal postcondition). As a result, Hoare logic reduces to classical logic since the verification of $\{P\}\pi\{Q\}$ reduces to the validation of either $P \Rightarrow wlp[\pi; Q]$ or $slp[P; \pi] \Rightarrow Q$.

Computer programs are built incrementally and the most elementary statement is the assignment. The semantics of an assignment like $x := f(x, y)$ is easily summarized into the triple

$$\{x = x_0 \wedge y = y_0 \wedge z = z_0\}\, x := f(x, y)\, \{x = f(x_0, y_0) \wedge y = y_0 \wedge z = z_0\}$$

but there is a more general, and therefore more helpful triple about this assignment, that is,

$$\{P(f(x,y), y, z)\}\, x := f(x, y)\, \{P(x, y, z)\} \qquad \textit{(Hoare Axiom)}$$

This triple is true for each ternary predicate P and is therefore adopted as the *assignment axiom.* The elementary language we will use to write the insert-sort algorithm consists of assignments, conditional statements ("if-then-else") and loops ("while-do"); a sequence of statements is also a statement. More formally, let us assume that S, S_1 and S_2 are programs, and B is a condition; we have four kinds of programs:

- An assignment is a program.
- $[S_1; S_2]$ is a program (sequential composition).
- [if B then S_1 else S_2] is a program (conditional statement).
- [while B do S] is a program (loop).
  (B is the *condition* or the *guard* of the loop; S is the *body* of the loop.)

The semantics of assignments is given by the Hoare Axiom; the semantics of sequential compositions, conditional statements, and loops are given by three inference rules:

$$\frac{\{A\}\ S_1\ \{B\}\ ,\ \ \{B\}\ S_2\ \{C\}}{\{A\}\ S_1; S_2\{C\}} \tag{R1}$$

$$\frac{\{A \wedge B\}\ S_1\ \{C\}\ ,\ \ \{A \wedge \neg B\}\ S_2\ \{C\}}{\{A\}\ \text{if } B \text{ then } S_1 \text{ else } S_2\ \{C\}} \tag{R2}$$

$$\frac{\{I \wedge B\}\ S\ \{I\}}{\{I\}\ \text{while } B \text{ do } S\ \{I \wedge \neg B\}}\ . \tag{R3}$$

Rules $R1$ and $R2$ are self-explanatory but rule $R3$ is a bit more subtle. Running "while B do S" means running S repeatedly (never, once or several times), until B becomes false.[2] Assertion I is an *invariant.* A last inference rule establishes a link between the Hoare system, used to reason about programs themselves, and the underlying logic, used to reason about program data (numbers, sets, lists, trees, graphs, ...):

$$\frac{\models A \Rightarrow B\ ,\ \ \{B\}\ S\ \{C\}\ ,\ \ \models C \Rightarrow D}{\{A\}\ S\ \{D\}}\ . \tag{R4}$$

[2] If B is always true, then the computation goes on forever. Recall that Hoare logic is about finite computations only.

This rule expresses monotonicity: we can always make a precondition stronger and a postcondition weaker; rule $R4$ clearly formalizes the link between proper Hoare logic and the logic used to reason about program data. Note that writing $\models A \Rightarrow B$ means only that the (classic) formula $A \Rightarrow B$ is valid; we do not assume a specific axiomatic system, or decision procedure, to determine if $A \Rightarrow B$ is valid or not. One can prove [4] that the Hoare axiomatic system (Hoare Axiom and rules $R1$ to $R4$) is sound and relatively complete, which means it is complete if we assume that any valid sentence about program data can be deduced. For instance, if the validity of any sentence of full number theory could be decided, than every true Hoare triple about every program involving unbounded integers and arithmetical operations could be deduced within Hoare axiomatic system.

1.3 Formal Verification of the Sorting Algorithm

In practice, however, the only difficult point in a Hoare-based correctness proof is rule $R3$ or, more precisely, the discovery of an adequate invariant I. This is not surprising since invariants really are the formal counterpart of algorithmic ideas; we illustrate this with the insert-sort example. The left part of Fig. 1 suggests the following program scheme:

$$
\begin{array}{l}
i := 1; \\
\{\, \text{Inv-Ext} \ \wedge\ i = 1 \,\} \\
\text{while } i \ \leq\ n \text{ do} \\
\qquad \{\, \text{Inv-Ext} \ \wedge\ i \ \leq\ n \,\} \\
\qquad \text{“execute macro-step } i\text{”;} \\
\qquad \{\, \text{Inv-Ext}\,[i+1\,/\,i] \ \wedge\ i \ \leq\ n \,\} \\
\qquad i := i + 1 \\
\qquad \{\, \text{Inv-Ext} \,\} \\
\{\, \text{Inv-Ext} \ \wedge\ i = n + 1 \,\}\,.
\end{array}
$$

If Φ is a formula, then $\Phi[e\,/\,x]$ denotes the formula obtained by replacing all free occurrences of x by e in Φ. The notation “Inv-Ext” stands for the invariant of the external loop; according to Fig. 1 (left part), it can be written as

$$
\text{Inv-Ext} \;:\; \begin{array}{l} 1 \ \leq\ i \ \leq\ n+1 \ \wedge\ \mathit{perm}(A, 1, i-1, A_0, 1, i-1) \ \wedge \\ \mathit{ord}(A, 1, i-1) \ \wedge\ \mathit{equal}(A, i, n, A_0, i, n)\,. \end{array}
$$

The index i corresponds to the double arrow in the figure. The invariant states that the left part $A[1 : i-1]$ of the array has been sorted, that is, it is an ordered permutation of $A_0[1 : i-1]$; besides, the right part $A[i : n]$ of the array has not been altered.

The macro-step (right part of Fig. 1) has now to be considered in a more detailed way. The role of this macro-step is to put element $A_0(i)$ in its correct place, within the left part of the array. Let x be the name of the variable used to record this element. The implementation will use an inner loop; each "micro-step" of this loop consists in shifting an (already sorted) element of the array to the right. A new index j, represented as a single arrow in the right part of Fig. 1, is used to refer to the next cell to be shifted. According to the figure, the invariant of the inner loop could be

$$\text{Inv-Int} : \quad \begin{array}{l} 1 \leq j \leq i \leq n \wedge x = A_0(i) \wedge \mathit{inf}(x, A, j+1, i) \wedge \\ \mathit{perm2}(A, 1, j-1, A, j+1, i, A_0, 1, i-1) \wedge \\ \mathit{ord2}(A, 1, j-1, A, j+1, i) \wedge \\ \mathit{equal}(A, i+1, n, A_0, i+1, n) . \end{array}$$

The predicate $\mathit{inf}(x, A, k, \ell)$ means that x is smaller than all the members of the set $\{A(k), \ldots, A(\ell)\}$; $\mathit{perm2}(A, i, j, B, k, \ell, C, m, n)$ means that $A(i : j)$ concatenated with $B(k : \ell)$ is a permutation of $C(m : n)$. Finally, $\mathit{ord2}(A, i, j, B, k, \ell)$ means that the concatenation of $A(i : j)$ and $B(k : \ell)$ is an ordered array. The index j divides the left part of the array into two subparts; the right one, that is, $A(j+1 : i)$, contains elements which have already been checked as greater than x.

We can now give a full program for the insert-sort (Fig. 2). The assertions between curly braces are a sketch of a formal proof of triple (1).

```
 1  i := 1;
 2  { Inv-Ext ∧ i = 1 }
 3  while i ≤ n do
 4          { Inv-Ext ∧ i ≤ n }
 5          x := A[i]; j := i;
 6          { Inv-Int ∧ j = i }
 7          while j > 1 ∧ A[j − 1] > x do
 8                  A[j] := A[j − 1]; j := j − 1;
 9                  { Inv-Int }
10          { Inv-Int ∧ (j ≤ 1 ∨ A[j − 1] ≤ x) }
11          A[j] := x; i := i + 1
12          { Inv-Ext }
13  { Inv-Ext ∧ i = n + 1 } .
```

Fig. 2. The insert-sort program with a proof sketch

We will not write a full proof but only outline an important fragment of it. When the execution of the inner loop terminates, the execution of the outer loops resumes. More specifically, the sequence $A[j] := x;\ i := i+1$ is executed (see Fig. 2, lines 10–12); the connection between loops is correct provided that the triple

$$\{ \text{Inv-Int} \wedge (j \leq 1 \vee A[j-1] \leq x) \}\ A[j] := x;\ i := i+1\ \{ \text{Inv-Ext} \}$$

is valid. According to the Assignment axiom and rule $R1$, we have to check that the (classical) formula

$$\text{Inv-Int} \wedge (j \leq 1 \vee A[j-1] \leq x) \Rightarrow \text{Inv-Ext}[i+1/i][A'/A]$$

is valid, where A' is such that

$$A'(k) = A(k)\,, \text{ for all } k \neq j\,;$$
$$A'(j) = x\,.$$

Lemma 1. Assertion $\text{Inv-Ext}[i+1/i][A'/A]$ is

$$1 \leq i+1 \leq n+1 \wedge \mathit{perm}(A', 1, i, A_0, 1, i) \wedge$$
$$\mathit{ord}(A', 1, i) \wedge \mathit{equal}(A', i+1, n, A_0, i+1, n)\,.$$

Proof. Elementary substitution.

Lemma 2. Formula

$$(\text{Inv-Int} \wedge j \leq 1) \Rightarrow \text{Inv-Ext}[i+1/i][A'/A]$$

is valid.

Informal proof. The internal invariant specifies $1 \leq j$, which, taking into account the additional formula $j \leq 1$, allows us to reduce the assertion under investigation to

$$1 = j \leq i \leq n \wedge x = A_0(i) \wedge \mathit{inf}(x, A, 2, i) \wedge$$
$$\mathit{perm2}(A, 1, 0, A, 2, i, A_0, 1, i-1) \wedge \mathit{ord2}(A, 1, 0, A, 2, i) \wedge$$
$$\mathit{equal}(A, i+1, n, A_0, i+1, n)\,.$$

The definitions of *perm2* and *ord2* allow a further reduction of this precondition into

$$1 = j \leq i \leq n \wedge x = A_0(i) \wedge \mathit{inf}(x, A, 2, i) \wedge$$
$$\mathit{perm}(A, 2, i, A_0, 1, i-1) \wedge \mathit{ord}(A, 2, i) \wedge \mathit{equal}(A, i+1, n, A_0, i+1, n)\,.$$

Besides, $A'[2:i] = A[2:i]$ and $A'(j) = A'(1) = x = A_0(i)$ and $A'(1)$ is smaller than $A'(2), \ldots, A'(i)$, hence the precondition can be rewritten as

$$1 = j \leq i \leq n \wedge x = A_0(i) \wedge \mathit{inf}(x, A', 2, i) \wedge$$
$$\mathit{perm}(A', 1, i, A_0, 1, i) \wedge \mathit{ord}(A', 1, i) \wedge \mathit{equal}(A', i+1, n, A_0, i+1, n)\,.$$

Due to Lemma 1, the postcondition is a logical consequence of the latter formula.

Lemma 3. Formula

$$(\text{Inv-Int} \wedge A[j-1] > x) \Rightarrow \text{Inv-Ext}[i+1/i][A'/A]$$

is valid.

The proof is omitted here.

Comment. Hoare logic can be extended to prove program termination. In the present case, proving termination is trivial. The body of the outer loop is executed n times, so the inner loop is also executed n times. Each execution of the inner loop involves a variable number of executions of the body of the inner loop, but this number is bounded by $n - 1$.

Hoare logic has been introduced in [16]; the weakest-precondition calculus is the basis of a design and verification method presented in [5]. The invariant method has been investigated, e.g., in [4] and many applications have been published.

Conclusion.

We observe that Hoare logic is useful to express properties of programs and also to write their proofs in a formal way. However, the root of the argument in such a proof is always the validity of classical formulas (lemmas 2 and 3, for instance), since the Assignment axiom allows us to reduce Hoare logic into classical logic. Furthermore, we notice that this kind of proof does not involve deep mathematical results but is nevertheless not trivial for two reasons. First, invariants have to be constructed, and this is not an easy task. In practice, rules $R3$ and $R4$ are usually combined into

$$\frac{\models X \Rightarrow I\,,\ \{I \wedge B\}\, S\, \{I\}\,,\ \models (I \wedge \neg B) \Rightarrow Y}{\{X\}\ \text{while}\ B\ \text{do}\ S\ \{Y\}}\,.$$

This rule is not analytical: I occurs in the premises but not in the conclusion; experience shows that the critical step when using it is the discovery of an appropriate invariant I.[3] The second difficulty is that correctness proofs of programs are typically very long. As a result, formal methods are, for now, seldom used in practice, except for small but safety-critical programs. The only way to deal with long programs is to use computer-aided verification.

In the next section we show how logic can be used to address the first problem, that is, invariant discovery.

2 Using Logic for Program and Invariant Construction

2.1 Mutual Exclusion

Concurrent programming is used in computer networks, distributed databases and many important applications. This kind of programming is rather difficult since concurrent processes cannot be investigated in isolation; it must be checked that a process does not alter the results produced by another process. For instance, it is possible to book a seat in a flight from Paris to New York at several places, but care must be taken that the

[3] This is the reason why writing proofs in Hoare logic can be difficult, just as using a sequent system with a cut rule in practice can be difficult.

same seat is not allocated twice. This is an instance of the "mutual exclusion" problem. An elementary instance is as follows. Two processes, say P and Q, perform independent computations but now and then print results on the same printer. Output should not take place for both processes at the same time, otherwise the printed document would be unreadable. The situation can be modeled as follows. Both processes endlessly switch between a "non-critical" state (say, p_0 and q_0, respectively) where they perform independent computations, and a "critical" state (p_c, q_c), where they print results. An appropriate mechanism must be used such that the formula

$$\neg(at\ p_c \land at\ q_c)$$

is valid, that is, it is never the case that both processes are simultaneously in their critical state. How can we do this ? A first solution consists in using a shared variable T (for "Turn") whose value (p or q) determines which process has access to the critical, printing section. This shared variable is handled in a "polite" way. When a process, say P, needs to print results, it switches the variable to the value q, allowing process Q to print. Process P waits until process Q switches the variable again (to p) before printing its own results. A drawback of this technique is that strict alternation is required; it is not adequate if, for instance, process P needs printing far more often than process Q. Another solution to the mutual exclusion problem is that both processes use their own variables, say inP and inQ, to signal their need of printing; inP becomes true when P signals its need of printing and is reset to false when the printing is completed. Process P effectively begins printing only when inQ is false, and conversely. The aforementioned drawback disappears, but there is a new one: if both processes attempt to print at the same time, there will be a "deadlock", that is, they will wait forever and no printing will be achieved.

2.2 Peterson's Algorithm

Peterson's solution [22] to the problem combines both mechanisms mentioned above and is illustrated in Fig. 3. Observe that process P waits (state p_w) until either the first mechanism or the second one allows it to proceed, so strict alternance is no longer required and there is no risk of deadlock.

The formal specification of the mutual exclusion property is the assertion

$$J =_{def} \neg(at\ p_c \land at\ q_c)\,. \tag{2}$$

It must be true throughout the computation. We view a computation as a temporal series of steps; each step consists in executing a transition, by process P or process Q. Initially, at time 0, we assume that both processes

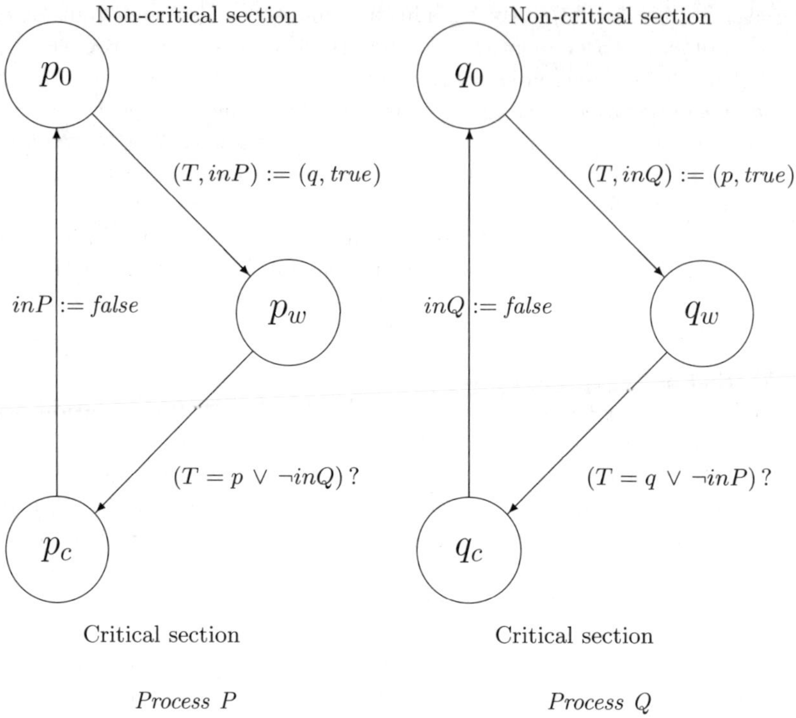

Fig. 3. Peterson's algorithm

are in their non-critical section, inP and inQ are false, and T is either p or q. This is formalized into the assertion

$$A =_{def} (at\ p_0 \land at\ q_0 \land T \in \{p, q\} \land \neg inP \land \neg inQ)\,.$$

The first step, leading to time 1, will be performed by process P or by process Q so, at time 1, either P will be in its waiting state, that is, the formula $(at\ p_w \land at\ q_0 \land T = q \land inP \land \neg inQ)$ will be true, or Q will be in its waiting state, that is, the formula $(at\ p_0 \land at\ q_w \land T = p \land \neg inP \land inQ)$ will be true. (We do not know which is the case; Peterson's algorithm, as most concurrent programs, is non-deterministic.) It is already clear that tracking computations is a lengthy and tedious work, so we need another method to establish that the property (2) is true at time n, for all n.

The standard way to prove a property for all $n = 0, 1, 2, \ldots$ is by induction. Clearly, mutual exclusion is satisfied at time 0: both $at\ p_0$ and $at\ q_0$ are true, so $(at\ p_c \land at\ q_c)$ is obviously false. However, assuming the property holds at time n, we cannot guarantee it will still hold at time $n+1$. If, say, $at\ p_c \land at\ q_w \land T = q$ holds at time n, property (2) is also true at time n, although the next step might lead to $at\ p_c \land at\ q_c$. In fact, property (2) cannot be verified by induction. However, we can find a stronger property that can be proved by induction, that is, an invariant of the program. In the present case, an appropriate invariant is

$$\begin{array}{rl} I = & (at\ p_c \Rightarrow (T = p \vee \neg inQ)) \\ \wedge & (at\ q_c \Rightarrow (T = q \vee \neg inP)) \\ \wedge & (at\ p_w \Rightarrow (T = q \vee at\ q_w)) \\ \wedge & (at\ q_w \Rightarrow (T = p \vee at\ p_w)) \\ \wedge & (at\ p_0 \equiv \neg inP) \\ \wedge & (at\ q_0 \equiv \neg inQ)\,. \end{array} \tag{3}$$

It is easy to check that the invariant is satisfied initially, and preserved by each transition, that is, $\{I\}\tau\{I\}$ holds for the six transitions of the system. The first step of the proof consists in reducing these triples into formulas of classical logic; as all variables of the program are Boolean, propositional formulas are obtained. The second step of the proof is the validation of these verification conditions. Both the production and the validation of verification conditions can be automated but, even for small programs, propositional verification conditions will likely be very big formulas, whose validation requires specific techniques (see [12]). Finally, we can check that property (2) is a logical consequence of the invariant (3), since $I \wedge at\ p_c \wedge at\ q_c$ implies inP, inQ, $T = p$ and $T = q$, which is inconsistent: variable T cannot have two distinct values at the same time.

Peterson's algorithm has been proved correct with respect to the mutual exclusion property, but there is an implementation problem. Both transitions $(p_0 \to p_w)$ and $(q_0 \to q_w)$ involve a double assignment: two variables are updated in one computation step. Usual hardware does not allow that, so assignment $(T, inQ) := (p, true)$, for instance, will have to be implemented either as $T := p;\ inQ := true$ or as $inQ := true;\ T := p$. It is possible to determine whether one or both decomposition(s) can be accepted, using Dijkstra's weakest precondition calculus. Only the results are given here.

The first decomposition consists in replacing $(p_0 \to p_w)$ and $(q_0 \to q_w)$ by

$$\begin{array}{ll} (p_0,\ T := q\ ,\ p_i)\,, & (q_0,\ T := p\ ,\ q_i)\,, \\ (p_i,\ inP := true\ ,\ p_w)\,, & (q_i,\ inQ := true\ ,\ q_w)\,. \end{array}$$

This is not appropriate, since the invariant cannot be updated in order to take into account the new control points p_i and q_i. It is easy to see that this "solution" does not ensure mutual exclusion. A counterexample is the following computation:

$$\begin{array}{ll} 0. & at\ p_0 \wedge at\ q_0 \wedge \neg inP \wedge \neg inQ \wedge T = p\,; \\ 1. & at\ p_i \wedge at\ q_0 \wedge \neg inP \wedge \neg inQ \wedge T = q\,; \\ 2. & at\ p_i \wedge at\ q_i \wedge \neg inP \wedge \neg inQ \wedge T = p\,; \\ 3. & at\ p_i \wedge at\ q_w \wedge \neg inP \wedge inQ \wedge T = p\,; \\ 4. & at\ p_i \wedge at\ q_c \wedge \neg inP \wedge inQ \wedge T = p\,; \\ 5. & at\ p_w \wedge at\ q_c \wedge inP \wedge inQ \wedge T = p\,; \\ 6. & at\ p_c \wedge at\ q_c \wedge inP \wedge inQ \wedge T = p\,. \end{array}$$

The second decomposition consists in replacing $(p_0 \rightarrow p_w)$ and $(q_0 \rightarrow q_w)$ by

$$\begin{array}{ll} (p_0,\ inP := true\ ,\ p_i)\ , & (q_0,\ inQ := true\ ,\ q_i)\ , \\ (p_i,\ T := q\ ,\ p_w)\ , & (q_i,\ T := p\ ,\ q_w)\ . \end{array}$$

If it is appropriate, we should be able to construct a new version I' of the invariant I. We cannot use $I' = I$ since I does not specify what is true when one or both process(es) are in the intermediate state(s) p_i, q_i. In order to find this information, we use weakest precondition and/or strongest postcondition calculus (details are omitted here). The result is

$$\begin{array}{rl} I' = & (at\ p_c\ \Rightarrow\ (T = p\ \vee\ \neg inQ\ \vee\ at\ q_i)) \\ \wedge & (at\ q_c\ \Rightarrow\ (T = q\ \vee\ \neg inP\ \vee\ at\ p_i)) \\ \wedge & (at\ p_w\ \Rightarrow\ (T = q\ \vee\ at\ q_w)) \\ \wedge & (at\ q_w\ \Rightarrow\ (T = p\ \vee\ at\ p_w)) \\ \wedge & (at\ p_0\ \equiv\ \neg inP) \\ \wedge & (at\ q_0\ \equiv\ \neg inQ)\ . \end{array} \tag{4}$$

It is also easy to establish that "deadlock" is not possible; this means, at every time point of the computation, at least one of the processes is allowed to proceed. In particular, when both processes are in their waiting state, the value of T decides which one will access its critical section first.

A final interesting point is that the algorithm behaves fairly: provided that, say, process P does not get stuck within its critical section forever, process Q will not be kept waiting indefinitely. When process P leaves its critical section, process Q is allowed to proceed since variable inP is reset to false.

The invariant method is frequently used to verify concurrent systems; see, e.g., [9, 11, 18]. See, e.g., [10, 13, 21, 24] for computer-aided verification methods. A full methodology for reasoning about concurrent systems is introduced in [2]. See also [6] for various logic-based proofs in several areas of computing science.

Conclusion.

We observe again that Hoare logic is appropriate to express properties and to write their proofs. For instance, the fact that I is an invariant is formalized into

$$\{I\}\ \tau\ \{I\}\ \text{holds for}\ \tau = (p_0 \rightarrow p_i), \ldots, (q_c \rightarrow q_0)\ .$$

Besides, the fact that Peterson's algorithm satisfies mutual exclusion is formalized into three more elementary propositions:

$$\begin{array}{l} \models (A \Rightarrow I)\ ; \\ I \text{ is an invariant}\ ; \\ \models (I \Rightarrow J)\ . \end{array}$$

However, as illustrated with the decomposition problem for Peterson's algorithm, the construction of proof arguments, i.e., the invariant, is done within classical logic.

Comment. Temporal logic is often used instead of classical logic for writing properties and proofs. The most usual temporal operators are

1. $\bigcirc$: "next time" ;
2. $\Box$: "always" ;
3. $\Diamond$: "eventually" ;
4. $\mathcal{U}$: "until".

These operators are unary except $\mathcal{U}$, which is binary; they are used to express properties of concurrent systems. For instance, we have

$$\begin{array}{ll} \text{Mutual exclusion :} & A \Rightarrow \Box J \ ; \\ \text{Invariant :} & (A \Rightarrow I) \wedge \Box(I \Rightarrow \bigcirc I) \ ; \\ \text{No lock :} & \Box(at\ p_w \Rightarrow \Diamond at\ p_c) \wedge \Box(at\ q_w \Rightarrow \Diamond at\ q_c). \end{array}$$

It is also possible to prove these properties within temporal logic; several axiomatic systems have been proposed. Algorithmic methods based on temporal logic have also been proposed for concurrent systems design and verification. (See, e.g., [8, 14, 17, 28].)

3 Logic Programming

In this section we show how the Horn fragment of classical first-order logic can be turned into a convenient programming language. The underlying principles are best understood within the propositional part of the logic; this is developed in the first paragraph. The second paragraph is devoted to the extension to first-order logic; the theory is illustrated with the classical example of list concatenation. The third paragraph shows how some second-order logic and extra logical features are used in Prolog[4] in order to turn logic programming into a practical computing and reasoning tool.

3.1 Propositional Logic Programming

The central problem in logic is, given a formula A and a set of formulas E, to determine whether A is a logical consequence of E (this is denoted $E \models A$) or, equivalently, if the set $S =_{def} E \cup \{\neg A\}$ is inconsistent. As every formula

---

[4] Prolog is the most popular programming language based on logic programming. The name Prolog denotes the language, but also the software system used to run programs written in this language. See, e.g., [26] for a detailed introduction to logic programming and Prolog.

is equivalent to a conjunction of clauses,[5] one can assume that S is a set of clauses.

Many algorithms exist for testing the inconsistency of a set of clauses. The most elementary one is to write a truth table; however, if the set S involves n distinct propositions, the truth table will contain 2^n lines, which is clearly intractable when n is even moderately large. No generally tractable algorithm is known for this problem, and maybe none exists.[6] As a result, automated deduction can be a challenging task, even in the propositional framework. Fortunately, there is a non trivial fragment of propositional logic for which the inconsistency problem is easy. A *Horn clause* is a clause containing at most one positive literal. An elementary but fast algorithm for testing Horn sets (i.e., sets of Horn clauses) is *unit resolution.* A (positive) unit clause is a clause whose only literal is a positive one. Let p be a proposition and therefore a unit clause, and c a Horn clause containing $\neg p$. The *resolvent* r of p and c is obtained by omitting $\neg p$ in c; it is noted $c - \neg p$. We observe immediately that the models of the clauses p and c are exactly the same as the models of p and r. The unit resolution algorithm for testing Horn set S_0 is as follows:

```
S := S₀;  { the initial value of S is S₀ }
while □ ∉ S do
        select p and c such that p ∈ S,  c ∈ S,  ¬p ∈ c;
        r := c − ¬p;     { clause r is the resolvent of p and c }
        S := (S\{c}) ∪ {r}.  { c is replaced by r in S }.
```

At every step, a unit clause p and a clause c containing $\neg p$ are selected in S, and clause c is replaced in S by the resolvent $c - \neg p$. So, at every step, a negative literal is suppressed in S. As there are only finitely many literals in S, the algorithm always terminates; the final value of S is denoted S_f. Also observe that S changes, but the set $\mathcal{M}(S)$ of S-models[7] does not change; the equality

$$\mathcal{M}(S) = \mathcal{M}(S_0)$$

is an *invariant* of the algorithm (cf. § 1.2). Termination can occur in two ways:

- The empty clause $\square$ is generated and becomes a member of $S = S_f$. As this clause is identically false, S_f is inconsistent and, due to the invariant, S_0 is also inconsistent.

[5] A clause is a disjunction of 0, 1 or several literals; a literal is an elementary proposition (positive literal) or the negation of an elementary proposition (negative literal). A clause is valid if and only if it contains a pair $\{p, \neg p\}$; it is inconsistent if and only if it is empty; the empty clause is denoted $\square$ and is equivalent to *false*.

[6] In technical terms, the satisfiability problem for propositional logic is NP-complete.

[7] An S-model is an interpretation for which all clauses of S are true.

- There is no further way to select appropriate p and c. This means that S_f and therefore S_0 are consistent; these sets admit at least one model, and the *minimal* model is the interpretation I such that, for each proposition q, $I(q)$ is true if q is a unit clause in S_f, and false otherwise.[8]

A Horn clause is *definite* if it contains a positive literal; it is *negative* otherwise. A definite Horn clause like $p \vee \neg q \vee \neg r$ is usually noted as $(q \wedge r) \Rightarrow p$ or, better, $p \Leftarrow (q \wedge r)$. In Prolog we write "`p :- q,r.`"; `p` is the *head* of the clause, `q,r` is the *body* of the clause. A *logic program* is a set of definite Horn clauses. A logic program L is always consistent; if I_L is its minimal model and if p is a proposition, then $L \models p$ if and only if $I_L(p) = true$. The minimal model can be determined by applying the unit resolution algorithm to L, but the full determination of I_L is usually not necessary; the standard problem of logic programming is, given a proposition q (the *query*), to determine whether $L \models q$ or not, that is, whether $L \cup \{\neg q\}$ is inconsistent or not. To answer this, we can run the unit resolution algorithm with $S_0 = L \cup \{\neg q\}$. We observe that the *goal*, that is, the negative clause $\neg q$, has a prominent role in any computation leading to the conclusion that S_0 is inconsistent, since L itself is always consistent. However, in the unit resolution algorithm, no specific role is devoted to the goal. For this reason, another approach, called the *input resolution* algorithm, is used instead:

$$
\begin{array}{l}
G := \neg q; \\
\text{while } \; G \neq \mathit{false} \text{ do} \\
\qquad \text{select } p \text{ and } c \text{ such that} \\
\qquad\qquad \neg p \in G, \\
\qquad\qquad c \in L \wedge p \in c; \\
\qquad G := (G - \neg p) \vee (c - p)\,.
\end{array}
$$

If the algorithm terminates successfully, i.e., with $G = \mathit{false}$, then $L \models q$ is proved. The input resolution is equivalent to the unit resolution (see [3] for a proof), but it turns out to be generally more efficient. The algorithm is highly non-deterministic since at every step several choices are possible for p and c. In practice, the goal G is a list of negative literals (and not an unordered set) and the selected literal $\neg p$ is always the first one. As all literals in G are negative, the negation is usually omitted; more precisely, instead of using a variable G (negative clause) one uses its logical negation, that is, a positive cube (conjunction of positive literals). Non-determinism is further reduced if the clauses in L are ordered, that is, a logic program is a *list* of definite Horn clauses. Let us immediately consider a toy example. The logic program is

[8] The model is called minimal since every proposition interpreted as true in this model is also true in any other model.

```
p :- q,r.
p :- s.
q :- r.
q :- u,s.
s.
```

Let p be the query. A possible computation generated by the input resolution algorithm is as follows:

```
0.  p.        (initial goal)
1.  q,r.      (clause  p :- q,r  has been selected)
2.  r,r.      (clause  q :- r    has been selected)
FAILURE       (no further progress achievable)
```

Another possible computation is:

```
0.  p.        (initial goal)
1.  q,r.      (clause  p :- q,r  has been selected)
2.  u,s,r.    (clause  q :- u,s  has been selected)
FAILURE       (no further progress achievable)
```

The last possible computation is

```
0.  p.        (initial goal)
1.  s.        (clause  p :- s  has been selected)
2.  .         (clause  s       has been selected)
SUCCESS       (goal is empty)
```

In Prolog, the search for a successful computation is organized as a search tree (see Fig. 4).

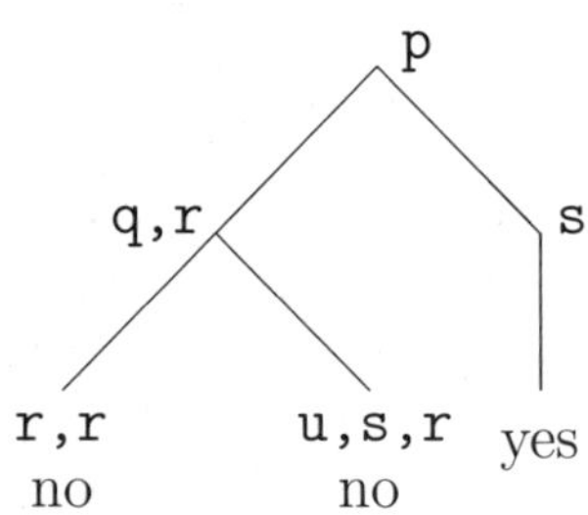

Fig. 4. A search tree

3.2 Predicate Logic Programming

Propositional logic programming is nothing more than automated tautology checking; no computation on terms can be achieved this way. Let us illustrate now how such a computation can be performed in the predicate framework.

A common structure in computing, the *list*, is used as an example. A list is simply an ordered sequence of objects; the number of objects is the *length* of the list. The empty list is denoted `[ ]`, the list whose elements are `a` and `b` (in this order) is denoted `[a,b]`. A list can be an element of another list, as in `[[a,b],c]`. Small letters denote constants, capital letters denote variables. The value of a variable can be any ground term, that is, any term without variable occurrences. (The only kinds of terms used here are constants and lists.) In the sequel, we use `X`, `Y`,... for variables with arbitrary values, and `Xs`, `Ys`,... for variables whose intended values are lists. In a non empty list `[a,b,c]`, the first element `a` is the *head* and the other elements form the *tail* `[b,c]`. The value of the non-ground term `[X|Xs]` is the list whose head and tail are the value of `X` and the value of `Xs`, respectively. For instance, the equation `[X|Xs]` = `[a,b,c]` admits only one solution, i.e., `X` = `a`, `Xs` = `[b,c]`.

A simple operation on lists is *concatenation*; the elements of the concatenation of two lists are the elements of the first list followed by the elements of the second one. For instance, the concatenation of `[a,b]` and `[c,d,e]` is `[a,b,c,d,e]`. The ternary predicate `append` models list concatenation; more precisely, "`append(Xs,Ys,Zs)`" means "the concatenation of `Xs` and `Ys` is `Zs`". A true instance of this predicate is "`append([a,b],[c,d],[a,b,c,d])`"; a false instance is "`append([a,b],[c],[])`". The predicate `append` can be defined as follows:

```
append([],Bs,Bs).
append([A|As],Bs,[A|Cs]) :- append(As,Bs,Cs).
```

Each clause has an implicit universal quantification on all variables. The first clause is a unit clause; it can be read as, "for each list `Bs`, the concatenation of the empty list and `Bs` is `Bs`". The meaning of the second clause is, "for each object `A` and for all lists `As`, `Bs` and `Cs`, the concatenation of the lists `[A|As]` and `Bs` is the list `[A|Cs]`, provided that the concatenation of the lists `As` and `Bs` is the list `Cs`".

Figure 5 illustrates how this definition (in fact, this logic program) can be used to find all true instances of the formula "`append(Xs,Ys,[a,b])`". Three solutions exist, which are

```
Xs = [] , Ys = [a,b] ;
 Xs = [a] , Ys = [b] ;
Xs = [a,b] , Ys = [] .
```

These solutions are found by input resolution. The principle is the same as in the propositional case, including the organization of the search as a tree, except that *unification* is used to match the terms in the clauses with the terms in the query; the purpose of the search is to obtain instances of the query "`append(Xs,Ys,[a,b])`", which are logical consequences of the clauses.

Let us see in detail how this tree is built. The root is labelled with the query. As the program contains two "matching" clauses, that is, two clauses

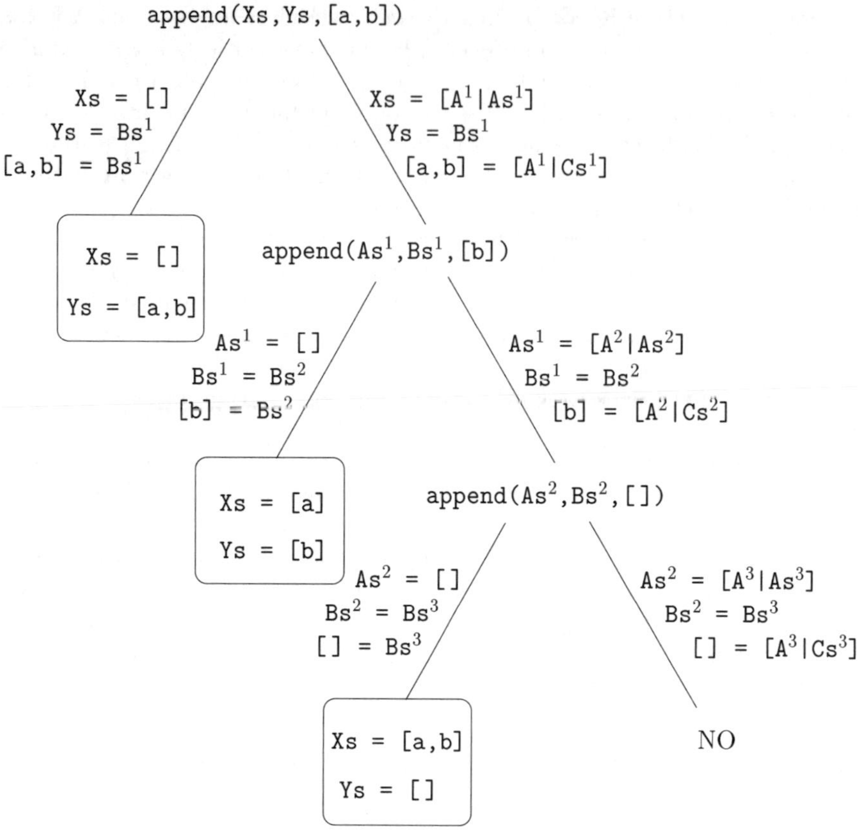

Fig. 5. Search tree for list concatenation and decomposition

whose heads share the same predicate as the query, the root has exactly two children. However, the query is `append(Xs,Ys,[a,b])`, whereas the head of the first clause is `append([],Bs,Bs)`. We have already mentioned that clause variables stand for arbitrary objects; for now, it is appropriate to consider the particular case `Xs` = `[]`, `Ys` = `Bs` = `[a,b]`, which leads to the first solution (see Fig. 5, left branch of the tree). In the tree we have written $\texttt{Bs}^1$ instead of `Bs`; this is to emphasize that the specific instantiations made with the matching clause will not prevent us making other instantiations (with the same clause) later. The superscript is conventionally chosen as the level in the search tree; we are at the first level, so now the superscript is 1. In the sequel, all variables issued from the program text will be superscripted in the search tree, whereas all variables issued from the original query will retain their name, without alteration.[9]

[9] This difference is quite natural. On the one hand, the name of clause variables does not matter, since "`append([],Bs,Bs).`" and "`append([],Ys,Ys).`", for in-

The second step consists in the alternate reduction of the query, using the second clause. The query and the head of the second clause become identical if some instantiations are made, i.e., `Xs` = `[A`^{1}`|As`^{1}`]`, `Ys` = `Bs`^{1}, `[a,b]` = `[A`^{1}`|Cs`^{1}`]`; they label the arc between the root node and its second child. The subquery associated with the latter is the body of the second clause, suitably instantiated, i.e., `append(As`^{1}`,Bs`^{1}`,[b])`. This subquery, just as the main query, can be solved in two ways, so the corresponding node has two children. The first one involves reduction with the unit clause, leading to an empty subsubquery, and therefore a second solution. The appropriate answer, as returned to the user, is the list of the values attributed to the variables contained in the user's original query. In the present case, these variables were `Xs` and `Ys`, so `Xs = [a]`, `Ys = [b]`, is returned to the user. The second child corresponds to the second clause, whose body is not empty. The appropriate instantiations label the arc between the node and this second child, which is labelled itself with a new subquery. This subquery is the body of the matching clause, that is, `append(As`^{2}`,Bs`^{2}`,Cs`^{2}`)`, maybe modified by the instantiations. From `[b] = [A`^{2}`|Cs`^{2}`]`, we deduce `[] = Cs`^{2}, so the subquery becomes `append(As`^{2}`,Bs`^{2}`,[])`.

The reduction of this last subquery gives rise to a third solution `Xs = [a,b]`, `Ys = []`; the alternate reduction fails: indeed, one of the required instantiation is `[] = [A`^{3}`|Cs`^{3}`]`, which clearly has no solution: the left-hand side is an empty list, whereas the right-hand side is a list with at least one element. As a result, the second possibility for reducing the last subquery fails, and this ends the search and the construction of the search tree.

It is clear that only two mechanisms are needed to construct search trees; input resolution provides node labelling, and unification, that is, the construction of appropriate variable instantiations, provides arc labelling. The instantiation problem has received a general, systematic solution, which is the unification algorithm (see [23] for the original paper; see also [3, 7, 19, 20] for more details).

We will not investigate first-order Horn logic programming further, but simply illustrate how it can be used to solve non-trivial problems in an elegant way. The example will be a classical riddle, à la Lewis Carroll.

---

stance, have clearly the same meaning. On the other hand, the name of query variables does matter, since it will be used by Prolog to communicate answers to the user. Observe nevertheless that, within a clause, variable names can be changed only in a consistent way. For instance, "`append([],Bs,Bs).`" cannot be rewritten as "`append([],Bs,Ys).`"; otherwise stated, the "scope" of a variable, viz. the text where it keeps its meaning, is the clause which contains it.

Five houses in a row, five nations, five colors,
Five drinks, five cigarette brands ... and five pets!

1. *Houses are numbered 1, 2, 3, 4, 5.*
2. *The English's house is red.*
3. *The Spanish's pet is a dog.*
4. *In the green house there is coffee.*
5. *The Ukrainian drinks tea.*
6. *The green house number is next to the ivory house number.*
7. *The Old Gold smoker raises snails.*
8. *Gauloises are smoked in the yellow house.*
9. *Milk is drunk in house number 3.*
10. *The Norwegian's house is numbered 1.*
11. *The Chesterfield smoker and the fox owner are neighbours.*
12. *The Gauloises smoker and the horse owner are neighbours.*
13. *The Lucky Strike smoker drinks orange juice.*
14. *The Japanese smokes Gitanes.*
15. *The Norwegian's house is next to the blue house.*

Who owns the zebra? Who drinks water?

There are $(5!)^5 = 120^5 = 24\,883\,200\,000$ ways of distributing the attributes among the five houses, but with a rather elementary logic program, the problem can be solved by computer in less than a second. This logic program will consist of two parts. The first one describes the concepts used to express the riddle, whereas the second one is the formal counterpart of the riddle itself, consisting in 15 "clues" and 2 questions.

The first relevant concept is membership: five houses are the "members" of a row. This notion is formalized by the following clauses:

```
member(X,[X|Xs]).
member(X,[Y|Xs]) :- member(X,Xs).
```

The predicate `member(A,As)` is true when the object `A` is an element of the list `As`. We observe that `As` must be non-empty, and therefore can be written as `[B|Bs]`. If `A` belongs to this list, then either `A` is `B`, or `A` is a member of `Bs`, hence the two clauses. As only five-element lists are considered, an alternate definition of `member` is

```
member(A,[A,B,C,D,E]).
...    ...
member(E,[A,B,C,D,E]).
```

Another relevant concept is neighbourhood:

```
neighbour(A,B,[A,B,C,D,E]).
...    ...
neighbour(E,D,[A,B,C,D,E]).
```

As number 1 and 3 are mentioned in the clues, we introduce

```
one(A,[A,B,C,D,E]).
three(C,[A,B,C,D,E]).
```

In these predicates, variables `A` to `E` represent the five houses. As each house is described by five attributes, it is convenient to model houses by compound terms `h(N,C,A,B,T)` where the arguments are the attributes. This is formalized as follows:

```
nation(h(N,C,A,B,T),N).
color(h(N,C,A,B,T),C).
animal(h(N,C,A,B,T),A).
drink(h(N,C,A,B,T),B).
tobacco(h(N,C,A,B,T),T).
```

The second part formalizes the fifteen clues and two questions. We first observe that solving the riddle means discovering the values of the nationality of the zebra owner, say `X`, and the nationality of the water drinker, say `Y`. The head of the clause defining the main predicate will therefore be `go(X,Y)`. Values of `X` and `Y` are adequate if they make true all clues and questions, so the body of the main clause will be the conjunction of seventeen parts. We only give the parts corresponding to clues 1, 2, and 15, and to questions 1 and 2:

```
go(X,Y) :- St = [h(N1,C1,A1,B1,T1),h(N2,C2,A2,B2,T2),
                 h(N3,C3,A3,B3,T3),h(N4,C4,A4,B4,T4),
                 h(N5,C5,A5,B5,T5)],                  % clue  1
member(X2,St),nation(X2,english),color(X2,red),       % clue  2
...
neighbour(X15a,X15b,St),nation(X15a,norwegian)
                        ,color(X15b,blue),             % clue 15
member(Q,St),animal(Q,zebra),nation(Q,X),         % question 1
member(R,St),drink(R,water),nation(R,Y).          % question 2
```

Clue 1 asserts the existence of a street (`St`) which is modeled as a list of five houses, each of them described as mentioned above. Clue 2 is *English's house is red*, which can be restated as follows:

> *There exists a house* `X2` *in the street* `St`,
> *whose owner's nationality is English and*
> *which is painted red.*

Clue 15 is *Norwegian's house is nearby blue house*, which can be restated as follows:

> *There exist two houses* `X15a` *and* `X15b`,
> *which are neighbours in the street* `St`*;*
> *The owner's nationality of* `X15a` *is Norwegian and*
> *house* `X15b` *is painted blue.*

Question 1 is *Who owns the zebra ?*, which can be restated as follows:

> *There exists a house* `Q` *in the street* `St`*;*
> *Its animal is a zebra and*
> *the owner's nationality is* `X`.

Question 2 is *Who drinks water ?*, which can be restated as follows:

> *There exists a house* `R` *in the street* `St`*;*
> *Its owner drinks water and*
> *the owner's nationality is* `Y`.

Now, let us run this program with the query

```
go(ZebraOwner,WaterDrinker).
```

Prolog starts building a (big) search tree and issues the solution, that is,

```
ZebraOwner = japanese,
WaterDrinker = norwegian.
```

If we ask Prolog for a second solution, the answer is `no`; indeed, the solution is unique.

Several textbooks are devoted to the applications of logic programming and Prolog to problems involving non-trivial reasoning; see, e.g., [1] and [25].

3.3 Beyond First-Order Logic Programming

The examples given above show that first-order logic or, more precisely, the Horn fragment of first-order logic, is an appropriate basis for logic programming. However, some second-order logic features, and also some extra-logical mechanisms, have been introduced in Prolog; they allow for more convenient solutions to many interesting programming problems.

A first useful extra-logical mechanism is writing. The special "predicate" `write(X)` has the effect of writing the value of `X` on the computer screen, and the predicate `nl` (new line) has the same effect as the "carriage return" key on a typewriter. Let us consider the following definition:

```
wlist([]) :- nl.
wlist([U|Us]) :- write(U),write(' '),wlist(Us).
```

This predicate is always true and writes the elements of its arguments, with spaces between them.

```
?- wlist([1,2,3,4,5]).
1 2 3 4 5
Yes
```

We use `wlist` to define a new version of `append`; this new predicate `wappend` issues the same results as `append`, but also writes the intermediate steps of the reasoning. Here is the definition:

```
wappend([],Bs,Bs) :- wlist([[],Bs,Bs]).
wappend([A|As],Bs,[A|Cs]) :- wappend(As,Bs,Cs),
                             wlist([[A|As],Bs,[A|Cs]]).
```

Here now is the computation corresponding to the query of Fig. 5:

```
?- wappend(Xs,Ys,[a,b]).
[] [a, b] [a, b]
Xs = [], Ys = [a,b] ;

[] [b] [b]
[a] [b] [a, b]
Xs = [a], Ys = [b] ;

[] [] []
[b] [] [b]
[a, b] [] [a, b]
Xs = [a,b], Ys = [] ;

No
```

The queries `append(Xs,Ys,[a,b])` and `wappend(Xs,Ys,[a,b])` admit exactly the same solutions but, in the second case, the structure of the search generated by Prolog is made apparent. We see that the first solution, corresponding to `(w)append([],[a,b],[a,b])` is a one-step reasoning. The second solution comes from a two-step reasoning: `(w)append([],[b],[b])` is established first, and then `(w)append([a],[b],[a,b])` is deduced. The third and last solution is a three-step reasoning: `(w)append([],[],[])` is established first, and then `(w)append([b],[],[b])` is deduced, and then `(w)append([a,b],[],[a,b])` is deduced.

Pure first-order logic programming allows automated reasoning, but the extra-logical predicate `write` provides an interesting addition: Prolog is now able to explain how results are generated,[10] that is, how reasoning is performed. This ability to provide explanations is very important for expert systems written in Prolog and for many applications in the domain of artificial intelligence (see, e.g., [1, 25, 26] for examples and more details).

The predicate `write` also allows simulation of second-order predicates. For instance, we have observed that the query `append(Xs,Ys,[a,b])` has three solutions, issued one after another. It is often convenient to collect all solutions to a query into a set (or a list). This can be simulated easily with the following definition:

```
append*(Xs,Ys,Zs) :- append(Xs,Ys,Zs), wlist([Xs,Ys,Zs]), fail.
```

[10] It is possible to obtain more detailed explanations; see, e.g., [26].

The predicate `fail` has no argument and is always false, since no clause with the head `fail` has been introduced. As a result, `append*`-queries will always fail, but the `wlist` predicate will issue the "solutions" nonetheless. Here is an example:

```
?- append*(Xs,Ys,[a,b]).
[] [a, b] [a, b]
[a] [b] [a, b]
[a, b] [] [a, b]
No
```

The three solutions to the problem of decomposing `[a,b]` are listed, although the system claims there is no solution. Clearly enough, for every predicate `pred`, we can easily define a corresponding variant `pred*`, that will issue all solutions in one step. In fact, Prolog provides a generic second-order predicate, `findall`, which collects in a list all solutions to a given query. Here is an example, giving the three decompositions of list `[a,b]`:

```
?- findall((Xs,Ys),append(Xs,Ys,[a,b]),Sols).
Sols = [([], [a, b]), ([a], [b]), ([a, b], [])]
```

The ability of Prolog to provide explanations for its own deductions is very helpful. Another feature which makes Prolog somewhat similar to a human expert is the ability to enrich or correct its own knowledge base. The second-order predicate `asserta` has one argument which is a clause. When "executed", this predicate places the clause at the beginning of the logic program. The special predicate `asserta` has many applications, but only one of them is mentioned here. A good human expert uses not only deduction abilities, but also memory: when some result has been deduced, it is stored in such a way that it can be reused later, without being deduced again. In mathematics, such an intermediate reusable result is a lemma, so we could define

```
lemma(Pred) :- Pred, asserta(Pred).
```

This is interpreted as follows: in order for `Pred` to become a lemma, first `Pred` has to be deduced, and then stored in the logic program. This definition, although correct, is not fully satisfactory. Indeed, when the lemma `Pred` is needed, is is first picked up in the logic program but, as Prolog usually searches for several ways to establish a goal, the lemma will also be reestablished again, which is exactly the redundant work to be avoided. So we use another definition instead:

```
lemma(Pred) :- Pred, asserta((Pred :- !)).
```

The cut (denoted "`!`") is a special predicate that cuts the search; here is has the required effect: when the lemma `Pred` has been established once, it will not be reestablished later.[11]

[11] The cut is a more subtle, and rather controversial extra-logical feature of Prolog, but its use cannot be avoided in most applications.

As an illustration, let us consider the classical case of Fibonacci numbers. The first Fibonacci numbers are $f_0 =_{def} 0$ and $f_1 =_{def} 1$; if $n > 1$ then $f_n =_{def} f_{n-1} + f_{n-2}$. This is translated easily into the following logic program:

```
fib(0,0).
fib(1,1).
fib(N,P) :- N>1, A~is N-1, B is N-2,
            fib(B,Q), fib(A,R), P is Q+R.
```

This program is a straightforward translation of the definition, but it is not satisfactory. If we ask the query `fib(30,A)`, a very long time is needed before the correct answer `A = 832040` is generated by the computer. The reason for this becomes apparent if the search tree is considered: intermediate results are computed several times. More precisely, in order to compute f_{30}, f_{29} is computed once, f_{28} is computed twice, f_{27} is computed three times and, for all i, f_{30-i} is computed f_{i+1} times. These redundancies are avoided with the following "lemmatized" version:

```
lfib(0,0).
lfib(1,1).
lfib(N,P) :- N>1, A~is N-1, B is N-2,
             lemma(lfib(B,Q)), lfib(A,R), P is Q+R.
```

Now, the computation takes nearly no time:

```
?- lfib(30,A).
A = 832040
```

4 Conclusion

Logic has various and deep connections with theoretical and applied computer science. Hoare logic is an example where logic is imported into the programming world in order to enhance the understanding of programs and their reliability. Prolog is quite different: Horn logic is turned into a programming language, that is, programming constructs have been imported in logic to allow automated deduction and computation. These examples also show that classical logic remains central in computing-related applications, but various logical systems used in computer science extend classical logic in a significant way.

References

1. I. Bratko: *Prolog – Programming for Artificial Intelligence* (Addison-Wesley, Reading, Mass. 1990 [2nd edition])
2. K.M. Chandy, J. Misra: *Parallel Program Design: A Foundation* (Addison-Wesley, Reading, Mass. 1988)
3. C.C. Chang, R.C. Lee: *Symbolic Logic and Mechanical Theorem Proving* (Academic Press, New York 1973)
4. J.W. De Bakker, L.G.L.T. Meertens: 'On the Completeness of the Inductive Assertion Method'. Journal of Computer and System Sciences **11** (1975) 323–357
5. E.W. Dijkstra: *A discipline of programming* (Prentice Hall, New Jersey 1976)
6. E.W. Dijkstra (ed.): *Formal Developments of Programs and Proofs* (Addison-Wesley, Reading, MA 1990)
7. K. Doets: *From Logic to Logic Programming* (MIT Press, Cambridge, Mass. 1994)
8. R. Gerth, D. Peled, M. Vardi, P Wolper: 'Simple On-the-fly Automatic Verification of Linear Temporal Logic'. In: *Proc. 15th Work. Protocol Specification, Testing, and Verification* (North-Holland, Amsterdam 1995)
9. P. Gochet, P. Gribomont: *Logique: méthodes pour l'étude des programmes* (Hermès, Paris, 1994)
10. D.M. Goldschlag: 'Mechanically Verifying Concurrent programs with the Boyer-Moore prover'. IEEE Trans. on Software Engineering **16** (1990) 1005–1023
11. E.P. Gribomont: 'Concurrency without toil: a systematic method for parallel program design'. Sci. Comput. Programming **21** (1993) 1–56
12. E.P. Gribomont: 'Simplification of Boolean verification conditions'. Theoret. Comput. Sci. **239** (2000) 165–185
13. E.P. Gribomont, D. Rossetto: 'CAVEAT: technique and tool for Computer Aided Verification And Transformation'. In: Lect. Notes in Comput. Sci. **939** (Springer, Berlin 1995) 70–83
14. P. Gribomont, P. Wolper: 'Temporal logic'. In: [27], vol. 2, ch. 4
15. D. Gries: *The Science of Programming* (Springer-Verlag, Berlin 1981)
16. C.A.R. Hoare: 'An axiomatic basis for computer programming'. *Comm. ACM* **12** (1969) 576–583
17. R.P. Kurshan, L. Lamport: 'Verification of a Multiplier: 64 Bits and Beyond'. In: Lect. Notes in Comput. Sci. **697** (Springer, Berlin, 1993) 166–179
18. L. Lamport: 'An Assertional Correctness Proof of a Distributed Algorithm'. Sci. Comput. Programming 2 (1983) 175–206
19. J.W. Lloyd: *Foundations of Logic Programming* (Springer-Verlag, Berlin 1987 [2nd edition])
20. J.-L. Lassez, M.J. Mahler, K. Marriott: 'Unification revisited'. In: *Foundations of Deductive Databases and Logic programming* (Ed.: J. Minker, Morgan Kaufmann, Los Altos 1988)
21. Z. Manna et al.: STeP: the Stanford Temporal Prover (Draft), Report STAN-CS-TR-94-1518, Stanford University, June 1994
22. G.L. Peterson: 'Myths about the mutual exclusion problem'. *Information Proc. Lett.* **12** (1981) 115–116

23. J.A. Robinson: 'A machine-oriented logic based on the resolution principle'. *Jl. ACM* **12** (1965) 23–41
24. D.M. Russinoff: 'A Verification System for Concurrent Programs Based on the Boyer-Moore Prover'. Formal Aspects of Computing **4** (1992) 597–611
25. Y. Shoam: *Artificial Intelligence Techniques in Prolog* (Morgan Kaufmann, Los Altos 1994)
26. L. Sterling, E. Shapiro: *The Art of Prolog* (MIT Press, Cambridge, Mass. 1994 [2nd edition])
27. A. Thayse et al.: *A logical approach to artificial intelligence* (3 vol.) (Wiley, New-York 1989, 1991)
28. M. Vardi, P. Wolper: 'Reasoning about Infinite Computations'. *Information and Computation* **115** (1994) 1–37

Stochastic *versus* Deterministic Features in Learning Models

Ion-Olimpiu Stamatescu

1 Introduction

This paper is not concerned with logics in the strict sense, it is, however, concerned with questions about the logical character of the scientific procedures in the wider sense of self-consistency, adequacy and interconnections between various schemes of thought. This is a reasonable point of view if we accept Peirce's claim that "according to its nature [logic] is forced to proceed with its research into the condition of reality itself and in doing that it cannot limit itself to language forms but must necessarily enquire about how and what we think." [6] But independently of patronage the above questions are important in investigating the main traits and alternatives in our scientific thinking. Here we shall try to open a discussion about the character and chances of statistical points of view in science.

Randomness and necessity is one of the conceptual pairs with a long history and unresolved argumentation. Although an axiomatic probability theory can be developed self-consistently as a mathematical system (Kolmogorov), the question of its connection to phenomena and thus of its adequacy and legitimacy for being used in the description of nature has never been an easy matter. Of course, statistical statements follow the rules of logics, nevertheless there seems to be in their basis a profound contradiction to (classical) logics.

An instance of this difficult relation is provided by the development of physical ideas. In classical physics the deterministic character of the physical laws appeared in some way as the evident realization of some intrinsic logic of scientific knowledge, and probability could be reduced to a measure of the lack of information about initial or boundary conditions. This already no longer works fairly in classical statistical mechanics. There we seemed to face a three–fold, unpalatable alternative: either we had to accept an irreducible concept of probability in the description of microscopic phenomena, or we had to renounce reducing thermodynamics to statistical mechanics, or we needed to deny an immutable character for the laws of the former. With the advent of quantum mechanics the argument acquires a new dimension and prompts one to ask whether it may not be meaningful to consider a basic statistical character for the laws of physics in general (Schrödinger). In this way, the question about "the logic character of statistical statements" becomes very acute in the epistemology of modern physics, as already evident in the discussion of the first half of the twentieth century (Cassirer, Reichenbach etc). As we know, this has led to a fullness of developments inside physics itself, from which I shall only mention two: 1) While there is no successful attempt to explain quantum effects systematically on a "lack of knowledge" basis[1] there is a successful program for explaining classical observations on the basis of quantum theory (the so–called "decoherence" program). 2) One has shown that one can reinterpret locally the formal derivations inside the quantum–mechanical theoretical scheme as statements in a logical scheme departing from the classical one.

The first remark very much weakens the claims of "naturalness", or "fundamentality" of many major features of classical physics, including classical determinism. The second one shows that the connection logics–statistics may go beyond the level of probabilistic statements.

In physics, statistic elements belong to the formal setting of certain theories or are introduced as approximations in building various models and the discussion takes place at the interpretation and epistemological level. In other sciences, however, like biology, cognition, economics, etc., we lack theories with precise mathematical schemes in which we can pinpoint the intervention of probabilities and ask about their epistemological status. On the other hand, most of the systems considered by these sciences are in some sense "open" and subject to uncontrollable influences. Randomness is therefore a natural element in building up models and the problem is not so much where it comes from but what it effects. One point of view is to consider randomness as noise and use statistical procedures to extract the underlying dynamics of the observed processes from the noisy observations. As in classical physics (but not classical statistical mechanics), statistics is here only a method to make sense out of "dirty" observations – with "sense" being the underlying, deterministic dynamics. The other point of view is to conceive of a special role of the stochastic element, determinant for the phenomena considered. Such is the case for evolution theories, whose dynamics itself is driven by random processes. In this case, and this is the interesting case, the question is to understand the interplay of statistic and deterministic elements and the relation and mutual determination between order and disorder.

Here I should like to discuss some aspects of this question in models for adaptive and learning processes. Learning and cognition are related to each other and therefore this discussion should also be of relevance in approaching the latter. Studying learning models for artificial systems is interesting for developing robust practical procedures for "Artificial Intelligence" and similar problems and for testing hypotheses concerning learning and cognition in general. In this frame we shall discuss the interplay of deterministic and

[1] Concerning initial conditions or other parameters, but without violating other principles, like locality.

random elements, in particular the capabilities of stochastic procedures in uncovering and building structures.

2 Approaches to Learning in Artificial Systems

Models of learning for artificial systems permit us to test quantitatively the interplay and effects of a few features, hypothesized to be active in learning. There are various perspectives on learning, however, and a large number of learning paradigms. Since we are only interested in a precise question we shall not attempt a systematic discussion but only mention some of the approaches. In one perspective, for instance, learning is seen as "improvement of behaviour over time". This perspective provides a good frame for discussing *learning from experience* (associative learning, reinforcement learning, individual versus multiagent learning, learning in fixed or variable environment, etc.), but also *learning by analogy, learning by developing and testing strategies,* etc. Another perspective views learning as "inductive structure finding": given a structure hidden in the "world" and for which a finite amount of data are available to the system, the latter has to find the best approximant to this structure from a reservoir of hypotheses made available to it (possibly generated by itself). In this frame one speaks of two major learning paradigms: *algorithmic learning* and *stochastic learning.* The typical opposition between these paradigms is: For the former: it uses enumerations, is based on orderings, exclusivity and uniformity, it implements discrete classes of algorithms, it provides unique (best) solutions, and is rigid. And for the latter: it uses statistic procedures both to obtain and to evaluate the solution, it implements basically continuous classes of algorithms, it does not necessarily provide unique solutions but allows the question of generalization, and is flexible. However, other perspectives define learning in accordance with the implementation frame rather than the goal: such are "soft computing learning", which attempts to exploit tolerance and imprecision to achieve robustness and tractability using logics of believing, fuzzy logics, randomness, probabilistic approaches, etc., or "neural network learning", where again we find associative and reinforcement learning, statistical approaches, etc., but redefined in the frame of these systems. As one can see, these considerations do not provide a true classification but only indicate different way of setting the accent or viewing the problems, therefore we preferred to speak of perspectives. From the point of view of our discussion – the role of stochastics in learning processes – the "learning from experience" and the "stochastic learning" paradigms are the most adequate frame.

3 Some Examples Concerning Stochastic Learning Procedures

We shall proceed in our discussion by just providing a number of examples as illustration for the capabilities of statistic considerations in the investigation of learning in artificial systems. They are meant to illustrate the questions of the possibility of perfect learning when only incomplete, statistical information is available, of the role of spontaneous stochastic behaviour in learning and of the quantitative characterization of learning involving stochastic procedures.

In the two examples below we shall consider learning from delayed, or unspecific reinforcement. One urgent problem with which a system, either natural or artificial, may be confronted when trying to improve its performance is to learn only from the final success/failure of *series* of consecutive decisions. This means that the system (agent) does not receive a feedback pertaining to each of its actions, but only one concerning the total result of a series of actions. Over repeated trials with random changes in both its actions and the environment the agent succeeds in tuning each of its actions although it never receives any feedback about their adequacy separately. We ask about the achievements and the stability of performances of such procedures.

The learning model we have devised [5],[7],[4] involves two steps:

1) *blind association*: assimilation of new relations without any control;
2) *unspecific reinforcement*: final control by feedback using only the global performance assessment.

Both steps implement stochastic elements in a significant way.

Illustration 1: *Unspecific reinforcement learning for perceptron.* The first example uses a neural network implementation (perceptron) to show that this algorithm leads to perfect learning in spite of unspecific feedback, and to make evident the learning dynamics. The typical learning problem is classification of patterns: the network is presented with patterns and learns (by modifying its internal structure) to classify them in a required way. In our case the network is presented with series of L random patterns at a time.
– At the presentation of each pattern the associated, free reaction of the network is learned by synapse updating.
– After each series of L patterns the previous associations are compared with the desired ones and the average error over the L steps is used unspecifically to "correct" the previous updatings equally. Thereby it is enough if only a random subset of these previous updatings is corrected (the agent "does not remember" everything it has done before it received the feedback).
The model shows very rich learning behaviour:
– perfect learning ("perfect generalization") is achievable,
– it requires both steps, more precisely: the relative strength of blind association/unspecific reinforcement is finite and bounded from below,

– the asymptotic behaviour depends on this ratio,
– the preasymptotic behaviour appears to be very important: the flow is controlled by a stable and an unstable fixed point and can go toward perfect generalization or toward complete confusion (depending on the parameters).
The fact that perfect learning *can be proved* to be achieved in this very simple model under very general conditions indicates that such a mechanism may be of major significance in natural learning processes. In [4] more details can be found, showing besides the capabilities of such models, also the analytic power of the approach. At a more general level we shall come back to this point in the third illustration.

Illustration 2: *Unspecific reinforcement learning and spontaneous random action.* The second example concentrates on the role of spontaneous stochastic actions of the agent in the learning process. It describes a robot which tries to go from one side to another side of a board with obstacles [5],[7]. The robot proceeds by spontaneous, random moves, with bias provided by learning:
– The robot is never told what it has to do; it just go and only the time it needs to arrive at destination serves as judgment *a posteriori* for all it did on the way (unspecifically).
– The robot recognizes its neighborhood and moves with higher probability in that direction which it remembers to have had used on successful paths.
The typical behaviour is a viable combination of stability and flexibility:
– after it has found a good path the robot begins to follow it;
– from time to time it makes random deviations, but it always returns after some trials to the known solution (it never loses it), *unless it has found thereby a better solution* to which it then switches.
This stability/flexibility combination is in some sense the key for "progress" in the agent's abilities. There are many similar problems, e.g., escaping from local optima, coping with changes, etc., Notice that:
– a "stupid" deterministic algorithm would need to repeatedly search the whole board to recognize new situations – it is therefore inferior;
– a "clever" deterministic algorithm can be superior, but it needs design: in principle problematic (it relies on the availability of a priori knowledge).

These have been illustrations of the statements that the statistical approach is both adequate and useful in learning. Stochastic and deterministic procedures can supplement each other. Another question is: which description of learning can be achieved in the frame of the statistical approach? Two observations are relevant here to:
– The typical continuity properties allow for dynamical descriptions: flows, fixed points, phase transitions, etc.
– Performance estimations are essentially statistical: average error, worst case bounds, confidence estimates, etc.
– The analysis uses state–space methods and differential equations, but also more specific statistical and computational physics methods.

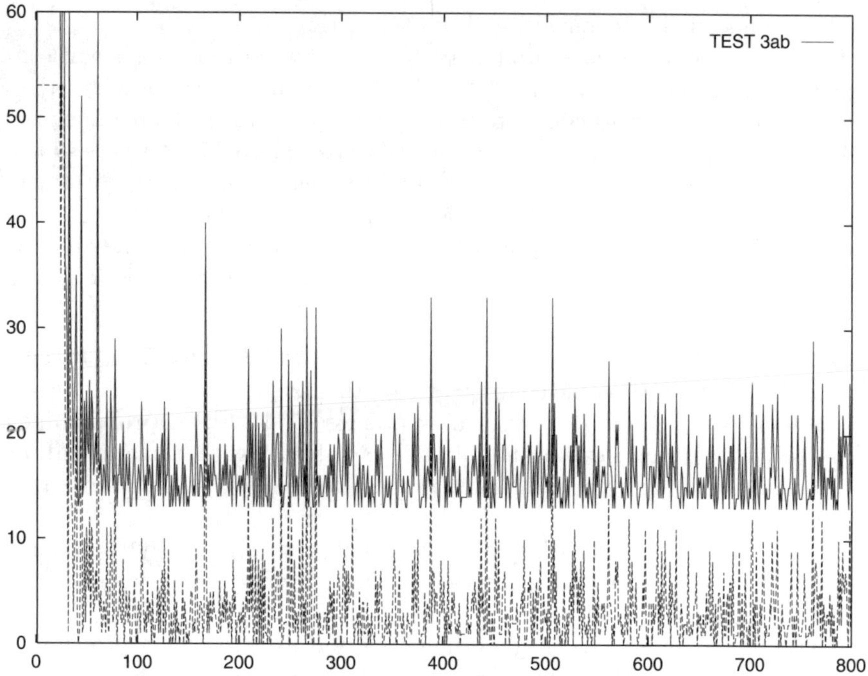

Fig. 1. *Performance* of the robot on a board with obstacles: Length of the path from the starting position to the upper edge of the board (full line) plotted run by run. After a period of about 50 runs during which it learns that it is expected to get to the upper edge (only in this case it receives a positive reinforcement) the robot finds the optimal path (13 steps in this example). From this moment on the robot will follow this path, with occasional fluctuations (suboptimal pathes). The fluctuations do not destabilize its behaviour, since an optimal path will never be forgoten, but they are useful if old obstacles disappear or new obstacles appear: in this case the robot switches rapidly to new pathes. The dotted line indicates the number of steps lost running against obstacles (the robot must learn by itself the impenetrability of the latter).

Illustration 3: *Worst case estimate in Stochastic Learning Theory.*
In this last example we shall briefly describe the Vapnik–Chervonenkis bound on the probability of poor generalization. We shall do this for neural networks, the theory, however, is general – see, e.g., [8]. Consider a network with N input and 1 output binary (e.g., Ising: $\pm$ 1) units. We produce inputs $\xi = \xi_\mathbf{i}(\mathbf{i} = \mathbf{1}, ..., \mathbf{N})$ from some distribution $P(\xi)$. We want to train the network to implement arbitrary Boolean functions f, i.e., such that on an input ξ the network must give as output $\zeta = f(\xi)$. Let us focus on a particular function of interest, f^*, implementable on our network. Then for any other function f

we can define the generalization ability of f (relative to the desired function f^*) as:

$$g(f) = \text{Probability}(f(\xi) = f^*(\xi) \text{ for } \xi \text{ randomly chosen with } P(\xi). \quad (1)$$

Notice that $g(f)$ does not depend on a particular ξ. We now produce training pairs $\{\xi^\mu, \zeta^\mu = f * (\xi^\mu)\}$, $\mu = 1, ..., q$. This defines therefore the information from which the agent learns, and we want to know how well it can do (using the best algorithm). Hence we ask: for an arbitrary function f, what is the generalization ability $g_q(f)$ on the training set q? The celebrated V-C bound is:

$$\text{Probability}(\max_f [g_q(f) - g(f)] > \epsilon \leq 4m(2q)\mathrm{e}^{-\epsilon^2 q/8}, \quad (2)$$

where the function $m(q)$ goes as 2^q up to some point $q = d_{VC}$ called the "V-C dimension". If d_{VC} is finite, then it can be shown that for $q \gg d_{VC}$

$$m(q) \leq q^{d_{VC}} + 1, \quad (3)$$

i.e., $m(q)$ does not continue to grow exponentially and the above bound is relevant: we can achieve ϵ as small as we please by just increasing q.

The "V-C dimension" d_{VC} is a measure of how well defined is the problem of implementing a certain function (i.e., of learning a certain task) by the best algorithm on a given network: d_{VC} is infinite for ill-defined problems. Typically, d_{VC} grows with increasing complexity of the network – on the other hand, of course, the network must be complex enough to solve a given problem (to implement a given f^*). In the appendix we will consider an explicit example.

Further results can be obtained, e.g., concerning bounds on confidence levels for automatized data processing in, say, classification of experimental data, etc. [1]. Again we see that we have strong instruments to characterize the capabilities of stochastic procedures in spite of their statistic character – sometimes, in fact, the latter allows us to make analyses which, because of lack of continuity properties, cannot be done for deterministic procedures.

4 Conclusion

Natural evolutionary processes are characterized by the prodigality of situations, by the complexity of interactions, by random "actions" and indirect feedback, by variable context, by statistical selection of performances involving multiple or global criteria, by the realization of apparently stringent development lines out of the interplay of seemingly random forces. These features pertain also to learning in general and to the development of intelligence.

The discussion of the previous section exemplified 3 important aspects of the intervening of statistical points of view in learning:

1) We have seen that exact solutions can be developed on the basis of only statistical information, in particular that accumulating random experiences allows the building of differentiate behavioural patterns using exclusively global information.
2) We have seen how a residual random element in the actions of agents with a superimposed bias developed by learning is profitable to further learning without destabilizing the behaviour: good solutions are only left after finding better ones.
3) We have seen that various kinds of learning dynamics allow a strict mathematical treatment and proofs of stability, performances, etc., and that also general characterization of learning performances can be obtained depending not of the particular learning algorithm but on the general agent architecture and data structure.

This discussion was only meant as an illustration for the adequacy and self-consistency of the statistical approach. Introducing statistical elements in modeling cognitive behaviour is one way to realize and control "softness". Other approaches proceed from the direction of "Fuzzy Logics", Fuzzy Set Theory", etc., which we did not touch upon here. Although some relations can be made between these approaches one should consider them independently. In connection with AI problems, "intelligent information processing", expert systems, and so on statistical method are used especially in problems where we gain by trading precision and exhaustivity for robustness and flexibility, etc.

In connection with cognition, neural networks models, e.g., may be understood as an attempt to describe formal aspects and to reveal general structural and functional features which develop depending on history, activity, experience, etc., of agents in the world. Learning processes in which stochastic elements play an important role could be expected to allow for such features to build up and thus to be involved in making the bridge between the particularity of "embodiments", the individuality of learning situations and the universality of cognitive achievements.

The main purpose of this paper, however, was to review and discuss features of the statistical approach to learning on the background of questions on the logic of scientific knowledge. We have seen that statistical elements can be rather successful in modeling cognitive behaviour and that we have adequate and powerful methods of analysis in this frame. The statistical approach to learning belongs to a certain way of thinking, in particular concerning cognition and artificial intelligence, with its special questions, methods, and achievements. This does not need to be taken as a new reductionist hypothesis, but needs nevertheless to be considered in its own right since it casts many aspects concerning the interplay between randomness and necessity which are specific to the natural processes of evolution and learning and which we may consider to be also relevant for us and for our knowledge.

5 Appendix

Consider a one-layer perceptron with N input binary neurons, then the number of possible inputs is simply 2^N, however, not all 2^{2^N} input/output association tables are implementable as Boolean functions – the simplest example is the XOR function ($N = 2$). In general, the number of Boolean function which a perceptron with N input units can implement to distinguish (classify) q examples is

$$m(q) = C(q, N) = 2 \sum_{k=0}^{N-1} \binom{q-1}{k}, \tag{4}$$

the number of dichotomies over q points in N-dimensional space [3]. For $q \leq N$ the sum stops at $k = q$ and we have $m(q) = 2^q$, for $q > N$, however, $m(q)$ grows more slowly. For $q \geq 2N \gg 1$ we have:

$$m(q) \simeq 2^q \left[1 + \mathrm{erf}\left(\sqrt{\frac{q}{2}}\left(\frac{2N}{q} - 1\right)\right)\right] \sim 2^q \mathrm{e}^{-q^2/4}. \tag{5}$$

Hence $d_{VC} = N$ and from the VC-bound (2) we see that for

$$q > 8N\ln(N)/\epsilon^2 \tag{6}$$

the training on the q examples practically ensures approaching the goal f^* within ϵ.

References

1. B.A. Berg, J. Riedler: Comp. Phys. Commun. **107** (1997) 39
2. B.A. Berg, I.-O. Stamatescu: 'Neural Networks and Confidence Limit Estimates'. In: *Field Theoretical Tools for Polymer and Particle Physics*, ed. by H. Meyer-Ortmanns, A. Klümper (Springer, Heidelberg 1998)
3. J. Hertz, A. Krogh, R.G. Palmer: *Introduction to the Theory of Neural Computation* (Addison-Wesley, Reading, Mass. 1991)
4. R. Kühn, I.-O. Stamatescu: J. Phys. A: Math. Gen. **32** (1999) 5749
5. L. Mlodinow, I.-O. Stamatescu: Int. J. Comp. Inform. Sci, **14** (1985) p. 201
6. C.S.S. Pierce: *Semiotische Schriften I.* ed. by Kloesel and Pape (Suhrkamp, Frankfurt 1986) p. 160 (Questions about Reality, 1868)
7. I.-O. Stamatescu: 'Statistical Features in Learning'., contribution to LEARNING'98, Madrid, (1998) cond-mat/9809135
8. V.N. Vapnik: *Statistical Learning Theory* (John Wiley and Sons, New York 1998)

Praxic Logics*

David Ritz Finkelstein and James Baugh

1 Empirical Logic

We and our species have learned most of our logic the hard way, from experience. Do new scientific experiences need new logics?

The question "Alternative Logics. Do Sciences need them?" suggests but two choices. We will encounter several, so we will speak of variant logics.

Von Neumann was probably the first to replace the singular form "logic" by a singularized plural "logics," to parallel "mathematics" and "physics." His answer to this question is clearly in the affirmative.

But if we do need variant logics, do we need them side-by-side, to use in different situations, or sequentially, as stages in the evolution of logic, perhaps towards some eventual Last Logic?

For example, quantum theory is sufficiently stable now that the quantum logic of Von Neumann, if it is not the Last Logic itself, is surely closer to one than (say) the logic of Boole, which it includes as a restricted case, adequate for certain quantum condensates like apples. Should we all dispense with pre-quantum logics and settle on Von Neumann's? If so, we have some work to do first, since the higher-order part of quantum logic is still under development.

The logic and set theory to be used in mathematical deductions are not at issue here. They must go on inside variant logics, much as Newtonian mechanics goes on inside relativistic mechanics. We discuss the logic of physical systems, not formulas, an extremely special case; the system logic, not the metalogic.

Then the existence of variant logics is not at issue. The question remains, how many do we need in science? We will restrict ourselves to physics. How many variant system logics does physics need?

First we must ask whether physics needs any system logics at all.

What we surely need is a way to estimate the value of our assertions about the system under study. On the other hand, logic is sometimes considered to be the theory of logical operations like AND, OR, NOT, and especially IMPLIES.

Predicates represent idempotent yes-or-no filter operations on systems. The natural way to combine filters is sequentially, $AB = A$ AFTER B. The

* Based on Finkelstein (1996, 1999a, 1999b).

usual logical operations of AND and OR are ways to combine such filters, assumed to commute, into a network of serial ∩-connections and parallel ∪-connections, which are then usable as compound filters for compound predicates.

We can go remarkably far with such series-parallel networks in the pre-quantum domain, though never nearly as far as rationalists like Laplace imagined, but that logic does not work well for predicates about photons or any other quanta. Two photon filters in series generally do not constitute a filter, because generally they do not commute, so a series-parallel network of photon predicates generally defines no predicate. If we restrict ourselves to any one family of mutually orthogonal simple predicates, then series-parallel networks of them do define predicates, but these networks fail to exhaust all photon predicates. They omit the quantum superpositions.

The most practical way to assign values to an assertion about the photon seems to be to compute its probability. This makes no explicit use of logic in the above sense today, and requires no series-parallel analysis, but uses only matrix algebra or operator algebra.

Boole assumed that predicates form a commutative semigroup under AFTER, recognizing at the same time that the question was an empirical one and that his assumption might be wrong. Photon predicates do not form a semigroup under AFTER, though of course they generate one; and the semigroup they generate is not commutative, and does not consist of predicates alone.

Matrix mechanics discomforts us as mathematicians seeking maximum mathematical generality, but the physical world is rather special, and mere logical consistency is a rather weak requirement on a physical theory. The only serious possibilities that we see for physically viable variant logics today are those with various matrix-dimensions and perhaps various coefficient rings. The main variability that remains concerns further structure that these matrix algebras should have to describe nature in more detail.

Every theory still has at least an implicit lattice logic for its system. The one implicit in quantum theory is a non-distributive variant of the pre-quantum Boolean lattice and is almost forced upon us by quantum experience. But this logic was not historically the foundation of quantum theory. On the contrary, Bohr was averse to using any formal logic at all.

People have set up new theories expressed in the language of logic, emulating Euclid's construction of plane geometry, but this has never led to a useful new quantum theory. The main function that we have found for such quantum logics has been to alert ourselves to outworn principles of pre-quantum thought that we might otherwise carry over into the quantum theory unthinkingly.

What took the place of the commutative predicate logic of AND and OR in setting up quantum theory is a non-commutative matrix algebra whose elements projectively represent actions on a system under study, including acts of selection; call it an action algebra to emphasize the physical interpretation. Instead of discussing the truth of propositions about the photon it deals with the probability of the photon passing certain test operations.

From the start, quantum theory dealt with modes of action where pre-quantum theories concerned themselves with states of being. It was pragmatic (deed-based) rather than ontic (being-based). This allowed it to be simpler (in the algebraic sense) and more general.

The action semigroup of quantum theories replaces the Boolean algebras of pre-quantum theories.

The classical logic of Boole is part of one kind of action semigroup, which is categorial. The quantum logic of Von Neumann came out of another kind of action semigroup, simple and not categorial. In both cases the action algebra is more important in physics than the small part of it comprised by predicates.

No static system of logic underlies physics as a kind of absolute foundation in the way that a certain "Aristotelian" logic underlies Euclidean geometry. Logic is not part of the "foundations of physics." Physics is an experimental science and can have no foundations. The creation of the world is still going merrily on and physicists have to be ready for whatever comes. What comes can overturn any alleged foundations.

Furthermore, physics no longer compartmentalizes itself into separate supposedly self-contained theories like logic, geometry and dynamics, each with its own meaning and validation, except approximately and transiently. There are no watertight compartments in physics. Each domain depends vitally on the whole for its meaning and validity. Logic is not some eternal Truth but only a high-order approximate language arising out of a certain useful dynamics, and it has been of limited utility in the quantum realm. The idea that we see things as they are, and deal with these valid perceptions by logic, died a slow death by parts as the contemporary theory of light and perception developed.

Quantum theory extends relativity to a deeper sphere of our mental habits than Einstein's relativity, which disturbed the tense-structure but not the logic of physics. Special-relativistic mechanics is a special case of pre-relativity mechanics, namely, with Poincaré-invariant action principle. No quantum physics is a special case of any pre-quantum physics.

A physical theory has a syntax to form its meaningful sentences about its system under study, has a semantics[1] to use these sentences in studies on

[1] The assertion that a physical theory needs a semantics may seem obvious in this community, so we confine the point to this footnote, but it is not universally accepted by physicists. Those who deny the need for a semantics often cannot regard the formulation of quantum theory by Bohr and Heisenberg as a step in the right direction, as far as it goes. This is only natural: Those who do not recognize the existence of their semantics cannot be ready to change their semantics. For example some otherwise reliable physicists hold that the right set of equations will declare its own meaning. They argue that in ordinary life and in prequantum physics we use no explicitly formulated semantic theory. My daughter Aria used "papaya" correctly without ever being told what it means, since it was her first

that system, and has a logic to deduce new useful sentences from old. Thus the logic of a physical theory presupposes a semantics that we use to decide the validity of some sentences by statistical inference from experiment.

Quantum physics needs a variant semantics and a variant system logic because it studies systems and processes in much deeper detail than pre-quantum physics. After we specify the variant semantics we can ask for the variant logics that work for dealing with quantum systems.

Since quantum theory is more truthful than pre-quantum physics, it is best to understand it in its own terms, to think quantum and speak quantum. We can usefully discuss macroscopic apparatus in pre-quantum language but not a photon. To deal comfortably with quanta, we first reason in the algebraic language of quantum theory, painfully slowly. Then with practice the language becomes transparent and we find the quantum logics intuitive, though never instinctive.

2 Variant Semantics

The difference between pre-quantum and quantum semantics is like the difference between the meanings of statues and of musical scores, or between the meanings of model boats and computer programs. Pre-quantum physics describes objects but quantum physics represents actions. In quantum semantics, we give operations first, and define quantum objects by giving the operations with which we create, transform, and register them. Therefore it is usually an improper mix of old and new philosophies to ask what these quantum operations act on; it is the operations themselves that define what they act on. We can throw our innate logic organ a sop by telling it that these operations act on "state vectors," but since there are no such things out there, this error eventually has to be compensated by another error,such as the "collapse of the state-vector," or the total renunciation of operational definition.

All quantum kinematics is a variation on the theme of Pierre-Etienne Malus, who first deduced the probability that a photon from one Iceland-spar polarizer will pass through a second similar polarizer rotated relative to

word. Why should aspiring quantum physicists not do the same for "$i\hbar\, d/dx$", which is another term for "momentum" in some contexts. Indeed we might, if we only spent enough time in appropriate quantum laboratories where this quantum phrase was actually used in successful practice, and were appropriately rewarded every time we used the language correctly, and harmed when we did not. But then it would not be the equations that are telling us their meaning but the use of these equations in practice. This usage is exactly what must be supplied for any new theory. In real life any word can readily be given any meaning by any user. There has to be a feedback to facilitate the formation and propagation of standard meanings; like delivering papaya when a child calls out "papaya" and Mother when she says "mama." A formal system by itself lacks this feedback.

the first about the ray axis. These modest gates are now the prototypes of all quantum analyses. Let us indicate the polarizer orientation by a unit vector ψ engraved on the polarizer. We indicate the orientation of the analyzer by a similar vector ϕ inscribed on the analyzer. With the arrow ψ goes a projection operator – projector for short – $\Psi := \psi\psi^{\dagger}$. Similarly, the analyzer has a projector $\Phi := \phi\phi^{\dagger}$.

Malus used no photon models but simply described the experiment and the experimental results, without intervening hypotheses, in accord with the philosophical precepts both of Newton and modern quantum theory. The probability that a random photon striking the polarizers will go through them both and be detected is half the square of the cosine of the angle between ψ and ϕ, according to Malus.

Actually Malus gave the conditional probability $p(\Phi, \Psi)$, the probabilty of Φ conditional on the photon passing Ψ. Called the transition probability today, this lacks the factor 1/2. But to measure a conditional probability we must effectively measure two unconditional probabilities and divide. The transmission probability $p(a)$, unconditional, involves one less experiment, and so is simpler.

To illustrate the general schema we write Malus' Law as

$$P = \operatorname{Tr} X X^{\dagger} / \operatorname{Tr} 1 \tag{1}$$

where P is the transmission probability, unconditional, and

$$X = \Phi \otimes \Psi \tag{2}$$

contains both polarizing filters. Max Born extended this law to a domain much vaster than the polarizer experiments of Malus. It is now the basic logistical formula of the quantum theory of any system, from quarks to quantum computers.

The skeleton of the Malus–Born formula is a sequence of operators $E = \Phi \otimes \Psi \otimes \ldots$ forming a flow chart, program or score for the atomic experiment or experience under study, read from right to left. The input of the random photon at the beginning and its outtake at the end are left implicit. We may represent sources and sinks of non-random photons by inserting suitable filter projectors into the program at the beginning and end.

In both pre-quantum and quantum kinematics we represent operations in series by a product. We represent operations in parallel by a special kind of sum. Actions Φ and Ψ are called orthogonal when $\Phi\Psi = 0 = \Psi\Phi$. Parallel operations are represented by orthogonal sums. It happens that in pre-quantum logic distinct sharp classes are orthogonal, and so the orthogonal sums that represent sharp operations in parallel are simply sums of distinct sharp classes.

In quantum theory there is also the new possibility of quantum superposition. We may form a linear combination of the vectors ψ and ϕ underlying the operators, and then forming the projector upon this linear combination.

This almost never can be expressed in terms of Φ and Ψ. It is not easy to give a general prescription for making such superpositions. In the two-slit interference experiment, the selective act using both open slits is approximately a quantum superposition of the two one-slit acts.

Some of the operators occurring in E may be projectors representing selective actions. Some may be unitary operators representing dynamical development between selective actions. In the polarizer example the program is a sequence of two actions upon the system, represented by Ψ and Φ. When we multiply these operators we lose information about the individual steps of the experiment. We represent the structure of the experiment in more detail by the above tensor product than by the operator product.

Then to find the transmission probability of the experiment E we contract the tensor product F representing the experiment in some detail to an operator product X representing only the resultant of all the steps, and then use the Malus-Born law.

The entities represented by ψ, ϕ, Ψ, and Φ are not to be measured on the quantum system. They are not like the vector for position that we measure on a planet. They describe what we do to the system. We measure them on our macroscopic apparatus, which they describe.

If we confused Ψ and Φ with variables of the system, then the fact that we choose Ψ and Φ independently, one to begin the experiment and one to end it, would become mysterious. We might find ourselves speaking of a "collapse" of Ψ to Φ to describe what are actually our own experimental decisions. But of course no one would make such a basic categorical mistake. Malus' meaning is perfectly clear. Ψ and Φ describe the polarizers, not the photon.

$P = 0$ means that the transmission is forbidden, that is, has never happened and is predicted not to happen. $P = 1$ means that the transmission is assured; that whenever E is carried out, a positive result is found.

Here, $\dagger$ is an antilinear mapping $\dagger : V \to V\dagger$ to the dual space, called the adjoint operation. The image $\dagger\psi$ is usually written $\psi^\dagger$. The implication of

$$\phi = \psi^\dagger \tag{3}$$

is that $\Phi = \Psi$ and that the experiment $\Phi\Psi = \Psi\Psi = \Psi$ has the same transmission probability as Ψ alone.

These input and outtake actions are theoretical elements of a kind alien to pre-quantum physics and pre-quantum logic. In using quantum theory we do not look at the system to determine these vectors but at the apparatus.

In the Malus experiment, we may print a unit vector ψ on the polarizer to define its orientation. There is no way to determine the orientation angle ψ of the polarizer by any action on one photon that it transmits. We may print a dual vector ϕ on the analyzer as a grating of parallel lines with unit line density. A is then the cosine of the angle between polarizer and analyzer. We read it by counting the lines of the grating ϕ that cut the vector ψ.

We do not look at the photon but at the optical apparatus to determine the input and outtake vectors. We have no choice in this matter. No experiment on the photon will tell us the ψ it came from or the ϕ it is going to.

Atoms make spontaneous transitions. They carry little memory of the past or precognition of the future. These input and outtake action vectors are not carried by an atom, in that they cannot be learned from it. They describe our actions far more than they describe the atom.

This is a good reason not to call them states; for in pre-quantum physics one imagined that the system under study truly carried a state, a complete determination of its responses to past and future actions, within itself.

Rumpelstiltskin was an imp who could be controlled by anyone who knew his exact and true name. Classical physics is Rumpelstiltskin physics. The pre-quantum physicist was convinced that each object has one true and complete description that controls its future actions. The first challenge to the imagination of the quantum physicist is unlearning this belief that things have complete names that determine the course of events. Names, as macroscopic objects, are subject to quite different laws than electrons, including different logical laws. They are products of a macroscopic condensation process. Physical entities come in and out of existence without true names. Things happen of themselves. Names, words, logos are given by us.

Each input vector ψ stands for how the metasystem (including ourselves) produces the system, much as a force stands for how the surroundings of a particle push on the particle.

3 Variant Epistemologies

What does it mean to know something about an atom?

Heisenberg's original insight was basic: Suppose an atom had a complete description or state, as Descartes thought, and perhaps Newton. Then even so, according to Bohr's provisional quantum theory we observe the atom only in a transition from one atomic state to another, when it emits a photon that produces such a transition in us. Even if states existed we would never see them.

This must change our concept of knowledge, our epistemology. In one ancient conception, knowledge was a state of the knower that correlates with the state of the known. Bohr raised the question: What can it mean to know something that is always in transition during the process of knowing?

The answer seems to be that knowing is a transition in the knower that produces or is produced by a transition in the known. We never experience a state but only a transition.

This answer also shows the two modes of knowing that we have mentioned, the input and outtake. "Non fingo hypothesis", Newton put it, or words to that effect. Einstein, following the long line of Newton, Francis Bacon, and Ockham before him, taught us yet again that we should not take for granted the existence of what we can never see, whether it be the ether, absolute time, or the Stoic pneuma. We think Heisenberg followed Einstein's line of thought one step further: Since we never see states, we should not invent them. What counts is what physicists do.

In pre-quantum physics we expected different atoms to do the same thing under the same circumstances. We thus implicitly assumed that individual behavior was already determined at the collective level. In pre-quantum physics, the input and outtake actions of an allowed transition uniquely determine each other and can be identified. The resulting entity is the state. Different atoms act alike when they have the same state.

In quantum physics, individual behavior is determined at the individual level. Different atoms generally behave differently. They have no state in the original sense of properties determining their behavior. In an allowed transition, the input and outtake actions are independent variables and cannot be identified.

Often the word "state" is applied to an input vector, and this has caused much confusion, for input vectors lack the most basic feature of the pre-quantum state. They are not there in the system.

This statelessness is the incompleteness of quantum theory. Quantum theory and arithmetic are both incomplete in the same sense though for different reasons: their postulates do not decide all their propositions.

We would rather speak of quantum spontaneity than quantum incompleteness, however. Calling quantum physics "incomplete" is a negative way to express its greater richness and openness, and contrasts it with a pre-quantum fantasy of a non-existent completeness. After all, pre-quantum mechanics too was incomplete. Its initial data were not even imagined to be specified by the theory. It is merely that it lumped this incompleteness at a beginning in the remote past, the creation of the universe. In quantum theory, spontaneity is distributed throughout all time and space.

The fundamental processes of quantum physics are stateless transformations, actions without passive objects. What Heisenberg discovered in 1925 he called a non-objective physics. The modern non-objective spirit expressed in mathematics by group theory, category algebra, and non-commutative geometry, and in physics by relativity and quantum theory, is expressed in logic by constructivist logic (Brouwer) and quantum or non-commutative logic.

4 Variant Logics

At first glance, quantization may not seem to simplify. The group of the projective transformations that preserve the transmission probabilities of the quantum theory seems no simpler than its limit as $\hbar \to 0$, the canonical group of the pre-quantum theory. To see the simplicity of quantum theory we must enlarge our transformations to include measurement actions, epistemic acts, and selective acts.

Classical mechanics evolved from astronomy. When we see a star we hardly change it. Pre-quantum thought tacitly supposes that in general the object of our study is unchanged by ideal observations. Pre-quantum physics naively omits measurement processes from the theoretical description of the experiment, as though we saw things as they "really are," and as though that expression had meaning. It asserts that the mass particle "is" at some point x.

During scientific revolutions, the invisible becomes obvious, and the obvious becomes false. For example, it is common sense that we see things as they are when we see them, and special relativity emphasizes the falsity of this common sense, by insisting on the finite speed of light. Seeing epistemic acts as full-fledged physical processes is another eye-opening process. Now it is not the time lag but the physical impact of the epistemic act that comes to the fore.

Selective actions generate a semigroup and not a group.

Call a subsemigroup $N \subseteq H$ of a semigroup H *normal* or *invariant* when every element of H commutes with N:

$$(\forall a \in H)(aN = Na). \tag{4}$$

Call a semigroup simple if it has no invariant subsemigroup but the trivial ones: the whole semigroup, the null set, the 1, and the 0.

Newton, the founder of pre-quantum mechanics, was already a pioneer of quantum theory, struggling with both the corpuscular and wave aspects of light. Indeed, he invented the ether mainly to carry the guide waves the he needed to account for the "fits and starts" of quantum interference of photons.

Similarly, Aristotle and Boole were already coping with the main problems of quantum logic even as they were inventing the pre-quantum logics that dominated medieval and Renaissance logic studies. For Aristotle, the theory of the syllogism was less interesting than the logic he called dialectical, required to cope with change. When you wake from sleep, according to his dialectics, there is a moment when you are both awake and sleeping, yet neither sleeping nor awake, "Aristotelian" logic to the contrary notwithstanding.

Similarly in Boole's epochal first publication on class algebra, he postulated in detail that our elective acts form a commutative semigroup with 0 and 1, not as a matter of necessity but as a matter of revisable empirical fact, which might be otherwise. To classify is to filter. Boole showed that the resultant of two filtrations is a filtration using his commutative postulate.

In physics we are more concerned with physical selection than mental election, which may be regarded as only one part of the whole process of filtering systems according to their properties.

Pre-quantum physics implicitly assumed what Boole made explicit: that selective acts form an invariant abelian subsemigroup of the semigroup of all actions on the system.

5 Simplicity

The existence of this invariant abelian subsemigroup means that the pre-quantum action semigroup is not simple. The classical logic, an invariant commutative subsemigroup, is exactly what complicates it. Because we believe in simplicity we must give up my old belief in logic as separate from dynamics.

In the quantum theory the selective actions, represented by idempotent projections, no longer commute. Boole's proof of their algebraic closure then fails. The resultant of two filtrations is not a filtration. For example, if Ψ and Φ represent two polarization acts with $\pi/4$ between their polarizations then $\Psi\Phi$ is not idempotent, hence not a filter. Quantum epistemic acts do not form a subsemigroup, let alone an invariant one.

Thus quantum theory simplifies the action semigroup when it relativizes the split between measurement (projection operators) and dynamics (automorphisms), knowing and doing. state of being and mode of action. By giving each observer her own subsemigroup of mutually commuting states, quantum logic relativizes states and classes within the action semigroup rather as special relativity relativizes space, an instant of time, within space-time.[2]

6 Relativity of the State

Ontism is the old doctrine that the world is to be analyzed into objects with absolute states of being and commutative classes. This is the realm of logics. Praxism, the doctrine that we am exploring, holds that the world is to be analyzed into actions. Objects are defined by the operations we can do on them. This is the realm of what may be called praxics.

Simplicity first becomes possible for a praxic theory. The simple semigroups that actually occur at the quantum level seem to be full matrix semigroups.

The primacy of actions, operations or processes is a crucial and salient feature of quantum thought. There have been many attempts to flatten it. Let me heighten it.

By an object (or being) we understand an entity that is completely defined by what it is, its state of being. The state of an object may be completely defined by a suitable input action upon the being, or also by a suitable outtake action upon it. The collection of all the possible states an object can have is called its state space.

Ontism presupposes that (sharp, or minimal) input and outtake actions of an allowed transition uniquely determine each other, and the allowed tran-

[2] We use the solidus for a non-semisimple fusion, for example for a fiber bundle F/B with fiber F fiber and base B. We reserve the hyphen for simple fusions. Aristotle had space | time, Galileo had space/time, and Einstein had space-time.

sition is in fact assured, compulsory. The input and outtake actions of an allowed transition define a state.

When one sharp input action allows many sharp outtake actions, as with quanta, such states do not exist.

Praxic entities are defined by a simple collection of actions, have no states, and have classes that do not commute.

Some go further and say that according to quantum theory atoms do not exist. This is in accord with the ontic doctrine that what exists must have a state. But, in fact, atoms do things and we act upon them. They therefore exist in the praxic sense. Quantum theory relativizes the state, keeping the atom as an absolute. In atomic physics we prepare and observe atoms, not states. Just as in pre-quantum physics, an atom is represented by the identity element 1 of its action semigroup, representing the class of all possibilities for that atom. As a predicate, 1 is the assertion of unqualified existence of the atom.

Whence the pre-quantum illusion of beings and objects, what Peirce called the suchness of things? For macroscopic systems, two randomly chosen quantum actions are orthogonal for all practical purposes. In pre-quantum physics, therefore, all the input and outtake actions under consideration are orthogonal (have vanishing mutual transmission amplitude) unless they are in fact one. To observe a transmission $\Phi\Psi$ on a macroscopic body in ordinary ambient conditions that is neither compulsory nor forbidden involves such precision and such a removal of entropy that it is practically impossible. Therefore macroscopic systems seem to have deterministic properties subject to pre-quantum physics. They seem to have states. This is what permitted us to erect and believe the pre-quantum logic. But there is no critical number of parts where suddenly an object and states materialize. It simply becomes harder and harder to notice and verify the absence of the state as the number of parts grows.

A maximal orthogonal collection of input actions is called a frame. Since transitions within each frame are either compulsory or forbidden, just as in pre-quantum physics, each frame may be regarded as defining concepts of being and state, relative to one experimenter. In pre-quantum physics there is only one frame for the system, composed of all its states, but in quantum physics there are many frames, each expressing some experimenter's choice of what experiments to do. Crescas and Giordano Bruno relativized the center of space, Galileo and Newton relativized the rest state, and Einstein relativized time; Heisenberg (and later, most explicitly, Dirac) relativized the frame, and thus the state. Quantum theory extended relativity to a domain where Einstein refused to follow.

7 Work to Be Done

This process is not over. The semisimplicity argument of Segal (1951) and of Inönü and Wigner (1952) is still a powerful one. It suggests now that all is quantum. Any element of quantum theory that seems to be an object with a state in the pre-quantum sense will eventually resolve under closer scrutiny into a complex of non-objective quantum actions. Each such element is a candidate for quantization, and the possible source of a variant action algebra, which may lead to a variant logic.

The most conspicuous pre-quantum objects left in quantum physics today are the observer, the space-time pointset underlying field theory, the physical law, and the imaginary constant $i\hbar$ that figures in the dynamical law. At the beginning of a calculation today, one usually stipulates implicitly that these are known in all their relevant properties, or may be known as completely as desired by physicists provided with enough time, will, and capital.

Our concept of a theory is still based on an essentially objective conception of a symbol and its states. When we must take into account the spontaneity of everything, surely we must allow some to the experimenters in it. Quantum theory mentions the metasystem (the system that studies the system under study) and the system-metasystem boundary but provides no way to move this boundary. A still more quantum theory would seem to call for an at least partially quantum experimenter using at least some quantum symbols.

We do not attempt this yet because the assumptions of the observer and the theory do not seem to us to be the cause of the difficulties of present physics.

Space-time complicates the field/space-time bundle of field theory just as time complicated the space/time bundle of Galilean physics. All field and string theories suffer from this fatal ailment.

One way to see this error is to look at our operations.

How do we know where an electron is? By detecting its electric field.

How do we detect the field of an electron? We measure the position of another electron, a test particle; for example by letting the first electron ionize atoms in an emulsion.

There is no hard-and-fast distinction between measuring position and measuring field at the particle level. We know each by the other. But field theory imagines an absolute distinction between the two concepts of field and space-time position, inherited from the macroscopic experiments of Coulomb.

To unify space and time meant to relativize the cut between them. Similarly, to unify space-time and field-matter means to relativize the cut between them [Finkelstein & Rodriguez (1984), Finkelstein (1996)]. This work is still in process.

While in pre-quantum physics there is a hard boundary between informing and transforming acts, between logic and dynamics, quantum theory relativizes this boundary. The product of two non-commuting projectors is not a projector, and thus represents a dynamical act, not an epistemic one.

One way to cope with this revolution is to build structures in quantum dynamics that resemble the pre-quantum logic. For example, since a Boolean algebra is a lattice, Von Neumann regarded the lattice of projectors of a quantum theory as a part of quantum logic.

But the operations of that lattice are remote from laboratory operations. $\Phi \cap \Psi$ for example is the infinitely iterated product $\Phi\Psi\Phi\Psi \ldots$. The product represents doing one thing after another, but a $\cap$ represents an infinite sequence of actions. The idea that this is fundamental is patently absurd.

The physicist might best regard logic as a high-level language supported by simpler ones of quite different structure. Boolean logics are a good tool for analyzing the behavior of machines that we have built precisely to carry out logical computations, such as digital computers. But they are not useful for systems studied at the quantum level of detail, where the simplicities of linear associative matrix algebra seem to govern.

8 Variant Second-Order Logics

Quantum theory led to several variant higher-order logics.

The theory of predicates about predicates has as its Boolean correspondent the theory of classes of classes, based on the power set function 2^S. This answers a question that arises in both pre-quantum and quantum physics. If we have a kinematical theory for an individual of some kind, how do we make the kinematical theory for an assembly of such individuals? How do we construct many out of one?

In pre-quantum thought, the passage from individuals to aggregates is quantification, so called because it converts "Yes-or-no?" questions into "How-many?" questions.

Quantification was called "second quantization" in the early quantum vernacular, but this is generally recognized to be a misleading name.

The mode of quantification proper to an individual is called the "statistics" of the individual.

In pre-quantum mechanics the aggregate — say the solar system — is traditionally represented as a sequence of individuals, not say a set of individuals. The difference is that a sequence is generally changed when we permute its elements, but not a set.

Objects whose aggregates are sequences are said to have Maxwell–Boltzmann statistics. Variant quantifications arose first in the quantum physics of photons, which take Bose–Einstein or bosonic quantification, and of electrons, which take yet another kind of quantification, with Fermi-Dirac or fermionic statistics. Only bosonic and fermionic statistics have definitely been observed for quanta so far.

In algebraic language, the action algebra of a fermionic assembly is the endomorphism algebra of the vector space of antisymmetric tensors over the space of input vectors ψ of the individual boson. The action algebra of a bosonic assembly is the endomorphism algebra of the vector space of symmetric tensors over the space of input vectors ψ of the individual fermion.

These are therefore genuine variant second-order quantum logics.

9 Cliffordonic Second-Order Logic

What are the predicates of quantum logic about? What are the elements of its classes? What should we be talking about?

Presently, quantum predicates describe the quantum system: the atom, the photon, the crystal. In present quantum physics the dynamical law is regarded as an endomorphism from the action algebra of the quantum now to the action algebra of the future quantum. This quantum law is supposed to be a pre-quantum object, in that it may be completely known. This absolute may also need to be relativized in a quantum cosmology. Simplicity and reciprocity require the law also to be a quantum variable. Just as our determinations of space-time geometry modify it slightly through gravity, our search for the laws of nature modifies it slightly as well. At least in a small degree we must be law-makers as well as law-seekers. Isaac Newton, Peirce, Mach and Wheeler, and lately Landauer [7], and Smolin [10] all agree that the law is variable.

But this understanding still leaves the split between law and system, which violates reciprocity and thus simplicity. The law acts on the system; surely the system must act on the law.

The only way to eliminate this split that we can see is to relativize the split between dynamical law and the system under study [Finkelstein and Rodriguez (1984), Finkelstein (1997)]. The two previous relativizations of Einstein and Heisenberg fused a static concept with a dynamic one: space with time, state of being with mode of action. This relativization is similar. The static element is now the system, and its dynamic element is the dynamics. Perhaps these earlier relativizations should both be regarded as special cases of the general process of relativizing the split between the act of change and the object of change.

Ultimately, therefore, the system that attracts my attention is the entire dynamical action, and the important quantification is the one that relates the elementary dynamical action to the total action.

In quantum mechanics today, dynamical development is presented as a sequence (or Maxwell-Boltzmann assembly) of infinitesimal transformations $1 - iH dt/\hbar$, suggesting that the chronon is a pre-quantum (completely describable) infinitesimal action with Maxwell-Boltzmann statistics. This suggestion is probably wrong on several counts.

First, the chronon is presumably a quantum action, not a pre-quantum one.

Secondly, the infinities of present quantum physics suggest a finite elementary dynamical action for the chronon.

Thirdly, we propose that the correct quantification for chronons, which are distinguishable, is not the Maxwell-Boltzmann, since that is not simple, but the following Wiman-Wilczek statistics, using a representation of permutations discovered by Wiman (1893, 1898) and applied to physics by Nayak and Wilczek (1996), Wilczek (1997, 1998, 1999).

Here the story circles back in time unexpectedly. Double-valued complex simple (irreducible) representations of the permutation groups were discovered by Wiman (1893, 1898) and completely catalogued by Schur (1911) before the discovery of quantum mechanics. In the process, Wiman discovered spinors in many dimensions well before Cartan did in 1913. Finkelstein and Rodriguez (1984) also developed a "Clifford combinatorics."

In the W-W statistics the action algebra of the aggregate is the Clifford algebra over the mode space V of the individual quantum entity. We call such entities *cliffordons*. The number of cliffordons in the aggregate is the dimension of V.

Bosonic and fermionic statistics represent two-quantum swaps trivially, by $+1$ and -1, respectively. But cliffordonic statistics represents the swap (ab) of any two cliffordons by the difference $i_a - i_b$ of real-Clifford-algebra units associated bi-uniquely with the cliffordons being permuted.[3]

Cliffordonic statistics provides a deeper interpretation of spinors. We conjecture that the spinors of physics represent permutations of quantum assemblies rather than rotations of individuals. We call such linear-algebraic objects *plexors*.

Here the story again circles back unexpectedly. Before digressing to the Lamb shift, Richard Feynman studied quantum spaces with non-commutative space-time coordinates of the form

$$x^\mu = \text{Constant} \times [\gamma^\mu(1) + \cdots + \gamma^\mu(N)], \tag{5}$$

where for each n the $\gamma^\mu(n)$ obey the usual Dirac-Clifford relations and for different n they commute [Feynman (1961)]. We therefore call quantum spaces with such Clifford-algebraic coordinates *Feynman spaces*.

Feynman spaces now arise naturally in chronon dynamics with cliffordonic statistics [Finkelstein (1999b)], and have natural momentum and i operators as well as coordinate operators. If the spinors of physics are truly plexors, as we conjecture, even an electron creation will resolve into swaps of at least four cliffordons under closer inspection; since fewer than four support no truly double-valued representations for the permutation group. We are currently examining details of the kinematics of fermionic fields in Feynman spaces, a pretty exercise in this promising variant of higher-order logic. At the same

[3] Here, James Baugh and we simplify slightly the algebra proposed by Wilczek and collaborators, who follow more closely the representation of the permutation group given by Schur (1911). Wiman, Schur, and Wilczek represent some swaps by sums $i_a + i_b$ and some by differences $i_a - i_b$ in a complex Clifford algebra. We represent all swaps by differences $i_a - i_b$, and in a real Clifford algebra.

time we are constructing cliffordonic scenarios for space-time and gauge processes.

The debate between Platonists and Aristoteleans goes on among us today. Our innate logic will always conflict with quantum logic. If we think that we carry eternally valid forms or ideas within us that need merely be remembered, we cannot accept quantum logics as anything but a mathematical game. On the other hand, if we realize that all knowledge derives from experiment and has restricted validity, we can see quantum logics as a magnificent opportunity to work ourselves into greater accord with nature.

10 Acknowledgements

This work is supported in part by the Monie A. Ferst Foundation and the Institute for Scientific Interchange. We thank Sukanya Chakrabarti, Shlomit Ritz Finkelstein, Andrej Galiautdinov, Michael Gibbs, and Tony (Frank) Smith for many helpful and enjoyable discussions. David Ritz Finkelstein had the pleasure of discussions with Frank Wilczek.

References

1. R.P. Feynman: Private communication. (1961) We could not found a published reference for his concept
2. D. Finkelstein, E. Rodriguez: 'Relativity of Topology and Dynamical Law'. *International Journal of Theoretical Physics* **23** (1984) pp. 1065–1097
3. D.R. Finkelstein: *Quantum Relativity* (Springer, New York 1996)
4. D.R. Finkelstein: 'Action physics.' *International Journal of Theoretical Physics* **38** (1999a) pp. 447–455
5. D.R. Finkelstein: 'Third Relativity.' *International Journal of Theoretical Physics* **38** (1999b) p. 2937
6. E. Inönü, E.P. Wigner: 'On the Contraction of Groups and Their Representations'. Proceedings of the National Academy of Sciences **39** (1952) pp. 510–525
7. R. Landauer: Physics Letters A **217** (1996) pp. 188–193 (The Physical Nature of Information.)
8. C. Nayak, F. Wilczek: Nuclear Physics. B **479** (1996) p. 529
9. I.E. Segal: 'A Class of Operator Algebras Which are Determined by Groups'. Duke Mathematical Journal **18** (1951) p. 221
10. L. Smolin: 'Did the Universe Evolve?'. Classical and Quantum Gravity **9** (1992) pp. 173–191
11. F. Wilczek: 'Some Examples in the Realization of Symmetry'. Talk given at Strings '97, 18-21 June 1997, Amsterdam, The Netherlands. Nuclear Physics Proceedings Supplement (1997); also hep-th/9710135
12. F. Wilczek: Private communication (1998)
13. F. Wilczek: 'Projective Statistics and Spinors in Hilbert Space'. Submitted for publication (1999); also hep-th/9806228

14. A. Wiman: 'Ueber die Darstellung der symmetrischen und alternirenden Vertauschungsgruppen als Collineationsgruppen von moeglichst geringer Dimensionszahl'. Mathematische Annalen **52** (1898) p. 243
15. A. Wiman: Mathematische Annalen **47** (1893) p. 531
16. H. Yukawa: 'Atomistics and the Divisibility of Space and Time'. Progress of Theoretical Physics, Supplement Nos. **37** and **38** (1966) pp. 512–523

Reasons from Science for Limiting Classical Logic

Paul Weingartner

1 Introduction

As is clear from the title, the paper defends the view that there *are* reasons from Science for limiting Classical Logic (CL). In this paper such reasons are especially given from physics; even if I think that other areas (Philosophy of Science, Epistemology, Action theory, Ethics... etc.) lead to similar or to the same limitations which are proposed here.[1] In general problems arise when logic is applied to fields outside logic and mathematics. These problems are of different kinds but many of them have the same source. The source is what I have called elsewhere "replaceable and reducible parts in the consequence class". In other words: Classical valid inferences, (arguments) which have replaceable or reducible parts in their consequence class lead to paradoxical or incorrect results when applied to the respective area. Replaceable parts are parts which can be replaced by an arbitrary part (sentence, predicate) *salva validitate* of the argument. And reducible parts are parts which can be reduced to smaller parts salva validitate of the argument. In this paper I shall concentrate on problems which arise when logic is applied to physics. The paper is divided into 4 chapters. In the first chapter I want to deal with the problems of commensurability and to show that this problem is not specific to physics, but is much more general. The second chapter deals with commensurability in Quantum logic. In the third chapter special restrictions (filters) on CL are proposed for a solution of the above mentioned problems. In the fourth chapter it is shown how these restrictions can solve the problems concerning commensurability and distributivity when logic is applied to Quantum Theory.

2 The Problem of Commensurability Is Not Specific To Physics

There is a quite general question underlying the problems of commensurability. It is this: We assume that propositions p, q, r, etc., represent and describe

[1] This is justified by previous research on solving paradoxes in very different areas with the help of relevance restrictions on CL. Cf. Weingartner-Schurz (1986), Schurz-Weingartner (1987), Weingartner (1994), (1997).

certain states of affairs. And more specifically we assume that they represent and describe certain, physical states. Thus, if p represents and describes the physical state P and q represents and describes the physical state Q, does it always follow then that the conjunction $p \wedge q$ also represents and describes a special physical state (which is some kind of fusion of states P and Q)? The answer to this question is: No, it does not always follow. But this fact, just stated, is not specific to physics. This can be seen from the following examples.

2.1 Logic and Mathematics

Assume p represents (describes) an important Theorem 1, and q represents (describes) a further important Theorem 2. Does $p \wedge q$ represent (describe) a further important Theorem 3? Of course not. Just the conjunction of two important theorems would not be called a new theorem. If p has been proved and q has been proved it would be redundant to mention that $p \wedge q$ has been proved and $p \wedge q$ would never receive the predicate "theorem". This tells us that the expression "theorem" is used in such a way that it means more than just: "proposition which is correctly derived from the axioms" or "proposition which is correctly proved."

2.2 Animal Behaviour

Assume the proposition S represents (describes) the observable state of affairs that sexual excitement obtains, the proposition A represents (describes) the observable state of affairs that agression obtains and the proposition F represents (describes) the observable state of affairs that fear obtains. Then research about animal behaviour shows the following facts: $S \wedge F$ does not represent (describe) an observable state in male animals but does represent (describe) an observable state in female animals. On the other hand: $S \wedge A$ does not represent (describe) an observable state in female animals but does represent (describe) an observable state in male animals.

2.3 Human Actions

Assume the proposition H_1 represents (describes) the (observable) action (state of affairs) of writing an essay on Alternative Logics and the proposition H_2 represents (describes) the (observable) action (states of affairs) of making a ski-tour. Then $H_1 \wedge H_2$ does not represent (describe) an (observable) action. Though H_1 at t_1 and H_2 at $t_2 (t_1 \neq t_2)$ may represent actions.

2.4 Social Help Among Animals

Consider the formula for social help (assistance) among (higher) animals: $I \cdot r > L$. It says that I, the increase-rate (growth-rate) for the propagation of genes (of an animal) times r, the degree of consanguinity must be greater than L, the loss-rate for the propagation of genes (of the same animal). These magnitudes can be only observed statistically. For instance, the time an individual spends (per day, i.e., statistical average over some time) to care for others or to save others from enemies is a measure for L. On the other hand, the time an animal spends for producing offspring is a measure for I. The magnitude r is the probability that those genes which are in the donor are the same in the receiver. For instance, if an animal cares for his child then $r = 0.5$.

Now, if L receives a relatively sharp value then the value of I will be very unsharp though, in general, it will not be zero. Because help given to a relative may in the long run result in a positive effect for I. But, in general, I and L are not measurable (observable) at the same time with respect to the same animal (or population) especially if r is not easy to determine.

Thus we have a similar situation here: If the proposition $p(I)$ represents (describes) the states of affairs that the measurable rate I has the value i and the proposition $q(L)$ represents (describes) the states of affairs that the measurable rate L has the value l then the proposition $p(I) \wedge q(L)$ does not represent the state of affairs that the measurable rate of both I and L has a certain value; because there is no such measurable rate: I and L cannot be measured simultaneously with a specific (sharp) value.

2.5 Incommensurable Pairs of Magnitudes in Physics

Last but not least I mention the incommensurable magnitudes in physics; not because they are less important but because they are well known. With these magnitudes i.e. the pairs position and momentum, energy and time, etc. the whole discussion about incommensurability began. And the basic logical situation is just the same as with the other examples mentioned above: if p represents (describes) the state of affairs that the measurment of the position of a particle has a certain value precise (sharp) enough and if q represents (describes) the state of affairs that the measurement of the momentum of that particle has a certain value precise (sharp) enough then it is not the case that $p \wedge q$ represents a measurable magnitude at all.

2.6 The Problem with Distributivity

In the famous paper of Birkhoff and von Neumann the following counterexample against the distributive law is given:[2] Assume a particle in a box. We are concerned with the complementary pair of position and momentum of the particle. For a rough estimate let the position be replaced by the determination of whether the particle is in the left part of the box (L) or in its right part ($R = L'$) and the momentum be replaced by the determination of

[2] Cf. Birkhoff and von Neumann (1936), p. 831. Cf. also Finkelstein (1979).

whether the particle is in an even (symmetric) state (E) or odd state (E'). This is determined by the parity of a certain quantum number n (even or odd) proportional to the momentum which means that the wave-function is either invariant or is changed. Then the respective instance of the distributive law is this:

$$E \cap (L \cup R) = ((E \cap L) \cup (E \cap R)).$$

Since $L \cup R$ means $L \cup L'$ and gets the value 1 in a Boolean Algebra the left side of the equation reduces to E. Translated into CL $L \cup L'$ means that the particle is either on the left or the particle is not on the left, which is a tautology. As to the right side of the equation evenness (for momentum) and leftness or rightness (for position) are incommensurable properties such that the subspace of the wave-functions of such a particle vanish, i.e., they get the value 0. That is $E \cap L$ and also $E \cap R$ get the value 0 and so the whole right side gets the value 0. Translated into CL this means that "particle s has evenness and particle s has rightness" is contradictory. Thus the left side of the distributive law gets the value E (i.e., a certain positive value) and the right side the value 0, which shows the violation.

3 Commensurability and Distributivity in Quantum Logic

3.1 General Considerations and Notation

For the language for QM we make the usual assumptions. The analogue to the global state in Classical Physics is the Hilbert Space in Quantum Physics. The elementary propositions $a, b, c \ldots$ say that the state vector of the physical system lies in the respective subspace A, B, C of the Hilbert Space H. The compound propositions can then be understood as follows: '$a \vee b$' means that the state-vector of the (physical) system lies in the subspace $A \oplus B$, where "$\oplus$" denotes the *linear span*, which is the least upper bound of any two subspaces (the set of all subspaces of H being partially ordered by set-inclusion). We can also think of it in the following way: if $S(a)$ is the subspace corresponding to proposition a and $S(b)$ for b then $S(a \vee b)$ is the linear span of the subspaces $S(a)$ and $S(b)$. '$a \wedge b$' means that the state vector of the system lies in the subspace $A \cap B$ (which is the usual intersection of A and B). Or again: if $S(a)$ is the subspace corresponding to proposition a and $S(b)$ for b then $S(a \wedge b)$ is the intersection of the subspaces $S(a)$ and $S(b)$. '$\neg a$' means that the state vector of the system lies in the orthogonal subspace $A_{\perp}$. The subspace consisting of all those vectors which are orthogonal to all vectors in the original subspace is called the orthocomplement of that subspace. Thus, in other words: if $S(a)$ is the subspace corresponding to proposition a then $S(\neg a)$ is the orthocomplement of the subspace $S(a)$. If there is such an orthocomplement defined, the respective lattice is also orthocomplemented. A proposition which is logically true then says that the state vector of the system lies somewhere in H.

3.2 Basic Definitions and Theorems of Quantum Logic

I am taking these definitions, conditions, and theorems from the system of Quantum Logic L_q developed by Peter Mittelstaedt[3].

Definitions

3.21 Quasimodularity:

$$b \leq a, c \leq \neg a \Rightarrow a \wedge (b \vee c) = (a \wedge b) \vee (a \wedge c).$$

3.22 Modularity:

$$b \leq a \Rightarrow a \wedge (b \vee c) = (a \wedge b) \vee (a \wedge c).$$

3.23 Weak Modularity:

$$b \leq a \Rightarrow (a \wedge (b \vee \neg a) = b).$$

3.24 Distributivity:

$$a \wedge (b \vee c) = (a \wedge b) \vee (a \wedge c).$$

From a comparison of the three one can see that modularity and quasimodularity are weakenings of distributivity.

3.25 Commensurability:

$$a \sim b \Leftrightarrow_{def.} \ a = ((a \wedge b) \vee (a \wedge \neg b)).$$

There is an important relation between commensurability and distributivity: L_q is not distributive (does not obey 3.24) but if the elements b and c are both commensurable with a then distributivity can be proved. This means that in L_q a weakening of distributivity (3.26) holds

3.26

$$b \sim a, \ c \sim a \Rightarrow a \wedge (b \vee c) = (a \wedge b) \vee (a \wedge c).$$

Theorems[4]

3.27 In the lattice L_q the relation of commensurability ($\sim$) is closed under the operations $\wedge, \vee, \neg$. Thus the following three theorems hold:

---

[3] Mittelstaedt (1978).

[4] Only some special ones related to our purpose are mentioned. For further theorems and for details (like proofs, etc.) see Mittelstaedt (1978).

1. $a \sim b, a \sim c \Rightarrow a \sim (b \wedge c)$
2. $a \sim b, a \sim c \Rightarrow a \sim (b \vee c)$
3. $a \sim b \Rightarrow a \sim \neg b$.

3.28 In L_q commensurability is symmetric, i.e.

$$a \sim b \Rightarrow b \sim a.$$

This does, however, not hold if the lattice is not quasi-modular, that is, without the additional axiom:

$$b \leq a, c \leq \neg a \Rightarrow a \wedge (b \vee c) = (a \wedge b) \vee (a \wedge c).$$

3.29 In L_q it holds that if two elements are partially ordered they are commensurable, i.e.

$$a \leq b \Rightarrow a \sim b.$$

In L_q $a \wedge b$ implies commensurability of the elements:

$$a \wedge b \leq a \sim b.$$

3.3 Implication

One type of reservation against Quantum Logic has been pointed out by Jauch and Piron: Since the lattice of QM lacks an implication connective or conditional operation and, in consequence, does not have deduction schemata it is "very questionable whether we may properly call the lattice of general quantum mechanics a logic".[5] But, as a matter of fact, implication can be defined in Quantum Logic by a partial ordering relation.[6] The partial ordering relation $\leq$ of L_q can be interpreted as the valid implication thus:

3.31 $a \leq b$ iff $\mathbf{V} \leq (a \rightarrow b)$

where $\mathbf{V}$ is the unit element of L_q. In order to interpret an orthocomplemented quasi-modular lattice as a logic it is necessary to introduce also *material implication*. But the important thing here is that material implication has to be introduced in such a way as to neither presuppose nor imply commensurability or distributivity. We cannot therefore have all the conditions for a Boolean lattice.
3.32 The following weaker conditions are suitable for a weaker kind of material implication:

---

[5] Jauch and Piron (1970), p. 176.
[6] This was shown by Mittelstaedt (1972) and (1978).

1. $a \to b = \neg a \vee (a \wedge b)$
2. $a \wedge (a \to b) \leq b$ Modus Ponens
3. $((a \wedge c) \leq b) \Rightarrow ((\neg a \vee (a \wedge c)) \leq (a \to b))$.

It can be proved then that an orthocomplemented lattice with the properties that for any two elements a, b, there exists an element $a \to b$ satisfying conditions (1)–(3) above, is quasi-modular.

Commensurability and Implication

On the basis of 3.32 the following relations between commensurability and implication hold in L_q:

1. $a \sim b \Rightarrow (a \leq (b \to a))$
2. $a \sim (a \to b)$.

4 An Alternative Logic for QM

Can there be reasons for adopting a new (weaker than classical) logic? I think there can but I agree with Suppes that "there are, however, a number of reasons for moving very slowly to the adoption of a new logic, especially a logic that is clearly weaker than classical logic."[7] Because of such reasons and several others[8] which are in some sense analogous, my general view is to preserve classical logic (usually First-Order Predicate Logic) with its concept of validity (since it is sound and complete: a sentence is valid iff it is provable) but make the changes as restrictions with some kind of filter (limitative criterion).[9] Thus, concerning Quantum Logic this restriction separates valid (derivable) and realizable conjunctions and distributions from valid and non-realizable conjunctions and distributions. Concerning Relevance Logic the same restriction separates valid and relevant (non redundant) formulas or derivations from valid and non-relevant (redundant) ones. Concerning Epistemic and Deontic Logic the same restriction separates valid unparadoxical from valid paradoxical statements, etc. The main point of this view is to leave classical logic as a basis unchanged concerning its concept of validity but to

[7] P. Suppes (1984), p. 91.

[8] A very general one was emphasized strongly by Popper on many occasions: Using weaker logics in science weakens testability strategies in a serious sense.

[9] The proposed restriction is analogous to a filter although it is not strictly speaking a filter, as it is usually defined. A filter is usually defined as satisfying the following criteria: (1) If a satisfies the filter and b satisfies the filter then $a \wedge b$ satisfies the filter. (2) If a satisfies the filter and $a \to b$ satisfies the filter then b satisfies the filter. We have already seen in Chap. 1 that condition (1) cannot be always satisfied when logic is applied. The "filter" proposed below does not, in general, satisfy (2) either.

restrict it by an analogon to a filter in an appropriate way when applied to certain fields of empirical science (natural sciences or humanities.)

Observe that the program of Anderson-Belnap and their followers is, in general, to really *change* logic, i.e., to change the validity concept. The view defended here, however, is based on the conviction that questions like that of the evaluation of conjunctions and distributions from the evaluation of their parts (cf. Quantum Logic) and like that of redundant elements in the consequence class (cf. Relevant Logics) or that of invariance of logical and mathematical results against language framework and coding (cf. Weingartner (1997)) go beyond validity questions (i.e., beyond questions of semantics in the narrow sense) of First-Order Logic. And therefore they cannot be solved by manipulating (weakening) validity in some way or other. Leaving validity unchanged and distinguishing it from additional properties like limitations for conjunctions, redundant parts, etc., which are handled by additional restrictions, seems to be more appropriate.

There is an interesting alternative approach by Hintikka[10] which concentrates on noncommuting (incommensurable) quantities; assuming they are represented by variables the relation of noncommutability is interpreted as dependence or independence of a variable on another one or more accurately by the dependence or independence of a quantifier on another one. For handling in a more general way the dependence or independence of quantifiers (which is fixed in an asymmetric way by the Standard First-Order Predicate Logic) Hintikka has developed a special logic (IF-logic) in Hintikka (1996).

It is not claimed here that the proposal given below is the only adequate one. There may be more than one to be sufficient to solve the logical problems coming up in Quantum Theory. But it can be claimed that the following proposal has the following three properties:

- It is very simple.
- It preserves classical logic as the underlying logic and rules out only the superfluous redundancies and reduces complex parts to simple elements in the restricted consequence class whereby the consequence class is (by CL) logically equivalent to the original one.
- It is applicable in a wide field of other areas to solve difficulties and paradoxes.

4.1 Preliminaries for the Alternative Logic

As is clear from what has been said the "Alternative Logic" proposed here is not alternative to Classical Logic (CL) with respect to its validity concept, it is rather based on CL. This will become clear from the definitions of R and RD (cf. Sects. 4.2 and 4.4). But it restricts the valid formulas of CL (First-Order Predicate Logic with Identity) with a kind of filter to a subset:

---

[10] cf. J. Hintikka (2000)

to those formulas which are both valid according to CL and relevant; or more accurately: CL-valid and obeying (restricted by) R (Replacement) and RD (Reduction). R and RD are applied to the consequences not to the premises or to the consequent of a valid implication, not to its antecedent. The first reason for doing so is that in all cases described in Chap. 1 and in most cases where paradoxes come up when CL is applied putting restriction on the conclusion (i.e., Conclusion-relevance) solves the problem, whereas restriction on the premises (premise-relevance) does not. The second reason is that the general principle that the premises are always permitted to be richer than the conclusion (consequences) seems very reasonable and should not be violated.

R and RD can be formulated in such a way as to apply to arbitrary formulas not only to the consequent part of an implication. But this has several drawbacks and will not be adopted here.[11]

Classical Logic (CL) - which is to be understood as classical two-valued Propositional Logic or Predicate Logic of First Order with Identity - has two important properties with respect to the consequence (conclusion) a of a valid inference $A \vdash a$:

P 1 Parts of α may be replaceable by arbitrary parts of the same category *salva validitate* of the inference.

P 2 Parts of α may be reducible to simpler (and usually shorter) parts salva validitate of the inference and by preserving the logical content of the conclusion.

Simple examples for P1 are that in $A \vdash (B \rightarrow A)$ or $\neg A \vdash (A \rightarrow B)$ or $A \vdash (A \vee B)$ "B" can be replaced by any arbitrary propositional variable *salva validitate* of the inference; examples for P2 are that $C \wedge C$, $C \vee C$, $\forall x(A \wedge B)$, can be reduced, respectively, to C and to: $\forall xA \wedge \forall xB$. (For a precise formulation see below.)

P1 and P2 lead to paradoxes. It can be shown that those inferences where α contains replaceable parts or reducible parts (or both) lead to paradoxes when CL is applied to fields outside logic or mathematics. And it can be shown that those paradoxes which come up in very different fields of application can be avoided, blocked or solved when the valid inferences of CL are restricted to those that do not contain replaceable parts or reducible parts. For the special application here: R and RD, which avoid P1 and P2, lead to those restrictions we need in Quantum Logic: Commensurability cannot, in general, be presupposed, adjunction is not generally available, distributivity is only available in one direction (from the more complex to the more simple).

4.2 Replacement Criterion (R)

R α is an R consequence of A (symbolically: $A^{\vdash}_{cr}\alpha$ iff the following conditions (1)–(3) are satisfied:

[11] For a discussion of several such possibilities see Weingartner (1994).

1. $A \vdash \alpha$.
2. It is not the case that a predicate (including propositional variable, the identity sign) is replaceable in α on some of its occurrences by any other predicate of same arity salva validitate of $A \vdash \alpha$.
3. There is no β such that $\beta \dashv\vdash \alpha$ and β is the result of the replacement of some occurrencies of $\tau_1 = \tau_2$ by t or f[12].

$\alpha \rightarrow \beta$ is an R implication (symbolically: $a \underset{cr}{\rightarrow} \beta$) iff $\alpha \overset{\vdash}{_{cr}} \beta$.

Condition 3 is necessary to rule out some subtle kinds of redundancies concerning identity formulas, though some can be ruled out already with condition 2. Thus, in $Fa \rightarrow [(Fa \wedge a = b) \vee (Fa \wedge \neg(a = b))]$ the identity sign can be replaced by any arbitrary two-place predicate. But consider $A \vdash a \wedge \gamma$ where γ is one of the formulas $a = b \rightarrow b = a$, $a = a$, $(a = b \wedge b = c) \rightarrow a = c$. Such cases require condition 3.

IR α is an irrelevant consequence of A (symbolically: $A \overset{\vdash}{_{cir}} \alpha$) iff R(1) is satisfied but either R(2) or R(3) are not satisfied.

4.3 Properties of R ($\overset{\vdash}{_{cr}}$)

T1 $\vdash_{cr}$ and $\underset{cr}{\rightarrow}$ are not transitive (hence do not satisfy the cut rule), nor monotonic, nor closed under substitution nor closed under modus ponens. T 1 shows very clearly why (at least stronger kinds of) relevance (non redundance) – here conclusion relevance – have to be distinguished clearly from validity (of inference or implication): For validity closure with respect to transitivity and substitution are very important. To keep it for relevance means to allow redundancies in a quite direct and straightforward way. By substitution one can enlarge the redundancies – by substituting tautologies, repetitions, etc., like the branches of a tree.

T2 $\vdash_{cr}$ and $\underset{cr}{\rightarrow}$ imply the Aristotelean criterion or variable-entailment criterion: All the propositional variables and predicate letters of the conclusion (consequent) are contained in the propositional variables and predicate letters of the premises (antecedent).

T3 $\vdash_{cr}$ ($\underset{cr}{\rightarrow}$) is not recursively axiomatizable. This is so only for propositional logic and for the decidable parts of predicate logic[13].

T4 $\vdash_{cir}$ ($\underset{cir}{\rightarrow}$) is transitive (and satisfies the cut rule), is monotonic and closed under substitution.

T5 $\vdash_{cr}$ satisfies only the left-to-right half and $\overset{\vdash}{_{cir}}$ only the right-to-left half of the deduction theorem. $A \vdash \alpha \rightarrow \beta$ iff $A \cup \{\alpha\} \vdash \beta$.

T6 1. $A \overset{\vdash}{_{cr}} (\alpha \wedge \beta)$ iff $A \overset{\vdash}{_{cr}} \alpha$ and $A \overset{\vdash}{_{cr}} \beta$.

---

[12] The "R" may stand for "replacement" or "relevance" or both; "cr" stands for "conclusion replacement" or "conclusion-relevance" or both. For properties of $\overset{\vdash}{_{cr}}$ see below.

[13] For the proof see Schurz (1991).

2. $\vdash_{cr}$ $and^{\vdash}_{cir}$ are preserved under applying the following equivalence transformations to subformula parts of the conclusion: $\vee \rightarrow$ transformation, double negation, De Morgan, $\vee\wedge$ commutation and association; $\exists\forall$ -exchange, quantifier scope reduction $\forall\wedge$-splitting and $\exists\vee$-splitting.

A very important question for R-relevance as a means to solve the problem of versimilitude and other problems in the application of logic is the following one: Is the set of all R-relevant consequences of a theory A always logically equivalent with the set of all consequences of A? Since this is so as theorem T7 says one does not lose anything by concentrating just on the R-relevant consequences of a theory rather than on the whole consequence class.

T7 1. The set of all consequences of A is logically equivalent to the set of all R-relevant consequences of A: For all A, $Cn(A) \dashv\vdash Cn_{cr}(A)$.
2. For all (formulas of predicate logic) α there exists an α' such that $\alpha \dashv\vdash \alpha'$ and $\alpha' {}^{\vdash}_{cr} \alpha'$.
3. For each consequence α of A there exists a β such that $A^{\vdash}_{cr}\beta$ and $\beta \vdash \alpha$.[14] Theorem 7 is a very important result and provides a necessary premise for further theorems.

4.4 Reduction Criterion (RD)

RD Let $A^{\vdash}_{cr}\alpha$. Then α is a RD consequence-element of A, iff it is not the case that there exist mutually distinct formulas $\beta_1, \ldots, \beta_n (n \geq 1)$ such that

1. each β_i is shorter than α,
2. $\alpha \dashv\vdash \beta_i \wedge \ldots \wedge \beta_i$,
3. for each β_i, $A^{\dashv}_{cr}\beta_i$.

By considering the conditions (1)–(3) it is easy to see that every R consequence α of A can be splitted up into a set of RD consequence elements $\beta_1, \ldots, \beta_n (n \geq 1)$ of A logically equivalent with the orignal α. For example:

[14] The proof of T7 for propositional logic was given in Schurz/Weingartner (1987), that for predicate logic was presented in Schurz (1998).

Formulas taken as R consequences of itself	RD consequence-elements
$A \wedge B$	A, B
$A \wedge (B \vee C)$	$A, B \vee C$
$A \vee (B \wedge C)$	$A \vee B, A \vee C$
$\forall x(Ax \wedge Bx)$	$\forall xAx, \forall xBx$
$\exists x(Ax \wedge B)$, x not in B	$\exists xAx, B$
$\forall xA, x$ not in A	A
$\exists xA, x$ not in A	A
$A \wedge A$	A
$A \vee A$	A
$\neg\neg A$	A
$\neg(A \vee B)$	$\neg A, \neg B$ etc.

The set of RD-relevant consequence elements obtained by these steps might not be unique, i.e. several $\alpha's$ with the same number of conjuncts might be possible. But here a further reduction is possible with the help of Tarski's concept of a "basis": A set of consequences is a basis of itself iff it is not equivalently replaceable by one of its proper subsets.

4.5 RD Solves Paradoxes

Already R solves well-known paradoxes which have been discussed for decades by logicians and philosophers of science. These paradoxes arise when CL is applied to different fields (outside logic and mathematics). The areas where R successfully solves paradoxes are the following: Theory of explanation, of confirmation of law statements, of disposition predicates; Epistemic Logic, Deontic Logic, and Value-Theory. This has been shown in detail elsewhere.[15] R, however, is not sufficient to solve paradoxes in connection with: (1) the concepts of versimilitude, (2) the noninvariance of important concepts with respect to logical equivalence transformations, (3) Quantum Logic, (4) Logic of Actions. For the solution of paradoxes in these four areas RD is successful (i.e., in addition to R one needs RD which includes R). (1) As to the first, it can be shown that Popper's original definition of verisimilitude (which was proved to be inadequate by Tichy and Miller) can be rehabilitated and revised by permitting only relevant consequence elements as logical consequences of a theory. By this method the inadequacy-proofs of Tichy and Miller are blocked and, moreover, correct results follow from the definition if it is applied to scientific theories.[16]

(2) Concerning the rather complicated problems of equivalence transformations it has been shown elsewhere how R and RD can be successfully applied.[17]

[15] Cf. Weingartner-Schurz (1986), Schurz-Weingartner (1987) and Weingartner (1988).

[16] Cf. Schurz-Weingartner (1987).

[17] Cf. Weingartner (1997).

(3) Concerning Quantum Logic it can be shown that the necessary restrictions for interpreting quantum-phenomena for which one of the usual lattices (like L_q described by Mittelstaedt) is a sufficient condition can also be obtained by filtering CL through RD. But RD put on CL does not produce the lattice L_q. There are several differences. For instance, L_q contains two axioms of addition ($a \vdash a \vee b$) which are ruled out by R ("b" is an arbitrary part). That RD is successful here is interesting because it and R are effective in solving difficulties also in so many different other areas (see Sect. 4.6 below).

(4) Some paradoxes coming up in the Logic of Actions are analogous to those in Epistemic and Deontic Logic. They are caused by arbitrary parts in the consequence and can be handled with the help of R. Other paradoxes, however, have their root in the principle of adjunction (cf. Sect. 4.6.) which is mistakenly viewed as harmless (though it is certainly harmless if only truth values are at stake). These latter paradoxes can be avoided by applying RD to CL, which is extended by operators for actions.

4.6 Application of R and RD to Quantum Logic

Commensurability

We investigate[18] first the definitions of commensurability which are frequently used. There are the following three (where the equality sign is already changed to an equivalence):

$$a \sim b \leftrightarrow df a \leftrightarrow ((a \wedge b) \vee (a \wedge \neg b)) \qquad \text{(cf. 3.25)}$$
$$a \sim b \leftrightarrow df a \rightarrow (b \rightarrow a)$$
$$a \sim b \leftrightarrow df b \rightarrow (a \rightarrow b)$$

(where "$a \sim b$" stands for "a is comensurable with b" and $a, b, c, \ldots$ are propositions representing states of affairs)

Since the right part of the three definitions is logically valid in CL, applying unrestricted CL means – when these definitions are reasonable on other grounds – that commensurability of any states of affairs is presupposed by logic. But this is unacceptable since commensurability holds only under special conditions. A little reflection shows that each definiens (right part) of the above definitions is ruled out as irrelevant or as having arbitrary (replaceable) parts (though classically valid) by R: b can be replaced by any arbitrary proposition *salva validitate* of the respective definiens. Thus, according to a "restricted logic", which results from CL by filtering it with R, commensurability cannot be presupposed in general. This means that commensurability has to be stated as an extra axiom wherever it can be assumed but cannot be presupposed as a part of the underlying logic.

[18] For an earlier version of such an application, see Weingartner (1993).

Adjunction

Next we investigate adjunction. In the Quantum Logic L_q the conjunction $a \wedge b$ implies commensurability of a with b:

$$a \wedge b \Rightarrow a \sim b$$

Thus the question arises: when is $a \wedge b$ given? The answer cannot be: "Whenever a, b are given". Because then commensurability would be given between all states of affairs which are given. Since this is not the case $a \wedge b$ cannot be given automatically when a and b are given separately (recall the counterexamples of Chap. 1). Now, in unrestricted CL the rule of adjunction holds: From a, b infer $a \wedge b$. $a, b \vdash a \wedge b$. Applying RD shows that the direction from the parts (a, b) to the whole $(a \wedge b)$ is forbidden. Only the other direction, i.e., from the whole to the parts $a \wedge b \vdash a$ and $a \wedge b \vdash b$ is permitted. Thus, RD rules out adjunction but permits simplification, i.e., splitting into (most informative) parts. By this also, all repetitions in conjunctions and disjunctions are reduced to simple parts. As the examples in Chap. 1 show unrestricted adjunction leads to absurd consequences when applied outside logic. Elsewhere I have shown that also the unrestricted rule of addition, i.e. $p \vdash p \vee q$ leads to absurd consequences when applied outside logic.[19]

These principles are usually understood as harmless. But on a closer inspection they are not. In the inference: from a contradiction $p \wedge \neg p$ infer any arbitrary statement q the rule of addition is the culprit (not as some claim: the rule of disjunctive syllogism). Because the rule of addition *introduces* the new (unrelated and arbitrary) variable q. In a similar way the rule of adjunction *introduces* the connective: conjunction. It introduces it without any condition on the parts. Certainly if these parts are only represented by the truth values T and F nothing wrong happens if the conjunction is introduced. But if the parts p and $q(a, b)$ represent observable states of affairs then there may not always be a conjunction available.

Distributivity

As is clear from the table in 4.4 $a \wedge (b \vee c)$ dissolve by RD to the parts $a, (b \vee c)$. By CL $a \wedge (b \vee c)$ is logically equivalent to: $(a \wedge b) \vee (a \wedge c)$. But RD rules out the direction from conjuncts to more complex disjuncts, i.e., from $a \wedge (b \vee c)$ to $(a \wedge b) \vee (a \wedge c)$. It permits only the other way from more complex to its parts: from $(a \wedge b) \vee (a \wedge c)$ to $a \wedge (b \vee c)$ and from the latter to $a, (b \vee c)$. Analogously $a \vee (b \wedge c)$ dissolves by RD into the parts $a \vee b$, $a \vee c$. By CL $a \vee (b \wedge c)$ is logically equivalent to $(a \vee b) \wedge (a \vee c)$. But RD rules out the direction from conjuncts like $(a \vee b) \wedge (a \vee c)$ to disjuncts $a \vee (b \wedge c)$. It permits only the other way: from $a \vee (b \wedge c)$ to $(a \vee b) \wedge (a \vee c)$ and from the latter to

[19] Cf. Weingartner-Schurz (1986) and Weingartner (1988).

the parts: $a \vee b, a \vee c$. Thus the implications of the distribution laws which are permitted by RD are:

$$((a \wedge b) \vee (a \wedge c)) \Rightarrow^{RD} (a \wedge (b \vee c))$$
$$((a \vee (b \wedge c)) \Rightarrow_{RD} ((a \vee b) \wedge (a \vee c)).$$

The respective right parts can be reduced by RD further to their conjuncts:

$a \wedge (b \vee c)$ is reduced by RD to $a, (b \vee c)$
$(a \vee b) \wedge (a \vee c)$ is reduced by RD to $(a \vee b)$, $(a \vee c)$. But the way back in the opposite direction (from the simplest parts to the more complex ones) does not hold in general.

Now, the above directions of the distribution laws which are permitted by RD are exactly those which are applicable in Quantum Theory. They do not cause any problems. Thus RD rules out exactly these directions of the laws (which are equivalences in CL) which are harmful in the application in Quantum Theory. The Quantum Logic L_q accepts the full distribution laws (in the form of equivalences) only under the condition that commensurability holds. Independently of the question whether one direction of the distribution laws should be forbidden in general (when logic is applied), or only under certain conditions we have shown elsewhere that the general idea of RD to split the consequence class of an inference into (non redundant and most informative) parts proves very successful in other areas, for example, in the theory of verisimilitude[20] and in general: always when we speak of the consequences inferred (with the help of logic) from scientific theories. Since these consequences have to be put to test they should always be the most simple, most informative, non redundant consequence elements. And when two scientific theories are compared with the help of their consequence classes then it is essential to take only the most informative and shortest non redundant consequence elements; otherwise one can easily show that redundancies lead to paradoxes in the comparison between the two theories. Thus RD a wider application than to serve just as a sufficient restriction for an alternative logic applicable to Quantum Theory.

References

1. G. Birkhoff, J. v. Neumann: 'The Logic of Quantum Mechanics'. In: Annals of Mathematics **37** (1936) pp. 823–843
2. D. Finkelstein: 'Matter, Space and Logic'. In: Hooker (ed.): **II** (1979) pp. 123–139
3. J. Hintikka: 'The Principles of Mathematics Revisited'. (Cambridge University Press, Cambridge 1996)
4. J. Hintikka: A Logic for Quantum Theory. Unpublished manuscript

[20] Cf. Schurz-Weingartner (1987)

5. J.M. Jauch, C. Piron: 'What is Quantum Logic?' In: *Quanta* ed. by P. Freund et al. (Univ. of Chicago Press, Chicago 1970) p. 166ff
6. P. Mittelstaedt: 'On the Interpretation of the Lattice of Subspaces of the Hilbert Space as a Propositional Calculus'. In: Zeitschrift für Naturforschung **27a** (1972) p. 1358 ff
7. P. Mittelstaedt: *Quantum Logic* (Reidel, Dordrecht 1978)
8. G. Schurz: 'Relevant Deduction. From Solving Paradoxes towards a General Theory'. In: Erkenntnis **35** (1991) pp. 391–437
9. G. Schurz: 'Relevance in Deductive Reasoning: A Critical Overview'. In: *Beyond Classical Logic* Conceptus-Studien. Ed. by G. Schurz and M. Ursic. (Academie-Verlag, St. Augustin 1998)
10. G. Schurz, P. Weingartner: 'Verisimilitude Defined by Relevant Consequence Elements. A new Reconstruction of Popper's Original Idea'. In: *What is Closer to the Truth?* ed. by Th. Kuipers. (Rodopi, Amsterdam 1987)
11. P. Suppes: *Probabilistic Metaphysics* (Blackwells, Oxford 1984)
12. P. Weingartner: 'Remarks on the Consequence-Class of Theories'. In: *The Role of Experience in Science* ed. by E. Scheibe. (de Gruyter, Berlin 1988) pp. 161–180
13. P. Weingartner: 'A Logic for QM Based on Classical Logic'. In: *L'art, la science et la metaphysique. Essays in Honor of André Mercier* ed. by M. de la Luiz Garcia Alonso, E. Montsopoulos, G. Seel (Peter Lang, Bern 1993) pp. 439–458
14. P. Weingartner: 'Can there be Reasons for Putting Limitations on Classical Logic?' In: *Patrick Suppes, Scientific Philosopher, Vol. 3.* ed. by P. Humphries. (Kluwer, Dordrecht 1994) pp. 89–124
15. P. Weingartner: 'Language and Coding Dependencies of Results in Logic and Mathematics'. In: *Philosophy of Mathematics Today* ed. by Agazzi, E. and György Darvas. (Kluwer, Dordrecht 1997) pp. 73–87
16. P. Weingartner, G. Schurz: 'Paradoxes Solved by Simple Relevance Criteria'. In: Logique et Analyse **113** (1986), pp. 3–40

The Language of Interpretation in Quantum Physics and Its Logic

Roland Omnés

1 About the Language of Interpretation

The present talk will be mainly devoted to the topic of language and more precisely the question: is there a sound and universal language for describing physical events? By "universal", I mean a language expressing macroscopic phenomena as well as microscopic (atomic) events; by "sound", I mean it obeys standard logic.

Historians tell us that Bohr, Heisenberg, and Pauli were asking that same question at the earliest stage of interpretation and they wondered what kind of language it would need (1, 2). If it exists, this language would have to express what is most obvious to an observer in an experiment, namely the experimental set-up, its various devices (such as an accelerator, a voltmeter or a Geiger counter) and their detailed working (which part is active at which time and how). It was obvious that classical physics is perfectly fit for such a purpose. The difficulties occurred, of course, when one attempted to deal on similar lines with the events happening among atoms and particles. Even the meaning of the word "event" became questionable after the discovery of uncertainty relations. The final outcome of the reflections in Copenhagen is well known and it can be grossly summarized as follows: the macroscopic aspects of physics are the only ones we can speak of with some sense of truth (or reality). The description of a microscopic system obeying quantum laws is, on the contrary, extremely limited and most treacherous because of its complementary aspects (an electron, for instance, may be described as a wave, or else as a particle, but certainly not as a wave and a particle). It is therefore wiser to abstain from speaking of such queer events.

There are at least two reasons for considering this traditional answer as unsatisfactory. Feynman made the first reason perfectly plain: One cannot *understand* the simplest interference experiment with a Young device if one cannot say: First, that the wave function consists at some time of two parts originating at the two holes; secondly, that what is detected finally is a particle. He did not say that this is how things happen, but that there is no other way to understand them and to convey them in plain words. We are all familiar with his famous sentence summarizing his puzzlement: "Nobody understands quantum mechanics." and indeed, nobody dared to take the challenge until very recently.

The second reason for doubting the necessity of the Copenhagen restrictions on language is that practically nobody respects them. You cannot read a textbook or an article describing an experiment, or listen to an experimental talk without noticing a superb disregard for the supposed awkwardness of quantum events. Physicists speak of them with practically the same words they use for speaking of an experimental set-up. They say "when the atom is in the cavity", after having just said "the atom velocity is controlled with a good precision", later they mention the time "when the two parts of the wave function are merging". And nobody seems to care, just nodding, except when some careless person goes a little too far. Then, and only then, Bohr's warning is quoted like an article of Law.

The point I want to raise with this example is not a pedantic criticism of too-frequent bad manners among physics circles. You may notice that the people using this apparently loose language are also often responsible for great advances in physics. Their way of understanding their experiment is extremely efficient (and who else understands their work better than they do?). I want rather to direct your attention to the efficiency of their language and ask a question: can it be that experimentalists have a better grip on quantum interpretation than the so-called specialists? Or in other words: is there a convenient language for expressing interpretation that was missed in the early times of quantum mechanics?

I intend to provide a positive answer for this question in the present talk.

2 Von Neumann's Predicates

The ancestors of the present language of interpretation were the famous "elementary predicates" by von Neumann. Let us then recall what they are.

A predicate is an elementary proposition about a quantum system. It says for instance that "the position of this electron is in this room". More generally, it would state that the value of some quantum obsevable A lies within some range Δ of real numbers.

Von Neumann observed that such a statement can be represented by (or associated with) a mathematical object that is directly rooted in the mathematical principles of quantum theory. It can be "translated" by a mathematical notion (like German can be translated into French) and the translation is called a projection operator. The predicate stating that "the position of an electron is in a space region V" can be associated, for instance, with a mathematical operation acting on the electron wave function ψ. Let us call this operation E. It clips off the wings of the wave function, i.e., puts it equal to zero outside V while it does not change it inside V. Intuitively, this operation discards the part of ψ. Corresponding to the electron outside V and it preserves ψ. in V. It turns out that this operation, or rather this operator, is a quantum *observable.* It can take therefore some values, which are its eigenvalues. It is a projection operator, by which one means that it repeats tself when acting twice $E^2 = E$. Accordingly, its eigenvalues are 0 or 1. They look very much like "truth values": 1 for "true" and 0 for "false", and in that sense, the predicates satisfy the fundamental logical rule of an excluded middle.

Similar results hold for more general predicates. One may specify at what time the predicate is supposed to express a physical property (the introduchtion of time making use of the Schrödinger equation). In the example of an interference experiment, one can therefore express the final position of a particle when it arrives on a detector. One may also express by means of another projection the earlier statement about the wave function (as originating from two holes). As a matter of fact, everything we might wish to say about a quantum system can be expressed by a projection operator.

It is well known that von Neumann met two very serious difficulties with his proposal: (i) he was not able to extend the equivalence of a statement with a projection in the case of a classical property. (ii) he found that, altogether, ordinary logic does not apply in the set of quantum predicates. His language therefore appeared as neither universal nor sound and it did not attract much attention among physics circles.

I shall now explain how these two difficulties can be alleviated.

3 Classical Properties

According to our definition, the language of predicates will be universal it it can encompass classical statements. let us first agree on what is to be meant by such a statement. it is concerned with some macroscopic object, or a part of such an object, for instance a pointer on the dial of an old-fashioned voltmeter. A classical statement about the pointer refers simultaneously to its position and momentum. Typically, it asserts some values for both quantities x, p within some error bounds $\Delta x, \Delta p$. It can be represented in the (x, p) plane by a rectangle with sides $2\Delta x, 2\Delta p$. the product $\Delta x, \Delta p$ must be, of course, much larger than Planck's constant in order to avoid a violation of the uncertainty relations. The question is: can one associate such a classical property with a projection operator in the space of wave functions? Von Neumann tried to find the answer but he did not succeed, because the necessary mathematical tools were not yet available.

These tools have been elaborated in the meantime (5,6). They have been worked out by mathematicians, as the so-called "microlocal analysis". To cut a long story short, one can say that a classical property is not exactly expressed by a unique projection operator but rather by a *set* of projections, which are equivalent and close to each other (in a well-defined sense I cannot describe here). This multiplicity of quantum translations for a unique classical property originates in the uncertainty relations, but it is well understood and it provides a basis for the derivation of clssical physics, with precise criteria for its validity. The conclusion is, in a nutshell: Von Neumann's language has now been shown to be universal and one may speak of macroscopic devices exactly as one does for quantum properties.

One may notice that this approach to the right language is exactly opposite to Bohr's since a language for quantum events is considere as fundamental because it is nearer the principles of physics whreas the language of classical physics is only derived from it.

4 Consistent Histories

The next step is best understood if it is meant as a precise justification for the trivial langugage one finds in physics books. As was the case for quantum predicates, the method is somewhat formal, which is the price to be paid for making its logical content explicit.

Let us start with the conventional description of some experiment. it begins generally with a description of the experimental set-up: it consists for instance of an accelerator, a vessel, and a detector. a good description must enter in these features in more detail; it says which kind of accelerator is used, how it performs, the matter, size and temperature of the vessel walls and everything else that is necessary for really understanding the set-up and reproducing it if necessary. then, one says something of the preparation: At some initial time, a particle 1 is injected in the accelerator and ther is another paricle 2 in the vessel. Then comes a description of som microscopic events: when going out of the accelerator, particle 1 has some velocity $v \pm \Delta v$. Some time later, it enters into the vessel. Ten, there is a collision between particles 1 and 2 resulting in a reaction $1 + 2 \to 3 + 4$. Then, particle 4 enters the detector and finally there is a measurement by the detector resulting in some datum, whch can be expressed as a classical property of some pointer belonging to the detector.

According to Robert Griffiths (10), this description should be translated by replacing every statement by a corresponding projection operator, each one of them referring to a definite value of time. This prescription includes quantum properties (for instance the occurrence of the reaction $1 + 2 \to 3 + 4$) as well as classical properties (for instance, the position of the pointer before and after measurement). The resulting sequence of properties (i.e., predicates or projections) is called a *history*. Because of the random character of quantum events, one history is not enough for an account of all possibilities. It might occur, for instance, that the collision between particles 1 and 2 results in a scatterin: $1 + 2 \to 1 + 2$ rather than the reaction $1 + 2 \to 3 + 4$. One takes all these possibilities into account by introducing a so-called complete family of exclusive histories involving different events.

Griffiths defined a family as consistent if one can write *bona fide* probabilities for the histories in it. Since the basic additivity of probabilities is not generally compatible with the additivity of quantum amplitudes, there must be some conditions for the existence of probabilities. These criteria, which can be written as some equations for the projections, are called *consistency conditions* and they are essential for the language of histories to be sound.

Two examples will be enough: Feynman's example of an interference experiment where one speaks of the wave function, and then of the detection of a particle in a definite place, is consistent: it makes sense. But if one says that the particle has gone first through a definite hole before being detected, there are no *bona fide* probabilities for the corresponding histories: it makes no sense to say that a particle has gone through only one hole, whichever it may be! Notice that this conclusion is reached without resorting to a measuring device which would check for the crossing of a definite hole. It is an intrinsic feature of the evolved language one is now using.

5 Recovering Standard Logic

I just said of two descriptions for the same experiment that one of them makes sense and the other does not. But what is menat by making sense or not? The answer is: the first description obeys the rules of conventional logic and the second one does not.

The point is that the standard rules of logic can be implemented in a consistent family of histories (8). As is well known, these rules involve three logical operations (not, and, or) and two relations (inference and logical equivalence). It is easy to define the operations by considering the histories as elements of a set and propositions as the corresponding subsets. It is easily shown that these propositions are the corresponding subsets. It is easily shown that these propositions involve everything one may find relevant for a discussion of the experiment. As fo inference, it is defined in a Bayesian form: As long as *bona fide* probabilities exist (in view of the consistency conditions), a proposition a implies another proposition b if the conditional probability $p(b \mid a)$ is equal to 1. As for the logical equivalence of two propositions a and b, it means simply that a implies b and b implies a.It is then easy to prove that all the rules of standard logic are valid.

To summarize, the two difficulties that had stopped von Neumann have now been solved. his language of predicates is universal and there is a clearcut method for selecting relevant predicates so that they obey standard logic. This selection agrees in most cases with everything that had been guessed painstakingly by Bohr, Heisenberg, and Pauli. One does not need, however, to be very clever or very learned to make sense of quantum physics or to point out a nonsense. Some straightforward calculations can do the job, and often once and for all. Is this by itself a progress? It looks so, at least in my opinion.

6 Verifying Logical Consistency

I said that verifying the consistency of logic involves some calculations, but this is not the end of the story. It turns out as a matter of fact that, when a description of an experiment makes sense and when the practical conclusions one may draw for it are correct (via implications), then in most cases no calculation is necessary. General results relying on the known range of validity of classical physics and on the now famous decoherence effect are enough. Rather amusingly, one gets embarked into heavy calculations only when something has gone wrong. a nonsense had been unknowingly or incautiously introduce: If one makes a blunder, one is punished by hard calculations on many sheets of paper that must be thrown finally into the waste basket, since they only acknowledge that a mistake was made.

The importance of decoherence for consistency was first pointed out by Murray Gell-Mann and James Hartle (9). It has far-reaching consequences in the derivation of the rules of measurement theory (including the rule of wave-function reduction, though there is no reduction effect and only decoherence). It also sheds light on one of the most intriguing problem in physics, namely the arrow ot time. Histories have a logical time ordering, which must coincide because of consistency with a direction of time existing in decoherence, and this is also necessarily the direction in the Second Principle of thermodynamics. This is however another story and I only mention it to stress a relation of logic with the arrow of time, which is more or less obvious when one comes to think of it, but is not too frequently noticed.

7 Complementarity

It is easier to understand the meaning of complementarity by means of an example. An old one, which was investigated by Bohr and Mott, is as follows: A radioactive source, located at the origin of space, emits an alpha particle at tim 0. The initial wave function of this particle is an isotropic S-wave. At a later time t, the alpha particle is detected by a geiger counter at some point x somewhat far away. It looks rather natural to assume that the particle had a straight-line motion between the origin of space to x. If so, the particle was certainly located at the midpoint $\frac{x}{2}$ at time $\frac{t}{2}$. Since stating an exact location for a particle might be asking too much of an analogy with classical physics, one may introduce some space volume V surrounding the point $\frac{x}{2}$ and state the expected quantum property as "the alpha particle position is inside V at time $\frac{t}{2}$".

But the alpha particle felt no external action before time t. One should therefore expect its wave function to have obeyed the Schrdinger equation. The initial wave function was an S-wave (with angular momentum zero). The same value of the angular momentum is expected to hold at time $\frac{t}{2}$, implying the existence of an isotropic wave function at that time.

We have here a paradigm of complementarity. Let us cast it into the language of histories for a thorough analysis. two histories are compared. Both of them involve the same initial and final events: the alpha particle emission at the origin in an S-wave and its detection at a point x at time t. the conflictin histories differ by their statement of the relevant event at time $\frac{t}{2}$: One history states that the particle was already going towards x in a straight-line motion; the other history says that, at that time, the wave function was still an S-wave. Which one makes sense really?

This is, of course, the old paradigm of wave-particle duality. It was partly solved by Mott, who made calculations with the following result: the detector at x is made of matter with which the alpha particle interacts. Let one assume that the intermediate space region V is also filled with some matter interacting with the charged alpha particle. then there is a straight-line track of the particle with one end in x and a segment in V. So far, so good, but the question of consistency we asked has only some interest when there is no matter in V. It is purely a matter of principle.

In the history approach, complementarity comes out directly. It means that (at least) two different families of consistent histories can describe the same physical events, even though they are incompatible, or more exactly complementary in so much as no wider consistent family of histories can include both of them. Both descriptions are logically consistent, but not consistent with each other!

As far as logic is concerned, the conclusion is therefore that non-standard logics do not enter quantum mechanics but exclusive logics do. Exclusive logics were only a curiosity in science, though international lawyers know them well: the same situation does not have the same interpretation according to the rules of Denmark or China. Complementarity does not mean, however, that paradoxes can occur, in view of a remarkable no-contradiction theorem. Anyway, because of complementarity and though the interpretation using the language of consistent histories and the decoherence effect has had many achievements, it cannot pretend to a *realistic* description of microscopic events by means of human words. I think I am in agreement with Bernard d'Espagnat if saying that the limitation is not in the existence of reality but in the limitations of a language when it does not express a plain perception (11).

8 Wider Issues

The present talk was limited to some logical aspects of interpretation. Much more, however, has been obtained with the help of consistent histories and decoherence (6). The Copenhagen rules for measurement theory, for instance, have been derived as so many theorems from more basic principles (Like the existence of non-commuting observables and Schrödinger's dynamics).

The derivation of classical physics from a quantum background has also explained the validity of classical determinism (which is a logical equivalence between two classical properties holding at different times, with a very smal *probability* of error). I even dare to say that one *can* understand quantum mechanics. The price to pay is, however, to take the basic principles of quantum theory as the right *philosophical* principles to be used. Other, more venerable Kantian principles, such as the category of causality, locality (which amounts to the existence of categorical *a priori* judgements) and separability (i.e., the existence of phenomena in the sense of Kant) are far from being so fundamental. They are valid in a rather wide macroscopic framework, with a domain that is quantitatively expressed by their respective amounts of error. By turning the tables of epistemology in this way and reversing premiss and conclusion, an interesting new field is opened for the philosophy of science, as I have tried to show elsewhere (8).

References

1. D. Murdoch: *Niels Bohr's Philosophy of Physics* (Cambridge University Press, Cambridge 1987)
2. C. Chevalley: *Ontologie et méthode dans la physique contemporaine. La physique quantique et la fin de la philosophie classique de la nature* Thesis, University Paris X (1997)
3. R.P. Feynman, R.B. Leighton, M. Sands: *The Feynman Lectures on Physics* Vol. III (Addison-Wesley, Reading 1965)
4. J. von Neumann: *Mathematische Grundlagen der Quantenmechanik* (Springer, Berlin 1932). English translation by R.T. Beyer: *Mathematical Foundations of Quantum Mechanics* (Princeton University Press, Princeton 1955)
5. R. Omnés: Rev. Mod. Phys. (64) 339 (1992); *The Interpretation of Quantum Mechanics* (Princeton University Press, Princeton 1994)
6. R. Omnés: *Understanding Quantum Mechanics* (Princeton University Press, Princeton 1999)
7. R. Omnés: J. Stat. Phys. **53**, 893 (1988)
8. R. Omnés: *Philosophie de la science contemporaine* (Gallimard, Paris 1994). English translation by A. Sangalli with additional material: *Quantum Philosophy, Understanding and Interpreting Contemporary Science* (Princeton University Press, Princeton 1999)
9. M. Gell-Mann, J.B. Hartle: In: *Complexity, Entropy, and the Physics of Information* ed. by W.H. Zurek (Addison Wesley, Redwood City 1991)
10. R.G. Griffiths: J. Stat. Phys. **36**, 219 (1984)
11. B. d'Espagnat: *Le Réel voilé, analyse des concepts quantiques* (Fayard, Paris 1994) English translation: 'Veiled Reality' (Addison-Wesley, Reading 1995)

Why Objectivist Programs in Quantum Theory Do Not Need an Alternative Logic

Alberto Cordero

Summary. Many different projects go by the name "quantum logic". The weakest and least controversial, aimed at clarifying superpositions and the structure of possibility in quantum physics, supplements but does not depose standard logic. At the other extreme, the strongest and most radical project seeks to completely supplant standard logic. In between there are intermediate projects aimed at empirically generalizing and revising standard logic in the spirit of the revision of physical geometry promoted by Einstein earlier in the century. Only the weakest program seems to retain some plausibility. This paper tries to further the critique of the logical turn in foundational quantum theory by questioning some key epistemological and metaphysical underpinnings of the project, in particular (1) the *a priori* exclusion of nonlinear stochastic transitions from quantum dynamics; and (2) the conception of physical properties as precise valued magnitudes. Reasons for moving away from these two assumptions are presented. The three most developed objectivist approaches to quantum theory to date (GRW), many decohering worlds, and Bohmian quantum theory) are then considered vis-à-vis the said underpinnings. While only one of the approaches in question takes the original linear dynamics as complete, all three incorporate "deviant" conceptions of physical property.

1 Introduction

Suppose someone asked something about don Quixote's hair. The availability of a sharp answer would depend greatly on the specific question posed. Queries which always have determined answers when asked about real human beings may not have such luck when asked about a fictional character. In the present case, Cervantes makes don Quixote sufficiently "normal" and gives enough information for us to know, for example, that during his fight with the windmill his head was not completely bald, that it had an exact number of hairs on it, and that this number was les than10^{100}. But now suppose we ask about the precise number. Relative to the novel, no assertion of the form "the number of hairs in don Quixote's head immediately before his encounter with the windmill was N" can be said to be true for any $N \epsilon [1, 10^{10}]$, even though it can be truthfully said that, at the moment in question, don Quixote's head did have less than 10^{100} hairs. Unlike their real counterparts, *the properties of fictional characters are indeterminate beyond a certain level of descriptive depth.*[1]

One could carry on along these lines and spell out a "Logic of Fictional Characters", centered perhaps on the following "deviant" notion of disjunctive property:

> For a fictional entity α, and a property A applicable to it and whose possible values in α are A_1 or A_2 or ... or A_n, the truth of the disjunctive proposition "$A_1(\alpha)$ or $A_2(\alpha)$ or ... or $A_n(\alpha)$" may be compatible with the falsity of each of the elementary propositions involved -i.e., $A_1(\alpha), A_2(\alpha), \ldots,$ and $A_{n(\alpha)}$. In this relaxed sort of way, therefore, discourse about fictional entities can be said to have "its own logic". Indeed, as many contributions in this volume make plain, virtually every special field can be claimed to have its own logic in a similar manner.

Not to be left behind, talk about "quantum logic" (QL) can be just as easy and uncontroversial. As is well known, for a quantum system σ in a state of superposition with respect to a property A applicable to it and whose possible values in σ are A_1 or A_2 or ... or A_n, the truth of the disjunctive proposition "$A_1(\sigma)$ or $A_2(\sigma)$ or ... or $A_n(\sigma)$" is indeed compatible with the falsity of each of the elementary propositions involved -$A_1(\sigma), A_2(\sigma), \ldots$ and $A_n(\sigma)$. Statements about a quantum-mechanical system are further complicated by Heisenberg's Principle, which denies that precise values can be determined for certain pairs of magnitudes (such "conjugate variables" as position and linear momentum). We could thus say that quantum theory follows "non-classical" rules of disjunction and implication, henceforth represented as v_q and $\supset_q$, respectively. Understood in this way, QL can be just a way of highlighting and making precise some core conceptual features of quantum theory, conspicuously superposition and possibility structure, by showing their import on how propositions about quantum systems relate to each other. Let us call this the "weak" QL program.

2 Logic and Empirical Science

Does contemporary physics actually *require* a shift from standard logic to a new logic proper? The present state of quantum physics makes such an idea scientifically very implausible. In actual physics mixed states are described using the standard concept of disjunction (represented by "v" in what follows). Only a system in an appropriate state of superposition is correctly described by means of the alternative disjunction. If so, however, the v_q connective supplements rather than replaces the standard connective. A similar

[1] An insightful contrast between the fictional and the real is found in Rescher (1999).

story can be told about the relation of implication and every other novelty brought by QL. The point is that the standard connectives in their traditional meanings still play a role in the quantum-mechanical description of the world. The quantum successors play a useful role, but it seems misleading to think of them as already replacing the standard classical connectives. There is, however, another older and deeper reason for not taking a radical QL too seriously. Given the state of the game, proposing too radical a move seems *logically incoherent*, as long as the argumentation used for it at the metalogical level remains attached to the rules of classical logic.

So, in what sense, if any, does quantum physics require a new logic? We need to examine the case more closely.

In the pre-quantum-theoretical description of an ordinary macroscopic body, all the magnitudes applicable to it are assumed to have precise values at all times. Objects like chairs and mountains, that is, are said to enjoy *complete determinateness* relative to each and all of their applicable properties. The systems dealt with by quantum theory, however, are not at all like this. In standard presentations of the theory (SQT) a magnitude has a definite value on a system only when the latter's quantum state corresponds to an eigenfunction of that magnitude. Quantum systems are peculiar in that, for any given magnitude $\mathbf{A}$, they naturally develop into superpositions of such eigenfunctions. However, when a system is subjected to measurement the outcome is always a definite value – even when the initial state is known to correspond to a superposition relative to the measured magnitude. SQT does not explain how this state "reduction" along the A variable occurs. The theory provides a linear dynamical account (QT_0) of how systems behave when they are not being measured. Their measurement behavior, on the other hand, is represented by a different dynamics, centered on a non-linear stochastic law (staged by the so-called "Projection Postulate" in conjuncion with Born's statistical rule). This second dynamics governs the fate of systems when external measurements are performed on them; it becomes operational only whenever a measurement takes place, bringing the state of the object system into one of the eigen-functions associated with the measured magnitude. But how does this happen? How does the system first pick the right family of eigen-functions and then select one of them? The process is left completely mysterious. SQT does not really tell us what a measurement amounts to physically. No wonder the standard Projection Postulate quickly came to be regarded by many thinkers as an infamous component of the theory.

One obvious first reaction to the resulting conceptual conundrum is to take quantum theory as a mere epiphenomenological statistical theory of atomic and sub-atomic particles. The findings of physics, however, make this line of thought extremely implausible. The properties of subatomic particles seem to be a great deal less determinate than the exact number of hairs on don Quixote's head, and yet those particles appear to be quite real. Understanding quantum superpositions has thus long been a pressing matter.

Birkhoff & von Neumann's logical-structural work[2] is most appealing in this regard because their representation of the structure of physical propositions in terms of modular lattices brings precision when talking about physical systems in general and quantum systems in particular. We thus get a "weak" QL program aimed at capturing such distinctly quantum-mechanical traits as superposition and property ascription by focusing on propositional relations that can be considered a kind of "logic". The set of propositions ascribing magnitudes to a system form a partial Boolean algebra, and this renders the resulting logic of property ascriptions plainly non-classical. Only one Boolean subalgebra applies to the system at any given time. The logic is classical within a Boolean subalgebra in the sense that, within it, ultrafilters can be used to assign standard (bivalent) truth-values to the propositions. Propositions outside this structure cannot simultaneously receive a determinate truth-value.

The question of probability assignments is, however, far from straightforward in any variety of QL. Suppose we apply standard probability theory to the new logic. Since, in QL, $[(p'_q r)v_q(p'_q s)]$ entails $[p'_q(rv_q s)]$, the probability of the latter should be no less intense than that of the entailing one. But consider a double-slit situation. The noted connection between entailment and probability seems to suggest that, when both slits are open, the probability that a particle lands at a point x on the screen should be at least as great as the probability that it lands at x, given that one or the other – but not both – of the slits is open. The phenomenon of interference, however, allows for *negative interference*. Accordingly, the probability of landing at x when both slits are open can be less than the simple sum of the probabilities derived from the two slits' individually being open. Sympathizers of the logical turn in foundational quantum physics quickly came to see that the non-Boolean character of the total lattice called for an amended account of conditional probability and proceeded accordingly.[3]

What I am calling "weak QL" does not have to challenge SQT. It does not, by itself, free the physics from mixed states but rather keeps them as developments generated by certain special interactions (possibly ones involving bodies with many degrees of freedom). So, no exclusivity is demanded for QT_0. The uses of this weak logical program are correspondingly uncontroversial – for example, in capturing the essential elements of the phenomenon of quantum superposition and linking lattice elements with propositions to which a probability value 1 can be assigned.

Not too long ago, however, a more exciting and ambitiously *stronger* project was pursued in foundational studies. Its aim was to free quantum theory of its deepest conceptual problems by both asserting the completeness and universality of the basic linear quantum theory (QT_0) and embracing an empirically learned deviant logic.

[2] Birkhoff & von Neumann (1936/1975).

[3] See, in particular, Bub (1977).

3 The Search for an Empirical Logic

The status of the outcome pursued by this stronger program was meant to be as *empirical and scientific* as that of the physical geometry spelled out by general relativity. Central to this program was the assumption that QT_0, the linear part of SQT, suffices to describe *all* physical systems completely, including ordinary macroscopic objects. No non-linear stochastic "reduction of the quantum state" needed to be included in the picture. As a universal theory, QT_0 was expected to apply to all material systems, including measuring devices, without any disruption of its laws at any level (classical, macroscopic, conscious observation, etc.). The wave function or state describing a system was taken to provide a complete description of the system's physics, as well as to evolve exclusively according to the theory's linear dynamics. In this stronger QL program the propositional lattice for a system in a general state of superposition *is* the properly fundamental lattice for *any physical system* (not just microscopic ones).

I am referring, primarily, to Hilary Putnam's valiant program of the late sixties, in which the principles of logic are presented as elements opened, in principle, to the possibility of critical generalization and revision *in the light of empirical reasons.* According to this point of view, the conceptual problems posed by quantum physics encourage one to adopt a logic along the lines of Birkhoff & von Neumann (1936/1975). Central to Putnam's project are the following three interpretive claims: (a) Physical systems are completely described by QT_0; (b) a physical system always has a precise value for every magnitude A applicable to it; and (c) a measurement of A conducted on a system reveals the A value that σ had prior to measurement.

Putnam's prototype of scientific revision here is Einstein's General Relativity. The proposal runs as follows: just as the challenge created by gravitational phenomena led Einstein to embrace a non-Euclidean geometry as the correct representation of physical space, now the challenge created by quantum phenomena leads us to embrace a new logic. This, it was further maintained, promotes a conceptually problem-free way of understanding quantum theory that allows us to hold onto a strong sense of dynamical magnitudes.

Unfortunately, if the proposed shift to a new logic eliminates some paradoxical elements, it introduces others which seem at least as bad. Suppose that a system σ that is initially in an eigen-function for the value X_i of a magnitude X applicable to it. The system is then subjected to a measurement aimed at finding the value of a second magnitude P, incompatible with X, with the result P_j. Are we to say that σ had this precise property value before the P measurement was conducted? If so, given that prior to this second measurement σ was in a state associated with the value X_i, how then can it be that during the period after the X measurement and before the P measurement the truth-value accorded to the proposition "$X_i(\sigma) \bullet P_j$"

is precisely false.[4] The program's emphasis on regarding quantum particles as things complete in themselves (i.e., having at all times every magnitude applicable to them), especially when compounded with its insistence that measurement reveals the pre-existing value of a magnitude on the relevant system, turns out to be problematic.

Again, the intended analogy with general relativity breaks down at a crucial point. The scientific acceptance of Einstein's theory was propelled by its manifest conceptual and experimental fertility. The strong QL program, by contrast, has not helped to discover any new phenomena. To be fair, it was never meant to do that – its aim was to deal with the measurement problem by means of a novel understanding of the concept of physical properties. Still, fertility of the noted kind has been the feature most prominent in the acceptance of all the great theories of 20th century physics.

It is largely because of critical observations like these that Putnam's program lost its original appeal. In addition, revising the logic was, at best, just one of many options. In particular, one can always place the revisionary burden at the level of the probabilistic rather than the logical level of relations.[5]

4 Lingering Classical Intuitions

Although the QL program as a whole is no longer pursued with anything like the intensity seen in the 1970s, some key components of it remain very much alive. There is, in particular, a lingering emphasis on precise-valued properties (PVP). At its most elegant and subtle, this is found, I think, in recent work by Jeffrey Bub. Consider, for example, his concern with *uniquely* **maximal** *sets of determinate observables* (i.e., sets that can be enlarged only at the cost of engendering Kochen-Specker contradictions). A pivotal element in Bub (1997) is a theorem characterizing such sets, which is offered as a consistency constraint on interpretations, along with the suggestion that to give an interpretation is to specify a preferred determinate observable. Thus, for example, the interpretation embodied by SQT links property determinateness of a magnitude A on a system to the latter being in an eigen-function of the operator associated with A. To Bub this agrees with specifying the identity operator as the preferred "observable". If, as Bub and many foundationalists favor, the completeness of QT_0 is to be maintained, this choice leads to the measurement problem. Hence, the charisma of Bub's theorem for seekers of "better" interpretations of quantum theory: a maximal set of properties determinate on a system in a given quantum state is uniquely fixed by specifying that state along with some *preferred determinate operator*. SQT

[4] Hughes (1989).

[5] See Professor Beltrametti's contribution to this volume.

picks identity as the preferred operator; Bohmian quantum theory[6] chooses position. Bub shows how to fit many historical interpretations into this framework, in particular Bohr's complementarity, Bohmian quantum theory, and modal interpretations.

However, I think the assumptions behind this persistent theme about PV properties in quantum physics need to be explored critically. Two related suppositions are (a) that *PV* determinateness has to be introduced somewhere into the picture, and (b) that to interpret quantum theory amounts to choosing a preferred basis. Bub suggests that any choice of a preferred determinate observable can, in consort with the appropriate dynamics, accommodate measurement outcomes. If so, however, an important question arises: why put determinateness in one place rather than another? What is to justify an interpretation of quantum theory? Bub favors Bohm's theory over other objectivist approaches. His specific choice here is most debatable (apart from yielding property ascriptions in a frustratingly tenuous sense, Bohmian theory does not take QT_0 as the total theory), but let us leave this for a moment. More important at this point, I think, is the Quinean suggestion that any choice of a preferred determinate observable can, in consort with appropriate theoretical revisions, accommodate measurement outcomes. Even if this were true, more would be needed to make a proposal serious. Extant scientific methodology, particularly regarding the acceptance and rejection of proposals, has to be brought to bear on grading possible interpretations. It may happen that *all the grand interpretations of quantum theory on display at a given time equally lack scientific seriousness.* In that case, however, the right move would not be to arbitrarily embrace an undeserving piece of speculation. Educated scientific sense would recommend one to remain uncommitted to any of the as yet insufficiently matured proposals.

Rather than elaborating on the specifics of theory choice, here I think it will be better to concentrate on a few questions about the very framework of *PV* properties in quantum physics. This framework, it seems, amounts to a lingering relic of classical physics via the *QL* program. In particular, I want to comment on the interrelated themes of interpretation, determinateness, and the advancement quantum theory.

5 On the Credibility of Property Determinateness

I take it to be agreed that preferential credibility should be granted only to claims that are so successful (intellectually and pragmatically) and so beyond reasonable doubt that they can be confidently held to be true independently of mere *personal preferences.* There is no shortage of such claims in the contemporary natural sciences. If, like everything else in science, interpretation is

---

[6] The seminal presentation is found in Bohm (1952). For a more contemporary presentation and discussion see Cushing (1994).

to be guided primarily by relevant current knowledge, then it would be unwise to embrace approaches that burden the foundations of science with arbitrary commitments. The only legitimate constraint on interpretations is whatever presently functions as objective knowledge about the relevant domains.

This has an immediate application to the project of universalizing quantum theory, for one of the greatest falsehoods of the twentieth century is that ordinary macroscopic objects are *known* to have PV property states. In fact, all that is known is that such objects have quasi-classical states (i.e., ones that are sharp by "ordinary standards"). So, to the extent that the notion of *PV* property determinateness lacks adequate scientific motivation, it cannot be maintained as a robust constraint on further theorizing in quantum physics.

If the above is accepted, classical determinateness need not be put anywhere in the contemporary scientific picture. All one is forced to assume is quasi-classical determinateness for ordinary macroscopic objects, i.e., sharp property states *within the limits imposed by the Heisenberg relations.* It follows that searches for any one naturally *PV* dynamical magnitude are not properly motivated by current knowledge. In this regard, we need to be particularly wary of the classical baggage that surrounds time-honored magnitudes like position, momentum, energy, etc. The Boolean sub-algebras of measurement results associated with such magnitudes seem to be exposed by quantum physics itself as being *approximate models* rather than idealizations. The actual dynamical workings of physical systems may tend to give more spontaneous sharpness to some particular observable, perhaps position (as variously suggested by the decoherence approach[7] and the *GRW* approach[8]). But this in no way requires *PV* magnitudes at any level. The point is simply that the search for maximal *PV* observables in foundational studies seems to be seriously undermined by background scientific and methodological considerations. *PV* magnitudes are not as theoretically sound as they once seemed to be.

How, then, should one go about interpreting and furthering quantum theory? The previous commentary recommends granting protagonic roles only to features made prominent by the reliable part of the theory. In this regard, it is a fact that scientific practice focuses on the quantum state field (somewhat misleadingly referred to as "the wave function") both because of this field's central role in the theory and because of the concept's exceedingly successful applications. I suggest that it is not an accident that the three most developed objectivist programs in foundational quantum theory today (Bohmian quantum theory, the *GRW* approach, and the decohering many worlds approach) all grant paramount importance to the wave function. In all these programs quantum theory is, at least in part, *about* this weird field that lives

[7] See, for example, Saunders (1993).

[8] The seminal ideas of this approach are presented in Ghirardi, Rimini & Weber (1986).

in configuration space. This theoretical element is picked out by the actual success of physics rather than by a whimsical choice among epistemically comparable options.

Is not the above move simply giving preference to the identity operator? If so, it would not be in the sense emphasized by defenders of PV property determinateness or of determinateness merely within the bounds of the Kochen & Specker strictures on property ascription. The *preference* for the quantum state field vindicated here has nothing to do with affirming the standard value-eigenvalue rule. That rule incorporates an unwarranted plea for forcing the spectrum of physical magnitudes to include PV determinateness as a physical possibility, whereas current background knowledge compels us to assume only that standard macroscopic bodies have approximately sharp property states. Thus, placing the state field at the interpretive center is not a matter of mere *preference* but something recommended both by current scientific practice and by the down-to-Earth notion that we should let only the most credible current information guide our search for further knowledge and understanding.

What are we left with? We end up with at least one robust objectivist core theory, as provided by the most warranted claims about the quantum domain.[9] This includes deep and detailed descriptions of individual quantum particles, ones capable of successfully guiding their actual manipulation in the laboratory, which lends further credibility to the quantum state as the theoretical counterpart of an actual physical field. Down-to-Earth realism about the state field or wave function is akin to this theoretical core, albeit just in a sense similar to ongoing ordinary realism about cats and such – no more metaphysical than that.

On the negative side, many questions are left unanswered. For example, the nature of the wave function is left opaque beyond its characterization as a field-like physical structure. The total dynamics may be linear and deterministic, or non-linear and stochastic, competing interpretations taking different lines here. Detailed explanatory speculations are made available by the GRW approach, Bohmian theory and the many decohering worlds approach, all of which take the wave function seriously and variously supplement the noted core theory. The GRW approach assumes the ontology of QT_0 but claims that the latter's linear dynamics provides only part of the story; Bohmian quantum theory supplements QT_0 both ontologically and dynamically; only the many decohering worlds approach builds entirely on QT_0. These theories predict different phenomena, but not different *observable phenomena in practice*; and they all seem otherwise comparably virtuous in terms of larger methodological considerations (see the previous reference). So, we are also left with an assortment of effectively speculative options. Still, this does not compromise the argued scientific preference for taking the

[9] For a discussion of objectivist explanatory theories and their present state of scientific equivalence, see Cordero (1998).

quantum state field seriously. This, I suggest, is a more perspicuous way of seeing what the *GRW* approach, Bohmian theory, and the many decohering worlds approach are trying to accomplish.

6 The Conception of Properties in Three Leading Objectivist Programs

As I said before, according to the standard quantum-mechanical concept of "having aproperty", a system has a property if and only if it is in a *PV* eigenfunction of the operator associated with that property. Which determinate properties a system has at a given moment can then be easily found by applying the value-eigen-function rule to its quantum state. But this easy way of thinking about properties has always been known to be unreal, especially for continuous magnitudes.

Both the *GRW* approach and the many decohering worlds approach eschew the standard rule as *bad idealization*, resorting instead to explicit "coarse graining" as an *ontically significant* feature of the physical world –one which turns out to be crucial to the objective rise of the quasi-classical world of chairs, mountains, and human beings. What this coarse graining amounts to, of course, is the introduction of *disjunctive properties* into the big picture. A way of representing such properties, favored by the two mentioned approaches is in modal terms: a system in state Ψ has a property A if and only if its probability of having that property within the interval Δ is close to 1 by a margin pragmatically set at ϵ_A, that is:

$$Prob_{\Psi}(A, \Delta) \geq 1 - \epsilon_A .$$

The Bohmian conception, in which reality is made up of the universe particle (in the multi-dimensional space appropriate for representing all the component particles that make up the world) and the universal quantum wave function, sticks to the *PV* model of properties but everything else in it deviates from the standard conception. For instance, a Bohmian sub-particle always has a well defined-position (and so, also trajectory), but all its other dynamical properties are not intrinsic but "contextual". The program's insistence on maintaining the category of particle in physics might lead one to expect a Bohmian particle to be the carrier of such intrinsic properties as mass, charge, and magnetic moment, but perceptive analyses of neutron interferometry experiments, especially by Brown et al., strongly suggest the opposite.[10] For example, experiments involving a tilted interferometer in free fall lead to phase shifts which seem to imply that the gravitational mass of neutrons is highly nonlocal. Experiments involving accelerated interferometers invite a similar conclusion about inertial mass. The neutron's magnetic

[10] Brown (1995), Brown (1996).

moment fares no better. How about charge? Does not a Bohmian electron have at least its charge localized on it? The answer appears to be "no", judging from studies of experiments of the Aharonov-Bohm type (where an electron passes on one or the other side of a magnetic flux), in which the relative phase shifts introduced by the process on the sub-trajectories turn out to depend on the charge. How can this be if, as Bohmian theory requires, the electron always has a well-defined trajectory? Like mass and magnetic moment, in Bohmian theory charge appears to be a property of the quantum field rather than the localized particle. Bohmian particles, in short, turn out to be a bit of a farce. They are initially introduced as things complete in themselves. Careful consideration, however, reveals them to lack intrinsic properties to such an extent that it is hard not to regard them as artificial constructs.

And so, none of the leading objectivist generalizations or revisions of quantum theory incorporates anything like the classical conception of properties. Moreover, two of the three reviewed programs turn their backs on the *PV* model. It seems therefore fair to state that, in quantum physics, appeals to classical-like properties are no longer supported by specific positive reasons. They have to do with some metaphysical nostalgia for the lost illusions of classical physics rather than scientific thought.

7 Conclusion

So, the quantum logical turn in foundational studies is now discouraged by challenges from many directions. Foremost among these are serious doubts about QT_0 as a complete theory of material systems. In addition, there are now doubts about the physical relevance of PV properties as proper idealizations. No presently forceful motivation exists for expecting ultrafilters to provide a physically meaningful representation of material systems.

This is not to say that the heroic paths explored in the sixties and seventies by Hilary Putnam and many others before and after, have not been of considerable intellectual service. For one thing, their explorations have helped us to envisage the kind of scientific developments that might conceivably lead us to revise our thought and understanding of the world at the deepest imaginable level – namely a scientific (as opposed to strictly logical) revision of the propositional structures relevant to physical discourse. It seems most doubtful that this could ever amount to a proper refutation of classical logic, just as it is doubtful that the Einsteinian revision of geometry can be taken as a refutation of Euclidean geometry. The envisioned revision could amount, however, to a wise shift to a "better logic" along the lines of extant scientific criteria of theory choice. What more proof can one have of the philosophical depth of science?

References

1. Birkhoff, G., J. von Neumann: 'The Logic of Quantum Mechanics' In: *The Logico-Algebraic Approach to Quantum Mechanics, vol 1: Historical Evolution* ed. by C.A. Hooker (Dordrecht, Reidel 1936/1975) pp. 1–26
2. Bohm, D.: *A Suggested Interpretation of the Quantum Theory in Terms of Hiddden Variables* Parts I and II. Physical Review **85** (1952) pp. 166–93
3. Brown, H.R: 'Bovine Metaphysics: Remarks on the Significance of the Gravitational Phase Effect in Quantum Mechanics'. In: Clifton (1996) pp. 183–193
4. Brown, H.R., C. Dewdney, G. Horton: 'Bohm Particles and Their Detection in the Light of Neutron Interferometry'. Foundations of Physics **25** (1995) pp. 329–347
5. Bub, J.: 'Von Neumann's Projection Postulate as a Possibility Conditionalization Rule in Quantum Mechanics'. Journal of Philosophical Logic **6** pp. 381–390
6. Bub, J.: *Interpreting the Quantum World* (Cambridge, Cambridge University Press 1997)
7. Cordero, A.: 'Physics and the Underdetermination Thesis: Some Lessons From Quantum Theory'. Forthcoming in: Proceedings of the Twentieth World Congress of Philosophy ed. by S. Dawson (Managing Editor). (Federation Internationale de Societes de Philosophie & Boston University, Boston 1998)
8. Cushing, J.T.: *Quantum Mechanics: Historical Contingency and the Copenhagen Hegemony.* (Chicago, University of Chicago Press 1994)
9. Ghirardi G., A. Rimini and T. Weber: 'Unified Dynamics for Microscopic and Macroscopic Systems'. Physical Review D **34** (1986) pp. 470–491
10. Hughes, R.I.G.: *The Structure and Interpretation of Quantum Mechanics.* (Cambridge, MA, Harvard University Press 1989)
11. Putnam, H.: 'Is Logical Empirical?'. In: *Boston Studies in the Philosophy of Science* **5** ed. by Cohen, Robert S., Marx W. Wartofsky (Dordrecht, Reidel 1969) pp. 181–241
12. Rescher, N.: 'How Many Possible Worlds Are There?'. Philosophy and Phenomenological Research **LIX** (1999) pp. 403–420
13. Saunders, S.: 'Decoherence, Relative States, and Evolutionary Adaptation'. Foundations of Physics **23** (1993) pp. 1553–85

Does Quantum Physics Require a New Logic?

Peter Mittelstaedt

1 Introduction

In order to answer the question whether quantum physics requires a new logic, we will at first briefly discuss the more general problem of a justification of the laws of logic. Here, we are not interested in the most general form of logic and its applicability to various fields. Instead, we will investigate propositions which assign properties to physical systems, i.e. to material objects of the external reality. On the basis of this kind of elementary propositions a language of propositions can be established by introducing a concept of truth, several connectives, and relations between propositions. Within the framework of this language we can find the logic of this language as an internal structure, which is justified by this way of reasoning.

The speaker of a scientific propositional language which is constructed in this way, is also an observer in the sense of physics who justifies or disproves propositions by measurement processes. The possibilities and abilities of the speaker-observer to verify or to falsify propositions by measuring the corresponding properties, have some influence on the most general rules of the syntax of this language, i.e., on the laws of logic. On the other hand, the possibilities of the speaker-observer to perform measurements will partly depend on the general structure of that domain of reality which is described by the scientific language. The observer is part of the world which he observes and subject to the law of the reality which he is investigating. Hence, the compatibility of the possibilities of the speaker-observer with the physical laws of the reality domain in question induces some unavoidable kind of self–referentiality into the physical theory.[1]

These more general remarks will be illustrated in the present paper by the well–known classical logic, which can be applied to the reality domain of classical physics (Sect. 2) and by the "new" quantum logic, which refers to the domain of quantum physics (Sect. 4). The main problem of this paper is, whether we are confronted here with a plurality of logical systems or whether there is a hierarchy with a priority of one of these logical systems over the other one. We show that there is no pluralism of various logical systems corresponding to different fields of experience but a strict hierarchy, such

[1] [Mi 86,98]

that there is only one "true" logic of propositions about physical systems. The reasons for this result are, on the one hand, the most general features of the quantum physical reality, and on the other hand the universality, and self–referentiality of quantum physics (Sect. 3).

2 The Logic of Classical Physics

2.1 The Language

We consider the object language S_C of classical physics, i.e., a propositional language with propositions $A(S,t)$ which assign properties $P(A)$ at time t to a physical system S. The domain of reality which is described by S_C is the macroscopic or classical physical reality. It is obvious that this language can also be applied to the apparatuses used by the speaker-observer.

The classical physical reality is characterised by the complete determination of its objects. This means that for all elementary propositions $A(S)$ it is objectively decided whether the property $P(A)$ or the counter–property $P(\bar{A})$ pertains to the system. These properties pertain to the system objectively and irrespective of the possibility of observation. However, we assume that for every property $P(A)$ there exists a process of observation or measurement, such that in a finite number of steps we can decide whether $P(A)$ or $P(\bar{A})$ pertains to the system.

The elementary propositions are logically independent and independent with respect to their verification and falsification. Elementary propositions pertain to an object irrespective of the tests of other elementary properties. Complete determination, finite decidability of elementary properties and their complete independence characterise the domain of classical reality. These most general "ontological" features have an essential influence on the language of classical physics, which is used for the description of this domain of reality.

2.2 Semantics and Syntax

The ontological preconditions mentioned guarantee the objective decidedness and finite testability of elementary propositions $A(S)$. Hence, $A(S)$ will be called to be *true*, $\vdash A(S)$, if and only if the system S possesses the property $P(A)$. According to the preconditions the truth of $A(S)$ can be shown by a finite proof procedure. In any case, either A or the counter–proposition $\bar{A}$ turns out to be true in this way. Hence, elementary propositions are value-definite. The semantics which is established by this concept of truth will be called "realistic".[2]

[2] [Mi 86]

On the basis of the set $S_C^{(e)}$ of elementary propositions we introduce the logical connectives by the possibilities to attack or to defence them, i.e., by the possibilities to prove or to disprove the connective. As an example we consider the two-place operation "sequential conjunction"

connective	denotation	attacks	defences
$A \sqcap B$	"A and then B"	$1.A?, 2.B?$	$1.A!, 2.A!$

where $A?$ means the challenge to prove A, and $A!$ the successful proof. This attack-and-defence scheme can be illustrated most conveniently by a proof-tree which is chronologically ordered.[3] The first branching point corresponds to the test of A at t_1, the second one corresponds to the B-test at t_2.

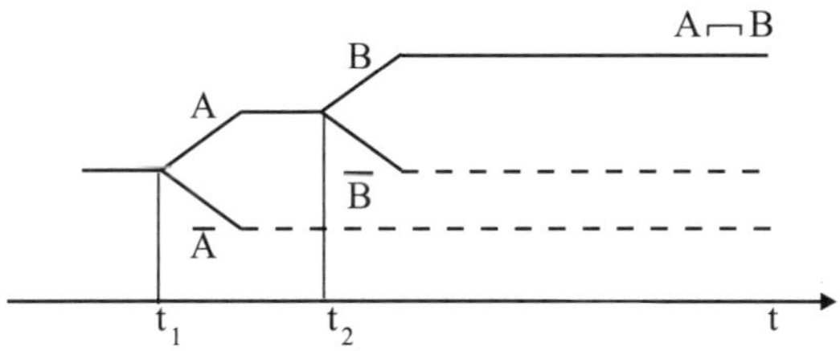

Fig. 1. Proof-tree for $A \sqcap B$ in classical logic

The temporal order is fixed here, but the time difference $\delta t = t_2 - t_1 > 0$ may assume arbitrary positive values. There is one branch of success. In a similar way the one place operation "negation" may be introduced by the proof tree for $\neg A$ (not A) with one branch for success and one branch for loss.

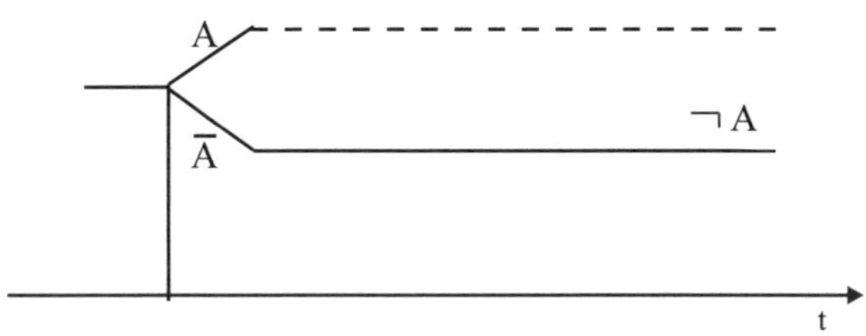

Fig. 2. Proof-tree for $\neg A$ in classical logic

The sequential conjunction (and the other sequential connectives) refer to two instants of time t_1 and t_2. The logical connectives refer to one common instant of time. The logical connective which corresponds to $A \sqcap B$ is the logical conjunction, which is defined by the attack-and-defence scheme

[3] [Mi78], [Sta 80]

connective	denotation	attacks	defences
$A \wedge B$	*A and B*	$A?,\ B?$	$A!,\ B!$

In contrast to the sequential conjunction we have here an arbitrary number of attacks and defences in arbitrary order. In this way it is guaranteed that the result of the A and B tests can be attributed to a common time value. If $A \wedge B$ is proved, then A and B are simultaneously true.

The assumed independent testability of propositions A and B implies that after a test of B the result of a preceding test of A is still valid and available without any restrictions. The *unrestricted availability* of the results of A- and B-tests implies that for the proof of $A \wedge B$ we need only two steps, provided their time difference $\delta t = t_2 - t_1$ is sufficiently small. There is no need to repeat the proofs.

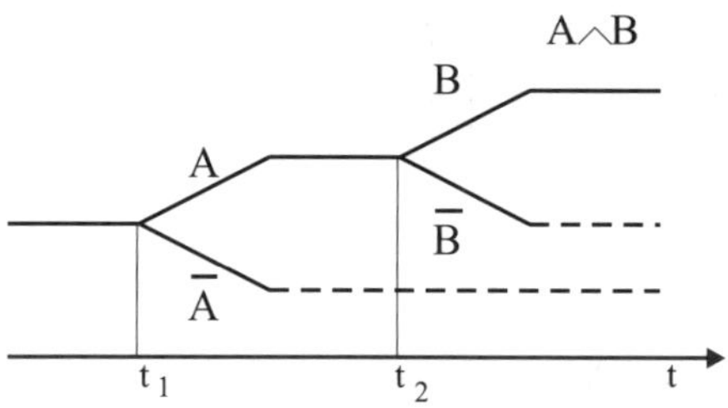

Fig. 3. Proof-tree for $A \wedge B$ in classical logic

In a similar way the other *logical* connectives, i.e. the disjunction $A \vee B$ (A or B) and the material implication $A \to B$ (if A then B) can be defined by the possibilities to prove or to disprove them, or by the respective proof-trees.

The full language S_C of classical physics can then inductively be defined by the set$S_C^{(e)}$ of value-definite elementary propositions and by all finitely connected compound propositions $A \in S_C$. The concept of truth can then be defined in the following way:

> A proposition $A \in S_C$ is said to be true if the proof-tree of A leads finally to a branch of success; it is called false if the proof-tree ends with a branch of loss.

Furthermore, on the set of propositions we introduce two binary relations,

1. the **value equivalence** $A = B, A$ is true if and only if B is true
2. the **implication** $A \leq B \Longleftrightarrow A = A \wedge B$

The full language can then be formulated as

$$S_C = \{S_C^{(e)}; \sqcap, \wedge, \vee, \to, \neg; =, \leq\}.$$

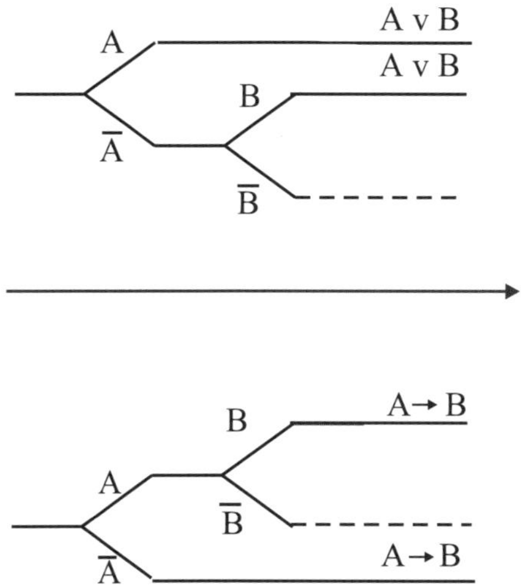

Fig. 4. Proof-trees for $A \vee B$ and $A \rightarrow B$ in classical logic

2.3 Classical Logic

The semantics described here is a combination of a realistic semantics (with respect to elementary propositions) and a proof semantics (with respect to compound propositions). Hence, the truth of a compound proposition depends, on the one hand, on the connectives contained in it, and on the other hand on the elementary propositions and their truth values. This leads to the following question: Are there finitely connected propositions $A \in S_C$ which are true in the sense of the semantics described, irrespective of the truth values of the elementary propositions contained in it? Propositions of this kind will be called *formally true*.

Examples for formally true propositions can easily be found. On account of the *value definiteness* of the elementary propositions we have the formally true proposition

$$A \vee \neg A \text{ (tertium non datur)}$$

and on account of the *unrestricted availability* of propositions in a proof tree we have the formally true proposition

$$A \rightarrow (B \rightarrow A),$$

which is true even without the assumption of value definiteness of elementary propositions. If both *value definiteness* and *unrestricted availability* of propositions are assumed, then we obtain the formally true proposition

$$A \rightarrow ((A \wedge B) \vee (A \wedge \neg B)).$$

There are many, even infinitely many formally true propositions. The totality of formally true propositions is called classical logic and the algorithm which provides propositions of this kind is the *calculus of classical logic*. Here we make use of a Bouwer-calculus for implications which turns out to be very convenient for our purpose.

For the formulation of the calculus of the classical logic L_C we make use of two special propositions,

- the *true* proposition (*verum*) V, such that $\vdash V$ and for all $A \in S_C$ we have $A \leq V$, and
- the *false* proposition (*falsum*) $\Lambda = \neg V$ with $\Lambda \leq A$ for all $A \in S_C$.

By means of these two propositions we have $\vdash A \Leftrightarrow V \leq A$ and, in particular, $\vdash A \to B \iff A \leq B$. The calculus L_C can be formulated as a calculus of implications with "beginnings" $\Rightarrow A \leq B$ and rules like $A \leq B \Rightarrow C \leq D$. The Lindenbaum–Tarski algebra of L_C is a complete, complemented, and distributed lattice L_B (Boolean lattice). If it is freely generated by a finite number of elementary propositions, it is also atomic and fulfils the covering law.

The calculus L_C of classical logic

1.1 $\Rightarrow A \leq A$

1.2 $A \leq B; B \leq C \Rightarrow A \leq C$

2.1 $\Rightarrow A \wedge B \leq A$

2.2 $\Rightarrow A \wedge B \leq B$

2.3 $\Rightarrow C \leq A; C \leq B \Rightarrow C \leq A \wedge B$

3.1 $\Rightarrow A \leq A \vee B$

3.2 $\Rightarrow B \leq A \vee B$

3.3 $A \leq C; B \leq C \Rightarrow A \vee B \leq C$

4.1 $\Rightarrow A \wedge (A \to B) \leq B$

4.2 $\Rightarrow A \wedge C \leq B \Rightarrow C \leq A \to B$

5.0 $\Rightarrow \Lambda \leq A, \Rightarrow A \leq V$

5.1 $\Rightarrow A \wedge \neg A \leq \Lambda$

5.2 $\Rightarrow V \leq A \vee \neg A$

The formal propositional logic does not depend on the elementary propositions which are contained in the formally true propositions. However, the logic depends on the general preconditions under which proof processes are possible. In the present case the most important preconditions are the finite decidability of elementary propositions and the unrestricted availability in a proof–process. Formally true propositions are not true in an absolute sense. Their truth follows from the pragmatic preconditions of proving or disproving propositions. Only in this transcendental sense they are a priori true.[4]

3 The Quantum Physical Reality

In the domain of quantum-physical reality the strong ontological preconditions of the classical language are no longer fulfilled.

3.1 Complementarity

Niels Bohr made the important discovery that it is, in general, not possible to observe simultaneously two elementary properties $P(A)$ and $P(B)$ of the same system. [5] The reason for this phenomenon, the *complementarity*, is, according to Bohr, that the measuring devices for $P(A)$ and $P(B)$ are mutually exclusive. The strong complementarity relation can be relaxed by Heisenberg's *uncertainty relation*: Two complementary properties $P(A)$ and $P(B)$ can be measured jointly on the system, if both properties are measured unsharply and if the product of convenient uncertainty measures, for $P(A)$ and $P(B)$ is always larger than a universal bound given by Plancks constant $\hbar$.[6] Here, we make use of the following terminology: Two properties $P(A)$ and $P(B)$ which can jointly and sharply be measured, will be called *commensurable*. If a simultaneous sharp measurement is not possible, then the two properties will be called *incommensurable*. There are various degrees of incommensurability. Properties which are maximal incommensurable will be denoted as *complementary*.

3.2 Nonobjectifiability

Complementarity and the *uncertainty relation* can be tested and confirmed experimentally. However, from these experiments it does not follow whether complementarity expresses merely the mutual exclusiveness of apparatuses or whether it refers to properties of the object system. If complementarity were merely describing some restrictions of measurement apparatuses, then it would still be possible after a measurement of $P(A)$ to assume that the system possesses in addition to $P(A)$ either $P(B)$ or $P(\neg B)$ – even if these

[4] [Mi 86]
[5] [Bohr 28]
[6] [Mi 98]

assumptions cannot be confirmed by experiments. The property $P(B)$ would then be objectively determined but subjectively unknown.

Within the framework of classical physics this (ignorance) interpretation can be justified by means of the logic of the language of classical physics. Indeed, if $P(A)$ was measured and thus the proposition A was shown to be true (in the sense of the realistic semantics), then for any other proposition B it follows from the logical equivalence $A = (A \wedge B) \vee (A \wedge \neg B)$ that also $(A \wedge B) \vee (A \wedge \neg B)$ is true. Hence, together with the property $P(A)$ the object system would also possess the property $P(B)$ – or it would not possess this property. In this case we call the property $P(B)$ "objective". The observer knows that an "objective" property is objectively decided even if it is subjectively unknown to him.

It is one of the most important results of quantum physics that this is, in general, not the case. If a property $P(A)$ is known to pertain to the object system, then one must not assume that, in addition, an arbitrary property $P(B)$ pertains to the system or not. If $P(B)$ is incommensurable with $P(A)$, then $P(B)$ cannot be tested by experiment. This is, however, less important. The essential point is that under the conditions described here the property $P(B)$ is not only subjectively unknown to the observer but it is *objectively undecided* whether the system possesses $P(B)$ or $P(\neg B)$.[7]

3.3 The Probability Argument

The justification of this nonobjectification argument makes use of the probability structure of quantum mechanics. Let us assume that the system possesses property $P(A)$. If, in addition, the complementary property $P(B)$ or its counter-property $P(\neg B)$ would pertain to the system, then the system would be in a mixed state of these alternatives with conditional probabilities $p(A; B)$ and $p(A; \neg B)$, respectively. The probability for any other property $P(C)$ which is neither commensurable with $P(A)$ nor with $P(B)$ is then given by[8]

$$p(A; C) = p(A; B)p(B; C) + p(A; \neg B)p(\neg B; C).$$

However, this formula does not hold in quantum mechanics. The correct formula in quantum mechanics reads

$$p(A; C) = p(A; B)p(B; C) + p(A; \neg B)p(\neg B; C) + p_{int}(A; B, C)$$

and contains the nonvanishing interference term $p_{int}(A; B, C)$. Hence, we find that the objectification assumption for property $P(B)$ is in disagreement with the well-established statistics of $P(C)$ outcomes. Experiments of this kind can easily be performed by a Mach–Zehnder interferometer.[9]

---

[7] [Mi 76, 98]

[8] This follows immediately from the logical equivalence $C = (C \wedge B) \vee (C \wedge \neg B)$ and the Kolmogoroff axioms.

[9] Cf. [Mi 98], p. 70 ff.

3.4 Universality

According to our present knowledge quantum physics is universally valid in all domains of the physical reality. Of course, this statement holds for the most general and most abstract laws of quantum mechanics but not for the large variety of special applications in nuclear physics, solid–state physics, etc. However, what is more important is that quantum mechanics is equally applicable to macroscopic physics as well as to microscopic phenomena. This somewhat surprising statement is confirmed by numerous macroscopic quantum effects. Consequently, quantum mechanics applies also to the macroscopic apparatuses which are used for the verification and falsification of quantum–mechanical propositions. Since the most general properties of the apparatuses belong to the pragmatic preconditions of our scientific language of physics, we arrive at the important conclusion that the laws of physics which describe quantum systems and their properties also determine the behaviour of apparatuses and hence the most general preconditions of the scientific object language of physics.[10]

This result has far–reaching consequences for the construction of a scientific language. If the speaker-observer takes into account the possibilities and restrictions which are known from the domain of reality of object systems, then the language which is constructed in this way will not contain more structure in its syntax and logic than the domain of reality that is described by it. This is by no means obvious. In the language of classical physics the strong pragmatic preconditions of value definiteness and unrestricted availability of propositions lead to classical logic, in particular to the equivalence $A = (A \wedge B) \vee (A \wedge \neg B)$. However, if this law is applied to quantum–mechanical propositions which are not commensurable, then one comes into conflict with the nonobjectifiability of propositions in quantum mechanics.

4 The Logic of Quantum Physics

4.1 Language and Reality in Quantum Physics

Since classical language and logic cannot consistently be applied to arbitrary quantum–physical situations we will make a new attempt at constructing a scientific language which can be applied to all quantum–physical situations. On account of the *universality* of quantum physics this language will be applicable also to macroscopic phenomena, which are usually treated within the framework of classical physics. For the construction of quantum language we will keep in mind the more general remarks about the interrelations between language and reality. Hence we make use only of those *restricted ontological preconditions* of the language which are compatible with the laws

[10] [Mi 98]

of quantum physics. On the basis of this new and weak quantum ontology a *quantum pragmatics* can be formulated which incorporates all restrictions for proving or disproving propositions which come from the incommensurability of quantum–physical properties. Since proof processes contain measurement processes, incommensurabilities restrict the possibilities of testing quantum–mechanical propositions.[11]

4.2 The Syntax of Quantum Language

For the constitution of a quantum language we begin again with elementary propositions A, which state that a property $P(A)$ pertains to the object system. Accordingly, the proof of the elementary proposition A consists of a measurement of property $P(A)$ with positive outcome. The possibilities for quantum measurements allow for the assumption that after the measurement of $P(A)$ we obtain either a positive or negative result.[12] Hence, an elementary proposition A can either be proved (result A) or disproved (result $\bar{A}$) and is thus value-definite. Furthermore, if after a successful proof of A a new proof attempt for A is made, then one obtains again the result A, if the applied measurement is repeatable. However, if after a successful proof of A another elementary proposition B is proved, then a new proof attempt for proposition A will, in general, not lead to the previous positive result. Hence, two propositions A and B are, in general, not simultaneously decidable. This is only the case if the corresponding properties $P(A)$ and $P(B)$ are commensurable. In this case we will also call the propositions A and B "commensurable".

Elementary propositions $A, B, \ldots$ are thus, in general, incommensurable, i.e. not simultaneously (jointly) decidable. If proposition A, say, was shown to be true, then after a proof attempt of B and irrespective of the result (B or $\bar{B}$), a new proof attempt of A will, in general, not lead to the previous result. Instead, this result is available after the B-test only if A and B are commensurable. In a sequence of proofs the results are only *restrictedly available*, where the restrictions are given by incommensurabilities. For the definition of the connectives the restricted availability is very important.

These restrictions do not invalidate the definitions of the negation $\neg A$ and the sequential conjunction $A \sqcap B$ which are defined here by the same proof-trees as in classical language. The negation is defined by one proof attempt and the sequential conjunction by two subsequent proof attempts. In both cases the restricted availability does not matter since repeated proof attempts do not occur here. However, the restrictions do matter if one tries to define the other connectives.[13]

[11] [Mi 78,86]

[12] [BLM 91]

[13] [Mi 78], [Sta 80]

The *logical conjunction* $A \wedge B$ is defined here by the same attack-and-defence scheme as in classical language. Since unrestricted availability is no longer given here, the proof-tree for

connective	denotation	attacks	defences
$A \wedge B$	A and B	$A?$, $B?$	$A!$, $B!$

$A \wedge B$ consists of an infinite number of steps and cannot be reduced to two steps as in classical language. If, however, A and B were commensurable, then it would again be possible to reduce the proof-tree to one A-proof and one B-proof. In order to arrive generally at a finite proof-tree we make use of the commensurability proposition $k(A, B)$, which is defined to be true if and only if A and B are commensurable. The counter–proposition is denoted here by $\bar{k}(A, B)$.[14] The logical conjunction $A \wedge B$ is then true if in addition to A and B also $k(A, B)$ is shown to be true. Hence, we have a proof-tree with three subsequent tests at time values t_1, t_2, t_3. Since the conjunction $A \wedge B$ is understood as a simultaneous connective, the time differences $t_3 - t_2$ and $t_2 - t_1$ must be sufficiently small.

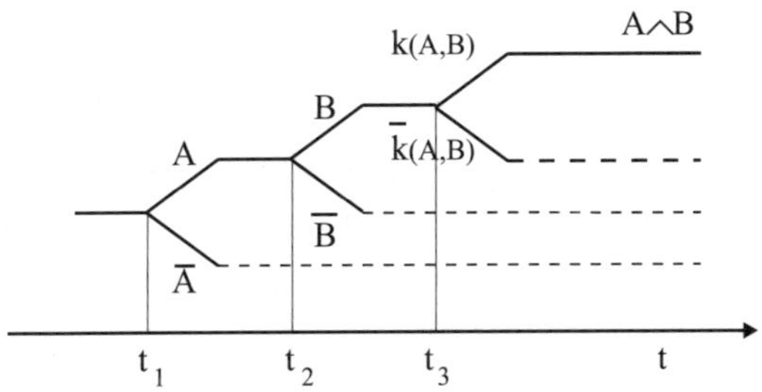

Fig. 5. Proof–tree for the logical conjunction

The commensurability propositions $k(A, B)$ and $\bar{k}(A, B)$ are contingent propositions whose truth must be shown by a convenient sequence of measurements. We will not go into detail here. By means of the commensurability propositions $k(A, B)$ and $\bar{k}(A, B)$ one can also define the logical disjunction $A \vee B$ and the material implication $A \to B$ by proof trees with a finite number of steps. Similarly as in classical language, we can define here binary relations between propositions. The proof equivalence $A \equiv B$ means that A can be replaced in any proof–tree by B without thereby changing the result of the proof–tree. The binary relation of value equivalence $A = B$ means that A is true if and only if B is true.[15] The relation of implication $A \leq B$ can

[14] [Mi 78,86]

[15] If two propositions are *proof equivalent*, then they are also *value equivalent*. The inverse is not generally true. However, in classical language the two equivalence relations coincide.

then be defined by $A \equiv A \wedge B$. Finally, we mention that again $A \rightarrow B$ is true if and only if $A \leq B$ holds and that the commensurability proposition is true if and only if $A \leq (A \wedge B) \vee (A \wedge \neg B)$ holds.

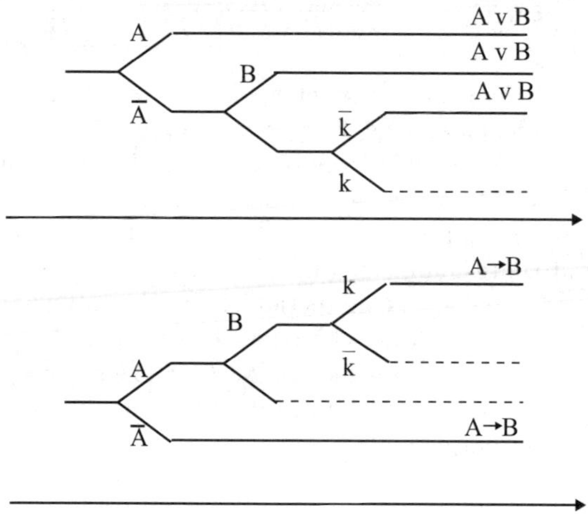

Fig. 6. Proof–trees for the logical disjunction and the material implication.

The full language S_Q of quantum physics can then inductively be defined by the set $S_Q^{(e)}$ of elementary propositions, the commensurability propositions k and $\bar{k}$ and the connectives mentioned. Together with the relations "$\equiv$", "$=$",and "$\leq$" the language S_Q reads

$$S_Q = \{S_Q^{(e)}; k, \bar{k}; \sqcap, \wedge, \vee, \rightarrow, \neg; \equiv, =, \leq\}.$$

4.3 Quantum Logic

Quantum logic is the formal logic of quantum language and its syntax. The reduced possibilities of proving propositions are, in particular, important for those propositions which are true, irrespective of the elementary propositions contained in them, i.e., for formally true propositions. It turns out that in quantum language there are less formally true propositions than in classical language. In order to make this more preliminary information more precise we will express the totality of all formally true propositions of quantum language by a calculus, the calculus of quantum logic.

There are, first of all, many formally true propositions of classical language which are also formally true in quantum language. The value definiteness of elementary propositions implies that also all finitely connected propositions are value–definite, i.e., the proposition $A \vee \neg A$, the *tertium non datur law*, is formally true. The precondition that measurements are repeatable in principle implies that $k(A, A)$ is always true and hence $A \rightarrow A$, the *law of identity*, is formally true. In a similar way, it follows that $\neg(A \wedge \neg A)$, the *law of contradiction*, is formally true in quantum logic. The three cases mentioned are not very surprising since these formally true propositions contain only one proposition A. Hence, commensurability problems cannot appear. There are, however, also formally true propositions in quantum logic which contain two ore more elementary propositions, where nothing is presupposed about their mutual commensurability. An example of this kind is the proposition $(A \wedge (A \to B)) \to B$, the *modus ponens law*, which is formally true in quantum logic irrespective of the truth or falsity of the commensurability proposition $k(A, B)$.

More important for the characterisation of quantum logic are those propositions which are formally true in classical logic but not in quantum logic. The shortest and in addition most important proposition which is formally true in classical logic but not in quantum logic is the proposition $A \to (B \to A)$. In classical logic the proof-tree for $A \to (B \to A)$ contains only branches of success.

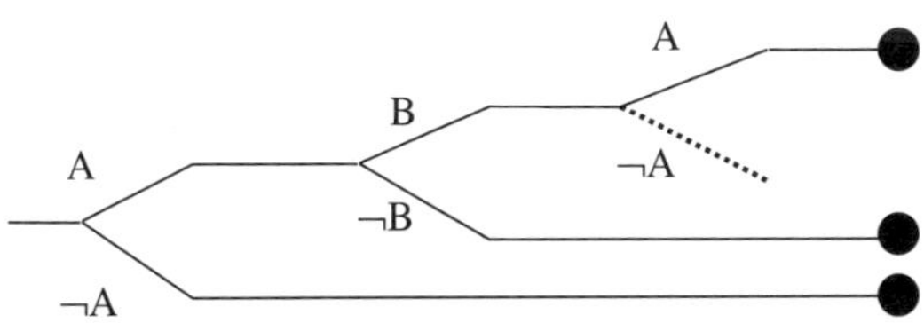

Fig. 7. Proof–tree for $A \to (B \to A)$ in classical logic

In quantum logic the situation is more complicated since the proof–tree for the material implication contains also the test of commensurability propositions $k(A, B)$. For this reason, the proof-tree for $A \to (B \to A)$ contains 5 branches, but only 3 branches of success. Only if the commensurability of A and B were presupposed would the proof-tree contain only successful branches. This means that, in general, the proposition $A \to (B \to A)$ is not true and thus not formally true.

The totality of all propositions which are formally true even under the restrictions that are provided by the commensurability tests is called quantum logic. There are –as in classical logic– infinitely many propositions which are formally true in the sense of quantum logic. They can be summarised in a quantum–logical calculus L_Q, which contains "beginnings" $\Rightarrow A \leq B$ and "rules" of the form $A \leq B \Rightarrow C \leq D$. For the formulation of this calculus we make again use of the two special propositions V (*verum*) and Λ (*falsum*) such that for all propositions $A \in S_Q$ the relations $\Lambda \leq A \leq V$ hold. If $A \to (B \to A)$ is true then the relation $A \leq (B \to A)$ holds. $A \leq B \to A$ implies $B \leq A \to B$ and *vice versa* and $A \leq B \to A$ holds if and only if $k(A, B)$ is true. Hence, in a calculus of quantum logic the commensurability

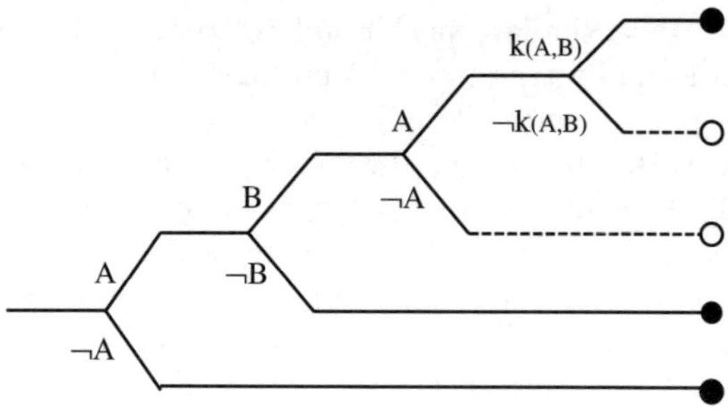

Fig. 8. Proof–tree for $A \to (B \to A)$ in quantum logic

propositions $k(A, B)$ can be eliminated by this implication and will no longer appear in its final formulation. The calculus L_Q of quantum logic reads:

The calculus of quantum logic

1.1 $\Rightarrow A \leq A$
1.2 $A \leq B; B \leq C \Rightarrow A \leq C$

2.1 $\Rightarrow A \wedge B \leq A$
2.2 $\Rightarrow A \wedge B \leq B$
2.3 $C \leq A; C \leq B \Rightarrow C \leq A \wedge B$

3.1 $\Rightarrow A \leq A \vee B$
3.2 $\Rightarrow B \leq A \vee B$
3.3 $A \leq C; B \leq C \Rightarrow A \vee B \leq C$

4.1 $\Rightarrow A \wedge (A \to B) \leq B$
4.2 $A \wedge C \leq B \Rightarrow A \to C \leq A \to B$
4.3 $A \leq B \to A \Rightarrow B \leq A \to B$
4.4 $B \leq A \to B; C \leq A \to C \Rightarrow B^\star C \leq A \to B^\star C$
$^* \in \{\wedge, \vee, \to\}$

5.0 $\Rightarrow \Lambda \leq A, \Rightarrow A \leq V$
5.1 $\Rightarrow A \wedge \neg A \leq \Lambda$
5.2 $A \wedge C \leq \Lambda \Rightarrow A \to C \leq \neg A$
5.3 $A \leq B \to A \Rightarrow \neg A \leq B \to \neg A$
5.4 $\Rightarrow V \leq A \vee \neg A$

Comparing classical logic and quantum logic we find that a proposition which is formally true in quantum logic is also true in classical logic. The inverse relation is, however, not true. There are infinitely many formally true propositions in classical logic which are not formally true in the sense of quantum logic. The reason for this important result is the restricted availability of quantum propositions or the general incommensurability of two arbitrary quantum propositions. Indeed, if the mutual commensurability of all pairs of propositions were presupposed, then the calculus L_Q of quantum logic would agree with the calculus L_C of classical logic. Hence, the quantum–physical restrictions lead to a logical system which is weaker than classical logic but obviously more general. Quantum logic is applicable to classical propositions as well as to quantum propositions. On account of the commensurability of classical propositions the laws of classical logic which hold for classical propositions are completely reproduced also in quantum logic. For this reason, quantum logic is the most general, universal logic of physical propositions.

The Lindenbaum–Tarski algebra of the calculus L_Q is given by a complete orthomodular lattice $\bar{L}_Q$. Subsets of mutual commensurable propositions constitute Boolean sublattices $\bar{L}_B^{(i)} \subseteq \bar{L}_Q$ of the lattice $\bar{L}_Q$.[16] From an algebraic point of view, the orthomodular lattice $\bar{L}_Q$ is much more complicated than the corresponding Boolean lattice $\bar{L}_B$. The orthomodular lattice of quantum logic can be further specified. It is freely generated by a finite (or infinite) number of elementary propositions. Moreover, if the entire quantum language S_Q refers to one quantum system, then the lattice $\bar{L}_Q$ is atomic and the atoms which correspond to pure states provide a maximal information about the system. In addition, the covering law is alsofulfilled by the lattice $\bar{L}_Q$.[17] Furthermore, the *Hilbert lattice* $\bar{L}_H$ of projection operators in Hilbert space[18] can be obtained from this lattice $\bar{L}_Q$ by adding the *Solér law*, the operational meaning of which is, however, still open.[19]

5 Conclusion

Quantum logic is weaker than classical logic and process semantics is weaker than a realistic semantics. These new structures do not contain new experimental information. On the contrary, for the formulation of quantum logic we left out all those supposed empirical results of classical physics which cannot be justified by quantum physics. Hence, quantum logic could have been discovered even without any quantum–physical knowledge merely by very cautious argumentation. Of course, there are situations which allow for classical logic. This is the domain of classical physics. However, the application of classical logic requires a justification in every individual case. Neither the realistic semantics nor the classical logic may be considered as the normal case. Hence, the relaxation of this "normal case" must not be justified

16 [BN 36]
17 [Sta 84]
18 [Neu32, BN 36]
19 [Sol 95]

but the “normal case” itself requires a legitimation. Quantum logic is the most general structure which is free from special presuppositions of classical physics.

References

1. P. Busch, P. Lahti, P. Mittelstaedt: *The Quantum Theory of Measurement* (Springer, Heidelberg 1991) (2nd edn. 1996)
2. N. Bohr: ‘The Quantum Postulate and the Recent Development of Atomic Theory’. In: *Atti del Congresso Internationale del Fisici, Como, 11-20, September 1927*, (Zanichelli, Bologna 1928) pp. 565–8
3. P. Mittelstaedt: *Philosphical Problems of Modern Physics* (Reidel, Dordrecht 1976)
4. P. Mittelstaedt: *Quantum Logic* (Reidel, Dordrecht 1978)
5. P. Mittelstaedt: *Sprache und Realität in der modernen Physik* (BI-Wissenschaftsverlag, Mannheim 1986)
6. P. Mittelstaedt: *The Interpretation of Quantum Mechanics and the Measurement Process* (Cambridge University Press, Cambridge 1998)
7. J. von Neumann: *Mathematische Grundlagen der Quantenmechanik* (Springer Verlag, Berlin 1932)
8. G. Birkhoff, J.v. Neumann: ‘The Logic of Quantum Mechanics’. Annals of Mathematics. **37** (1936) pp. 823–43
9. E.-W. Stachow: ‘Logical Foundations of Quantum Mechanics.’ Int. Journ. of Theoretical Physics. **19** (1980) pp. 251–304
10. E.-W. Stachow: ‘Structures of a Quantum Language for Individual Systems’. In: Recent Developments in Quantum Logic ed. by P. Mittelstaedt and E.-W. Stachow (BI-Wissenschaftsverlag, Mannheim 1984)
11. M.P. Solèr: ‘Characterisation of Hilbert Spaces by Orthomodular Lattices’. Communications in Algebra **23(1)** (1995) pp. 219–243

Experimental Approach to Quantum-Logical Connectives

Ernst-Walther Stachow

1 Introduction

This paper concerns an alternative logic, called "quantum logic", the general framework of which is presented by Peter Mittelstaedt in his contribution "Does Quantum Physics require a New Logic?". Starting with the pragmatic possibilities of the speaker of a language about the properties of a quantum-physical system who is also an observer of the system, a proof process semantics can be developed and a formal language can be constituted, the internal structure of which gives raise to the "new" quantum logic.

The pragmatic possibilities of the speaker-observer include the possibilities to prove or disprove elementary propositions of the language by experiments, as well as the possibilities of experimental setups in space-time. In the following, we shall analyse experimental proofs of elementary propositions in more detail, but still in a very schematic framework, and show that by means of particular setups of experiments connectives between propositions may be defined leading to a language which requires the "new" quantum logic.

2 Experimental Proofs of Elementary Propositions

Although this will not be necessary for the following experimental approach, we adopt a "realistic" point of view implying that the propositions A of the language to be established relate to properties $P(A)$ of a physical system S. Elementary propositions $A(S)$ are verified or falsified by yes-no experiments. If the result of such an experiment is "yes", we say that the proposition $A(S)$ is *true* and that the property $P(A)$ pertains to the system S. Whenever the result is "no", we say that the proposition $A(S)$ is *false* and that a counter-property $P(\bar{A})$ pertains to the system S.

A yes-no experiment for an elementary proposition $A(S)$ may be considered to comprise two experimental acts on a quantum-physical system S. The first experimental act is schematically shown in Fig. 1.

The incoming system S interacts with an experimental device illustrated by a rectangular box which "splits" the path of S into two alternative paths x and y [Sta 81, Al 93]. The device acts on the system S in such a way that path x is associated with a property $P(A)$ and path y is associated with

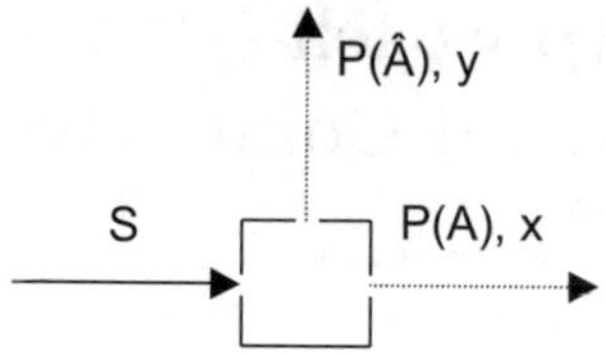

Fig. 1. First experimental act

the counter-property $P(\bar{A})$. What "$P(A)$ is associated with x" and "$P(\bar{A})$ is associated with y" mean is explained in the following.

By means of a subsequent experimental act, schematically illustrated by the round box in Fig. 2, the quantum-physical system S is registered to be in path x or in path y.

If the answer of the registration act in path x is "yes", as shown in Fig. 2, the proposition $A(S)$ is *true* and, hence, the property $P(A)$ pertains to S. If the answer is "no", as illustrated in Fig. 3, the proposition A is false and the counter-property $P(\bar{A})$ pertains to S.

It turns out that it makes no difference whatsoever in an arbitrary experimental context to perfom the registration act in path y. If the result of this registration is "yes", we can say that $A(S)$ is *false* and $P(\bar{A})$ pertains to S. If the result is "no", we can say that $A(S)$ is true and $P(A)$ pertains to S. Both experimental setups with registration in x and registration in y, respectively, are *proof-equivalent* with respect to the proposition $A(S)$.

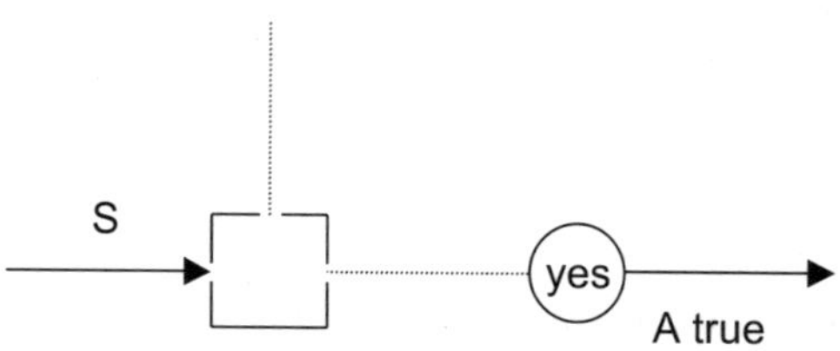

Fig. 2. Subsequent experimental act, A true

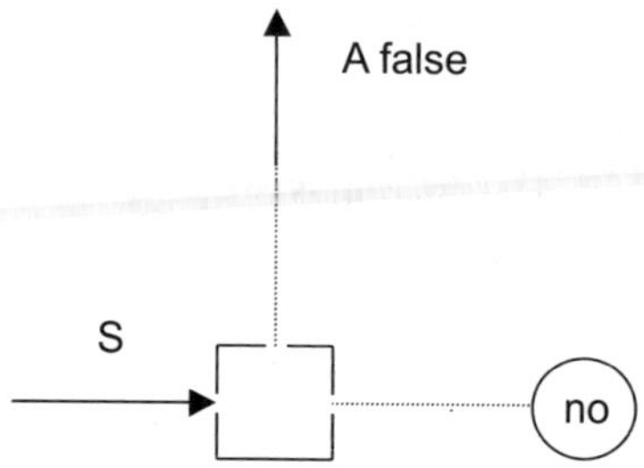

Fig. 3. Subsequent experimental act, A false

Yes-no experiments schematically described above may be realized by various devices for many kinds of quantum-physical systems, like electrons, neutrons, and atoms. For a spin-1/2-system, the first experimental act can be realized by means of an inhomogeneous magnetic field in a Stern–Gerlach–experiment. Here, the paths x and y are associated with the spin-properties of the system "spin-up" and "spin-down", respectively, in a certain direction. In order to determine the truth-value of such a spin-proposition, a registration-act is performed in one of the paths x or y.

Yes-no experiments of the kind described above are (nearly) ideal in the sense that they are proof-equivalent to an arbitrary repetition of themselves. *Proof-equivalence* here again means that no difference whatsoever can be established between one experiment and an arbitrary repetition of it in any experimental context. Concerning the first experimental act leading to the split-paths, a repetition of this act in path x, for instance, reproduces path x but does not produce path y. The second experimental act of registration is assumed to localize the system S in the path x or y without destroying it.

Yes-no experiments which have been considered above and will be used in the following are not restricted to path-split experiments. Instead of alternative paths x and y, other registrable alternatives, for instance energy levels, may be established by appropriate devices. Schematically, this does not lead to any difference to our above representations.

After registration of the respective alternative, for instance two alternative paths used as an example throughout this paper, the proposition $A(S)$ is either true or false. Classical language and logic can then be used for describing the alternative properties $P(A)$ and $P(\bar{A})$. In particular, the negation $\neg A$ can be defined to be true if and only if A is false and *vice versa.* Using the (classical) disjunction and conjunction we have that, after a yes-no experiment for "A", the connected proposition $A \cup \neg A$ is always true and $A \cap \neg A$ is always false. Furthermore, the double-negation law holds.

When, after performing a yes-no experiment with respect to the property $P(A)$ including registration, the speaker-observer does not know the result of the experiment, i.e., whether the elementary proposition $A(S)$ is true or false, we represent this situation by the diagram shown in Fig. 4.

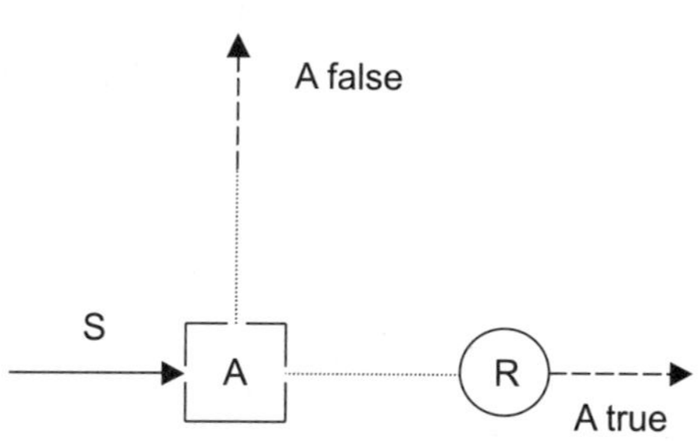

Fig. 4. Yes-no experiment with unknown result

3 Experimental Proofs of Connected Propositions

Let us now consider the experimental setup shown in Fig. 5.

In this setup, the direction of the paths x and y resulting from the proof device for $A(S)$ is changed, for instance by means of mirrors, and the deflected paths are joined to the single path x'. The whole setup is represented in Fig. 5 by a box with one outlet for path x'. The "elementary" proposition defined by this box not including any registration act and associated with path x', which is the join of the two original paths x and y, may be denoted by $A \vee \neg A$. Since the box has only one outlet and it is assumed that a deflection and rejoining of the original paths x and y do not lead to a loss of the system S, the proposition $A \vee \neg A$ is always true. This means that the negation $\neg(A \vee \neg A)$, which we may use to define the "elementary" proposition $A \wedge \neg A$, is always false.

In the same way we may, of course, also deflect and join the paths x and y in Fig. 4 and associate the proposition $A \cup \neg A$ with path x'. This experimental setup including a registration act and the one shown in Fig. 5, however, and this is the crucial point of our following results, are not proof-equivalent in the sense defined above.

This can be seen by means of the probability argument considered in the contribution by Peter Mittelstaedt in Sect. 3. We will be able to make use of this argument after having introduced experimental setups defining sequential propositions as follows.

Let us consider the experimental arrangement schematically shown in Fig. 6.

Here a proof of the proposition A including registration is followed by a proof of another elementary proposition B in path x. By means of this experimental setup according to Fig. 6, the sequential conjunction $A \circ B$ and its negation $\neg(A \circ B)$ are defined. $A \circ B$ and $\neg(A \circ B)$, however, do not represent a "pure" alternative, since $\neg(A \circ B)$ includes by itself the alternatives $\neg A$ and $A \circ \neg B$ one of which being true if and only if $\neg(A \circ B)$ is true.

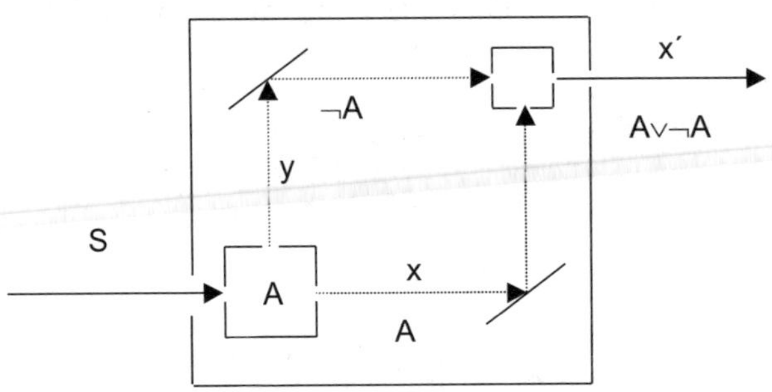

Fig. 5. Experimental setup for $A \vee \neg A$

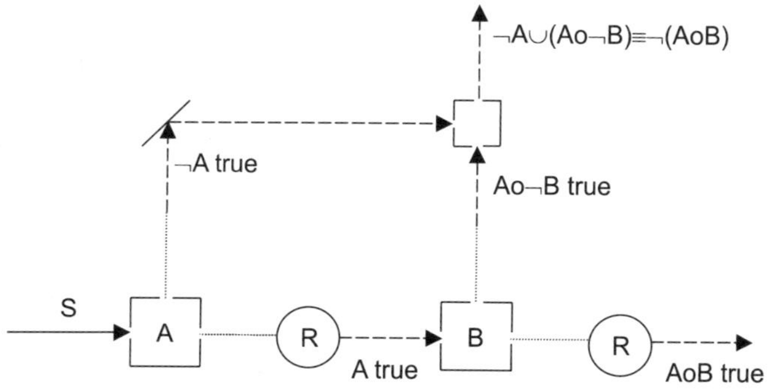

Fig. 6. Yes-no experiment for $A \circ B$

A subsequent proof of A does not, in general, reproduce the truth of A but generates the possibility of A to be false, i.e., of the y path. This is because of the general incommensurability of quantum-physical propositions, which we will consider below.

Starting from the set $S_Q^{(e)} = \{A, B, \ldots\}$ of elementary propositions about a quantum-physical system S, the always-true "elementary" proposition V, the always-false "elementary" proposition Λ and the (classical) logical disjunction $A \cup \neg A$ for elementary propositions A, a first set of sequential propositions can recursively be defined as $S_Q^{(s1)} := \{S_Q^{(e)}, V, \Lambda, A \cup \neg A$ with $A \in S_Q^{(e)}; \neg, \circ\}$.

With this language we can now describe the slightly more complicated experimental setup shown in Fig. 7.

If A and B are not commensurable, there is the possibility that the result of the proof of B in the right part of Fig. 7 is "no". The conditional probability

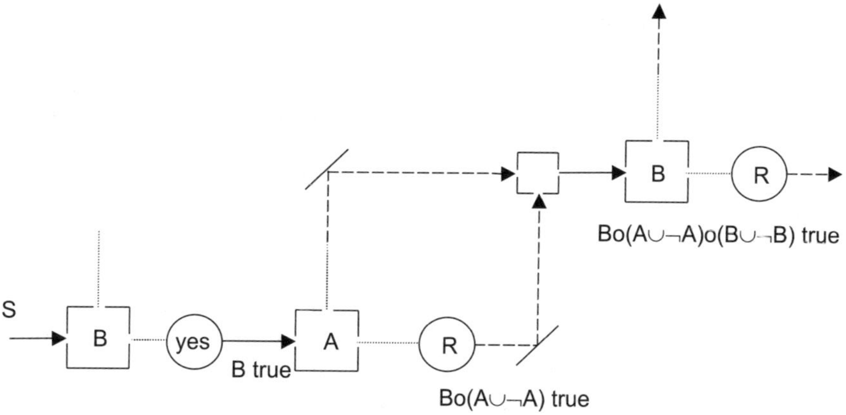

Fig. 7. Experimental setup for $B \circ (A \cup \neg A) \circ (B \cup \neg B)$

that, if B is true, the system S running through the experimental setup of Fig. 7 has the property $P(B)$ after the second proof of the proposition B is given by

$$p(B;(A \cup \neg A) \circ B) = p(B;A) \cdot p(B \circ A;B) + p(B;\neg A) \cdot p(B \circ \neg A;B)\,.$$

If B refers to the property of a spin-1/2-system to have "spin-up" in a first direction and B refers to the property that the spin of the system is "spin-up" in a direction perpendicular to the first direction, the probability is $p(B;(A \cup \neg A) \circ B) = 1/2$.

Now comes the probability argument demonstrating that the proposition $A \cup \neg A$ is not proof-equivalent to the always-true "elementary" proposition $V = A \vee \neg A$. Namely, if we replace the middle part of the experimental setup in Fig. 7 establishing the (classical) disjunction $A \cup \neg A$ by the arrangement shown in Fig. 5 establishing the "elementary" proposition $A \vee \neg A$, the result "no" of the second proof of B on the right side of Fig. 7 is no longer possible. The respective conditional probability is

$$p(B;(A \vee \neg A) \circ B) = 1\,. \tag{1}$$

This means that, in the particular context of the whole experimental setup shown in Fig. 7, replacement of the proof procedure for $A \cup \neg A$ by the proof procedure for $A \vee \neg A$ makes a difference. Therefore, both propositions $A \cup \neg A$ and $A \vee \neg A$ must be distinguished, although they are always proved to be true and therefore locally (this means not taking into account possible experimental contexts) indistinguishable. In this case we say that two propositions like $A \cup \neg A$ and $A \vee \neg A$ are value-equivalent.

We have seen that, after a proof of $A \cup \neg A$, either the proposition A or $\neg A$ is true. This leads to the question whether in the two-paths diagram of Fig. 5 it can also be *assumed* that either A associated with path x or $\neg A$ associated with path y is true, i.e., either the property $P(A)$ or the counter-property $P(\bar{A})$ pertains to the system S. Generally speaking the question is whether an elementary proposition $A(S)$ can be associated with a truth-value before the second experimental act of registration in a yes-no experiment. The answer is *no* for the above-stated reasons, if we want to describe a quantum-physical system S by means of a history of observable properties pertaining to the system S during particular time periods [Sta 84]. Since the experimental setups defining $A \vee \neg A$ and $A \cup \neg A$ are contextually distinguishable, i.e., not proof-equivalent, it would not be sufficient, in general, to characterize a system S during a particular time period by objective properties $P(A)$ or $P(\bar{A})$. It would rather be necessary to take into account the experimental setup of the situation in which $A(S)$ is assumed to be true as an additional parameter. An objective description of a quantum-mechanical system S independent of the choice of observation of a particular property $P(A)$ would not be possible. It should be noted that the objective language considered here is constituted on the basis of observation and experimental setups. After its constitution, the language, however, no longer requires additional references to experimental situations. If, on the other hand, one were not reluctant to give up an objective description, for instance by considering "preferred" observables [Bub 97], a truth-value assignment to the proposition $A(S)$ in an experimental situation as shown in Fig. 5 would be possible.

We also assume that a quantum-mechanical system S is exclusively characterized by propositions $A(S)$ defined by the above-discussed yes-no experiments. If one is willing to allow further parameters which are not observable in the above sense to determine a system S, as for instance Bohmian mechanics does [Bohm 52, BH 93], these parameters could also be used to distinguish the situation of S in an experimental setup for the proposition $A \vee \neg A$, on the one hand, and in an experimental arrangement establishing the proposition $A \cup \neg A$, on the other hand, such that a truth-value could be associated with A in the large box shown in Fig. 5.

An objective description of the history of a quantum-physical system S, which can be extended to a relativistic description of S in space-time [Mi 83], also excludes a "delayed-choice" interpretation [Whe 94] according to which by means of the decision of the observer to perform a registration act in the middle part of the experiment shown in Fig. 7 "after the system S has passed the box" it "shall have come by one of the paths x or y", with $P(A)$ or $P(\bar{A})$ pertaining to S "before registration".

The question whether the proposition A can be associated with a truth-value already before the registration act in a yes-no experiment, however, has no impact on the above conclusion that the always-true "elementary" proposition $A \vee \neg A$ has to be distinguished from the (classical) logical disjunction $A \cup \neg A$.

The result that the truth of a proposition defined by the join of two alternative paths does not imply that one of the propositions associated with the paths is true already indicates that the logic of such propositions including the connective "$\vee$" is not classical logic but a "new" logic. In the following we will develop this logic.

As a further step to this end we introduce another sequential proposition defined by the experimental set-up in Fig. 8 and denoted by $A \sqcap B$.

If $A \sqcap B$ is true, we can say that A was true and B is true. However, if $A \sqcap B$ is false or $\neg A \sqcup \neg B$ is true, we cannot say that $\neg A$ was (and is) true or $\neg B$ is true for the same reasons as they apply to the connective '$\vee$'. Since no registration act is performed between the boxes for A and B, the sequential conjunction $A \sqcap B$ is defined by a "pure" alternative like an elementary proposition and is represented by a box in Fig. 8, as distinguished from the sequential conjunction $A \circ B$.

If we join a further box for A after $A \sqcap B$ in the x path, this box, in general, generates an x and an y path associated with $P(A)$ and $P(\{A\})$, respectively, like a proof of A subsequent to $A \circ B$.

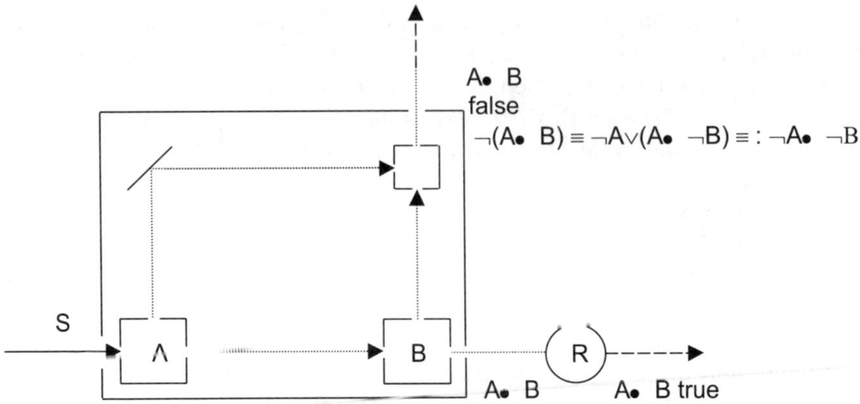

Fig. 8. Yes-no experiment for $A \sqcap B$

After having established the sequential conjunctions $A \circ B$ and $A \sqcap B$, the subpropositions A and B, which necessarily refer to subsequent time periods, we are now interested in the conditions under which the subpropositions A and B can be referred to the *same* time period so that it can be said that A and B are *simultaneously* true. Since a subsequent proof process of A and B, i.e., of $A \sqcap B$, in general is not sufficient to establish the simultaneous truth of A and B, we have to repeat the proof processes for A and B in order to test whether the proof results of A and B will remain the same or may change. If the truth values remain the same in an arbitrary number of alternating proof processes for A and B, we say that A and B are *commensurable.* If the proof results may change, we say that A and B are *incommensurable*. This means that, after a first proof process for A and a subsequent proof process B, the commensurability of A and B has to be tested to define a conjunction $A \wedge B$ which is true if and only if A and B are simultaneously true. This conjunction, which we call "quantum–logical conjunction", is defined by the experimental setup shown in Fig. 9, which includes an infinite number of proof processes for A and B in order to test their commensurability.

The alternative outcome of the proof process for $A \wedge B$, associated with path y, which defines $\neg(A \wedge B)$, is the join of paths associated with $\neg A$ and $\neg B$, respectively, for which we use the symbol "$\vee$" already introduced in Fig. 5. We call this connective the "quantum logical disjunction". If $A \vee B$ is true, it cannot be concluded that either A or B is true. Therefore, the proof processes for $A \wedge B$ and $A \vee B$ lead to "pure" alternatives and may be represented by a box characterizing such "pure" yes-no experiments.

The infinite proof process for $A \wedge B$ may be abreviated by a finite proof process if there exists an elementary commensurability proposition $k(A, B)$ such that $A \wedge B \equiv A \sqcap B \sqcap k(A, B)$, or if there even exists an elementary proposition C such that $A \wedge B \equiv C$.

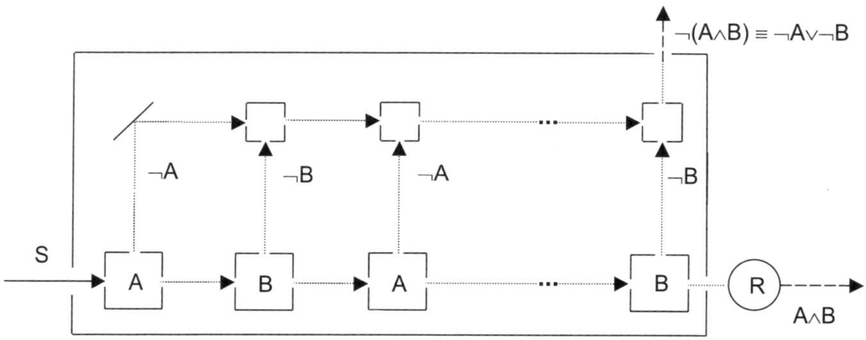

Fig. 9. Yes-no experiment for $A \wedge B$

Starting from the set $S_Q^{(e)}$ of elementary propositions, defined by elementary boxes according to Fig. 1, the set of *quantum-logically connected propositions* can be recursively defined to be the set $S_Q = \{S_Q^{(e)}; \neg, \wedge, \vee\}$.

For establishing the logic of the set S_Q we introduce the relation of implication $A \leq B$ by means of the experimental situation shown by Fig. 10, in which the proof process for B leads to an outcome along path x only but not to the alternative outcome along path y.

It can be shown that, if $A \leq B$ and $B \leq A$ for A, $B \in S_Q$, A and B are proof-equivalent. The following proof sketch makes use of the sequential connective "$\sqcap$". If A and B are commensurable, it makes no difference whether at first the proof process for A or the proof process for B is performed. This means that $A \sqcap B$ and $B \sqcap A$ are proof-equivalent.

Furthermore, if $A \leq B$, the proof process defined by the large box in Fig. 10 corresponds to the proof process for $A \sqcap B$ shown in Fig. 8. Since the proof process shown in Fig. 10 cannot be distinguished from a simple proof process for A, A and $A \sqcap B$ are proof-equivalent whenever $A \leq B$. The same argument for $B \leq A$ then implies that A is proof-equivalent to B. If A and B are proof-equivalent, we write $A \equiv B$.

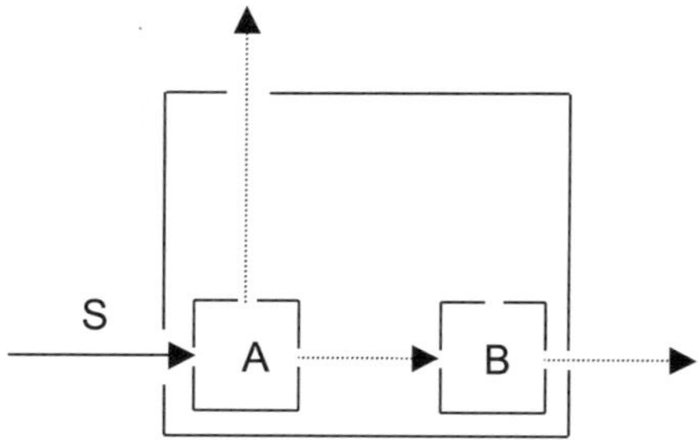

Fig. 10. Experimental setup for $A \leq B$

By means of arguments used in the above proof it can easily be shown that the following implications and rules between implications hold:

1.1 $\Rightarrow A \leq A$
1.2 $A \leq B;\ B \leq C \Rightarrow A \leq C$

2.1 $\Rightarrow A \wedge B \leq A$
2.2 $\Rightarrow A \wedge B \leq B$
2.3 $C \leq A; C \leq B \Rightarrow C \leq A \wedge B$

3.1 $\Rightarrow A \leq A \vee B$
3.2 $\Rightarrow B \leq A \vee B$
3.3 $A \leq C; B \leq C \Rightarrow A \vee B \leq C$

4.1 $\Rightarrow A \wedge \neg A \equiv L$
4.2 $\Rightarrow A \vee \neg A \equiv V$
4.3 $\Rightarrow \neg(\neg A) \equiv A$
4.4 $A \leq B \Rightarrow \neg B \leq \neg A$

The proof of the last rule deserves the following remarks. If $\neg B \leq \neg A$, we have the experimental situation shown in Fig. 11.

If the proof process for $\neg A$ had an A-true-outcome, we could join a commensurability test to this outcome. Since $A \leq B$, the commensurability would be verifiable with A-true-outcomes and B-true-outcomes. This, however, would contradict the previous B-false-outcome of the first proof process for $\neg B$. Therefore, there is no A-true-outcome of the second proof process for $\neg A$ in Fig. 11.

The above calculus would lead to classical logic if the distributivity axiom $A \wedge (B \vee C) \leq (A \wedge B) \vee (A \wedge C)$ were added. However, this axiom cannot be verified in quantum logic, as we have already seen in the example $A \sqcap (B \vee \neg B) \equiv A \wedge (B \vee \neg B)$. Only if A and B as well as A and C are commensurable does the distributivity axiom hold.

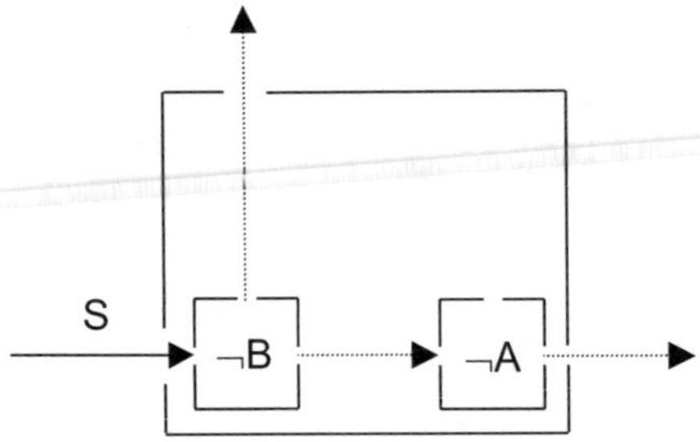

Fig. 11. Experimental setup for $\neg B \leq \neg A$

In order to show this, it will be sufficient to prove the rule

5. $A \leq B \Rightarrow B \wedge (\neg B \vee A) \equiv B \wedge A$.

The proof can be seen from the setup shown in Fig. 12.

The proof process for $\neg B \vee A$ in the particular setup is reduced to the box shown in Fig. 12, since A and B are commensurable because of $A \leq B$. The outcomes of this proof process are $B \wedge A$ and $B \wedge \neg A$, respectively. Since B and $B \wedge A$ are commensurable, the proof process for $B \wedge (\neg B \vee A)$ reduces to the large box shown in Fig. 12. Since the positive outcome of this box is $B \wedge A$, the negative outcome is $\neg(B \wedge A)$. This outcome is the join of $\neg B$ and $B \wedge \neg A$, hence, $\neg B \vee (B \wedge \neg A) \equiv \neg(B \wedge (\neg B \vee A))$. Therefore, it follows that $B \wedge (\neg B \vee A) \equiv B \wedge A$.

It immediately follows that the Lindenbaum-Tarski algebra of the above calculus, including rule 5, is given by a complete orthocomplemented quasi-modular (orthomodular) lattice. The lattice properties are given by axioms and rules 1.1 to 3.3. Axioms and rules 4.1 to 4.4 establish the orthocomplement whereas rule 5 establishes the quasi-modularity.

As mentioned in Peter Mittelstaedts contribution in paragraph 4.3, the Hilbert-space representation of the above calculus of quantum logic can be obtained by adding atomicity, the covering law, and the Solèr law. Whereas the atomicity and the covering law can be obtained by the possibility to describe the history of a quantum-physical system S by a sequence of states

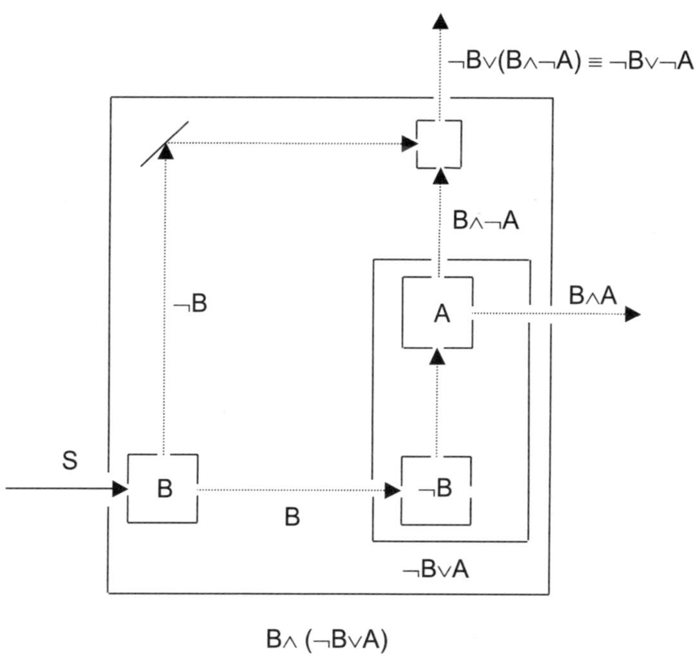

Fig. 12. Experimental setup for proving rule 5

[Sta 85], the meaning of the Solèr law within the quantum-logical approach is still open.

Although we have shown that the above calculus of quantum logic is consistent with respect to the semantics of experimental setups defining proof processes for elementary propositions and quantum-logically connected propositions, a proof of its completeness, i.e., that all implications defined by Fig. 10 can be derived by means of the calculus, is still outstanding.

We will conclude this paper with a few remarks concerning extensions of quantum logic to sequential propositions and proof processes including registration acts.

Starting from the set S_Q of quantum-logically connected propositions, the set of sequentially connected propositions can be recursively defined to be the set $S_Q^{(s)} = \{S_Q; \neg, \sqcap, \sqcup\}$. The language and logic of the set $S_Q^{(s)}$ has been investigated for a semantics which uses attack-and-defence schemes for the definition of the connectives in [Sta 80]. The Lindenbaum–Tarski algebra of this sequential quantum logic can be characterized by a Baer*-semigroup.

In the Hilbert space representation, the sequential proposition $A \sqcap B$ is represented by the product $E(B) \times E(A)$ of bounded linear operators on a Hilbert space. The set S_Q of quantum-logically connected propositions which are idempotent in the sense that $A \equiv A \sqcap A$ is represented by the set of projection operators on the Hilbert space.

If registration acts according to Fig. 4 are included in the proof processes, connected propositions, the subpropositions of which having simultaneous truth values, can be defined by using commensurability tests as shown in Fig. 13.

If $A \cap B$ is true, then again A and B are simultaneously true, as this is the case for $A \wedge B$. When, however, $A \cap B$ is false, which means that $\neg A \cup \neg B$ is true, at least one of the two subpropositions $\neg A$ and $\neg B$ is true. In this respect a (classical) logical disjunction "$\cup$" differs from the quantum-logical disjunction "$\vee$" for the same reasons as discussed in connection with Fig. 7.

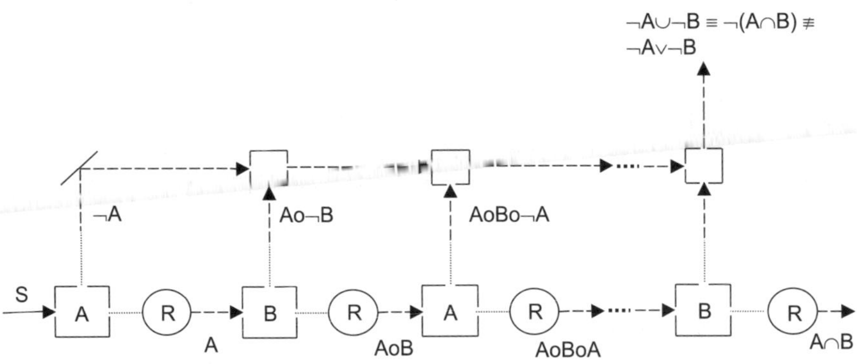

Fig. 13. Yes-no experiment for $A \cap B$

Although $A \cup B$ and $A \vee B$ are not proof-eqiuvalent, it can easily be verified that they are value-equivalent since $A \vee B$ can be proved to be true or false whenever $A \cup B$ is proved to be true or false, respectively, and *vice versa*. This means that $A \cup B \leq A \vee B$ and $A \vee B \leq A \cup B$ hold, which, however, does not imply $A \cup B \equiv A \vee B$ since $A \cup B \notin S_Q$.

The set $S_Q^{(c)} = \{S_Q; \neg, \cap, \cup\}$ can be recursively defined by means of proof processes shown in Fig. 13 on the basis of the set S_Q of quantum-logical propositions. The logical structure of the set $S_Q^{(c)}$ is classical in the sense that the distributivity axiom holds. It has, however, to be taken into account that, although the registration acts of each step of the proof process according to Fig. 13 force the propositions A and B to be simultaneously true or at least one of them to be false, truth values for A and/or B cannot be associated with the quantum-physical system S for every time period. Therefore, the axioms of contradiction and *tertium non datur* corresponding to axioms 4.1 and 4.2, respectively, of the above calculus of quantum logic do not hold. The Lindenbaum-Tarski algebra of this generalized (classical) logic turns out to be a generalization of a pseudo-Boolean algebra [Sta 84].

4 Conclusion

It has already be shown that attack-and-defence schemes used in constructive logics, in which defences of elementary propositions are performed by yes-no experiments with respect to quantum-mechanical systems, give rise to a "new" quantum logic, if it is assumed that every logically connected proposition is value-equivalent to an elementary proposition which is proved instead of the connected proposition [Sta 80]. In our present approach we have systematically considered experimental setups of yes-no experiments generating registrable alternatives associated with alternative properties of a quantum-physical system without registration (wherein the boxes illustrating such yes-no experiments). Starting from elementary yes-no experiments defining elementary propositions, subsequent arrangements of such elementary yes-no experiments and the possibilities to join the generated alternatives can be used to define yes-no experiments which define sequential propositions and finally, by taking into account commensurability tests, logically connected propositions the subpropositions of which can be referred to the same time periods. The "circuits" of quantum-mechanical alternatives establishing quantum-logical connectives can be compared with circuits used in classical computers to establish logical connectives. Irrespective of the possibility to use this "new" quantum logic for quantum computers, quantum mechanics is using this alternative logic as it is describing quantum-physical systems by means of the Hilbert-space formalism which, apart from some exceptional cases which are excluded by Solèr's law, is a realization of quantum logic. We therefore can say that quantum logic is a good example for an alternative logic which is needed in science.

References

1. D.Z. Albert: *Quantum Mechanics and Experience* (Harvard University Press, Cambridge, Mass. 1993)
2. D. Bohm: 'A Suggested Interpretation of Quantum Theory in Terms of "Hidden Variables" '. Parts I and II, Physical Review **85** (1952) 166–79, 180–93
3. D. Bohm, B.J. Hiley: *The Undivided Universe: An Ontological Interpretation of Quantum Theory* (Routledge, London 1993)
4. J. Bub: *Interpreting the Quantum World* (Cambridge University Press, Cambridge 1997)
5. P. Mittelstaedt: 'Relativistic Quantum Logic'. Int. Journ. of Theoretical Physics, **22** (1983) pp. 293–314
6. E.W. Stachow: 'Logical Foundation of Quantum Mechanics'. Int. Journ. of Theor. Phys. **19** (1980) pp. 251–304
7. E.W. Stachow: 'Sequential Quantum Logic'. In: *Current Issues in Quantum Logic* ed. by E. Beltrametti and B.C. van Fraassen. Ettore Majorana International Science Series, Vol. 8 (Plenum Press, New York 1981) pp. 173–191.
8. E.W. Stachow: 'Classical and non Classical Limiting Cases of Quantum Logic'. In: *Reduction in Science* ed. by W. Balzer, D.A. Pearce and H.-J. Schmidt (D. Reidel, Dordrecht 1984) pp. 397–418
9. E.W. Stachow: 'Structures of Quantum Language for Individual Systems'. In: *Proceedings of the Symposium on the Foundations of Modern Physics* ed. by P. Lahti and P. Mittelstaedt (World Scientific Publishers, Singapore 1985)
10. J.A. Wheeler: *At Home in the Universe* (AIP Press, The American Institute of Physics, Woodbury, NY 1994)

From Semantics to Syntax: Quantum Logic of Observables

Vladimir L. Vasyukov

An attempt of syntactical reconstruction of a logical counterpart of algebraic formulation of quantum theory is done (cf. [11]*). It includes R. Goldblatt's orthologic* [6] *as the two-valued case. An axiomatization of such quantum logic of observables (QLO) along with the set-theoretical semantics is proposed and soundness and completeness are proved. Also a normed quantum logic of observables (NQLO) is yielded. Exploiting von Wright's temporal operation two systems of temporal quantum logic of observables (TQLO and TNQLO respectively) are obtained for which soundness and completeness is proved. In conclusion, some perspectives for such an approach are considered.*

1 Introduction: Quantum Logic of Observables as a Converse Problem of Semantics

We start with the confusion concerning the notion of a logico-algebraic approach. As a rule a subject of the study in the framework of such an approach is propositional calculi and algebras of observables which are exploited in physical theories - that is commonly accepted formulation. Moreover, nowadays in the literature on quantum logics one observes the situation when a term "quantum logic" is applied tentatively both to some propositional calculus and to algebraic constructions - usually to orthomodular lattices. The most known case turns out to be Birkhoff-von Neumann's fundamental paper from 1936 entitled "The logic of quantum mechanics" [2] in which meticulous theorist cannot find any logical but only a "pure" algebraic construction.

Indeed, there is a tradition in algebra where "logic" might means a kind of an algebraic construction. Besides, it is commonly taken for granted that logical formulation always involves algebraic (semantic) considerations, e.g. a Boolean algebra in case of classical logic, a Heyting algebra in case of intuitionistic one etc. This practice early was based on Lindenbaum's theorem, describing the general method of obtaining an algebraic counterpart for logical calculus, and recently logicians always must be able to cope with issues of algebraizability of either calculus they works with.

Meanwhile, from the very beginning there was an asymmetry in this situation. Actually, one would be always able to pass from a respective quantum logical calculus to an algebra (e.g. even though to Lindenbaum algebra of this calculus) bur the converse passage more often then not is an attempt foredoomed to failure: there is any general method for such a passage. Hence, from a logical point of view the unity of logico-algebraic approach is a kind of illusion: a logical part is in most cases algebraizable while an algebraic part is hardly to be "logizable".

Beside, a interrelationship between logic and algebra in the framework of logico-algebraic approach has one more aspect concerning a problem which should be conditionally called a problem of direct and converse task of semantics. In a nutshell, it consists in follows. A direct task of semantics in mathematical logic is a quest for some semantics for previously yielded syntax of some formal theory. For instance, a possible world semantics was called into existence by issues of modal logics, a topological one was obtained for intuitionistic logic, T-F-semantics was constructed for many-valued logics etc. A converse task of semantics is in turn an yielding of some formal system (calculus), its syntax corresponded to available (nonformal in general case) system which is treated as some accomplished semantics. In effect, a quantum logic should be considered as an instance of the solution of this task.

Accepting such a awkward definition one would be tempted to consider an algebraic formulation of quantum mechanics as an accomplished semantics which is a good starting point for the reconstruction of syntactical formulation (e.g. proof-theoretical one). Unfortunately, there are many difficulties in this way making this enterprise a very problematic.

Observe, that common usage of the term "quantum logic propositions" purports that they are observables (projective operators) for which the name "propositions" is justified by their property (as observables satisfying the equality $P^2 = P$) to have in states with null dispersion a two-point spectrum, i. e. consisting from 0 and 1. In effect, this "two-valuedness" was precisely the reason for analogy with a logical propositional calculus. So one may naturally ask: would it be possible to search for such kind of an analogy in case of observables with more complex spectrum and many-valued logical propositions?

If we set our mind on maximality of description then it ought to consider observables with the spectrum of a mostly general kind, i. e. subsets of a real line. In this case the main barrier on the way would be that as a truth domain of many-valued logics the subset of (0,1]-interval is usually chosen (or even the whole interval for Łukasiewicz's infinite-valued logic $Ł_{\aleph_0}$). But in [12] it was shown that $Ł_{\aleph_0}$ has (in some specific sense) discrete model with the ordinal type $\omega + \omega*$. Since a Boolean algebra could be embedded in MV-algebra and an algebra of quantum logic propositions (an orthomodular lattice) according to the formulation of an algebraic approach in quantum mechanics could be embedded in algebra of observables having spectrum whatsoever then one should expect that syntactically it involves satisfiability of axioms and theorem of some quantum propositional calculus in the respective "spectrally many-valued" calculus of observables resembling satisfiability of two-valued propositions in $Ł_{\aleph_0}$.

Thus, the problem consists in yielding a syntactical system corresponding to algebraic structure of quantum theory which is treated as some algebraic semantics. Since an algebra of observables with spectrum whatsoever in contradistinction of quantum propositional calculus with two-valued spectrum is considered then hereinafter this syntactical system will be called a quantum logic of observables and will be denoted as **QLO**.

As a starting point of an algebraic approach in quantum theory usually the following serves as an axiom:

Any physical system Σ corresponds to an ordered triple $(\mathcal{A}, \mathcal{S}, \langle , \rangle)$ which is formed by a set $\mathcal{A}$ of all its observables, a set $\mathcal{S}$ of all its states and a map $\langle , \rangle : (\mathcal{A}, \mathcal{S}) \to R$ which associates with every pair (A, φ) from $(\mathcal{A}, \mathcal{S})$ a real value $\langle \varphi; A \rangle$. This we interpret as an average value of observable A in a state φ (cf.[5]).

What can we say concerning observables, states and average values of observable in some state from the point of view of a logic? Intuitively in physics one means under states of a physical system the whole complex of knowledge of average values of the system in question. But something similar we meet nowadays in mathematical logics too (see e. g. Belnap's notion of epistemic state in [1]).

If one take a system of Goldblatt's quantum orthologic [6] then its relational semantics is determined by a model $\mathcal{M} = \langle X, \perp, \xi, v \rangle$, where X is a non-empty set, $\perp$ is an orthogonality relation on X, (i.e. $\perp \subseteq X \times X$ is irreflexive and symmetric), ξ is a non-empty collection of $\perp$-closed subsets of X and v is a valuation. Since a proposition according to aforesaid is an observable with the two-point spectrum then it would be natural to consider a set X as a set of states. In this case a valuation assigns to each proposition a set of states in which measuring gives us 1 while a set orthogonal to it would be a set of states in which measuring leads to 0 as an average value. If a join of those does not exhaust X, then its complement be a state in which there is no such proposition. If one would endeavour to extend this picture in order to cope with observables of any kind then as truth-values one need to consider such subsets of X in which observable has values different from 0 and 1. But then the problem of negation arises - which relation would be counted as an orthogonality?

Let us return again to an algebraic approach. By virtue of the 2nd and 3rd axioms in [5] a set of all observables $\mathcal{A}$ of the system Σ is endowed with the structure of a real vector space. The third thesis of an algebraic approach claims:

1) there are two elements O, I in Σ such that for all $\varphi \in \mathcal{S}$ the equalities $\langle \varphi; O \rangle = 0$ and $\langle \varphi; I \rangle = 1$ are fulfilled;

2) for any observable $A \in \mathcal{A}$ and any real magnitude λ there is an element $\lambda A \in \mathcal{A}$ such that $\langle \varphi; \lambda A \rangle = \lambda \langle \varphi; A \rangle$ for all $\varphi \in \mathcal{S}$;

3) for any pair of observables A and B from $\mathcal{A}$ there is an element $(A + B) \in \mathcal{A}$ such that $\langle \varphi; A + B \rangle = \langle \varphi; A \rangle + \langle \varphi; B \rangle$ for all $\varphi \in \mathcal{S}$.

How could we interpret this axiom? Since all this operations are defined on average values of observables in some state then according to the analogy above those are, in fact, defined on truth-values, i.e. they ought to be defined as operations on a logical matrix for **QLO**.

As to the "+"-operation then it appears to be qualified as an algebraic counterpart of disjunction since it, incidently, assigns both observables a value being greater than any of them (maximum in that sense) - sum, which is not diverge with the usual definition of disjunction in many-valued logics (e.g. in $Ł_{\aleph_0}$-system). More complicated is a situation with the multiplication-on-real-value definition. Let us use with this aim a notion of J-function which is usually defined as $J_\alpha(x) = \begin{cases} 1, \text{if } x = \alpha \\ 0, \text{if } x \neq \alpha \end{cases}$, where α belongs to the domain of truth-values. But if to reformulate its definition as $J_\alpha(x) = \alpha x$, then it is easy to see that it gives us the result we need while being so unusual at first glance.

But it is not the end of the road. The 6th axiom of an algebraic approach requires an existence of symmetrical product of two observables where for those $\langle\varphi; AB\rangle = \langle\varphi; A\rangle\langle\varphi; B\rangle$ for all $\varphi \in \mathbb{S}_A \cap \mathbb{S}_B$ (here $\mathbb{S}_A$ is a set of all states with the null dispersion for observable A). Clearly, we need to introduce some operation on a logical matrix for attempt to construct a multiplication of observables through their addition would be failed. If one defines an operation of multiplication on matrix as a conjunction, i.e. $(x)\wedge(y) = xy$, then the only justification for the introduction of such a definition will be the fact that on the $[0, 1]$ this operation actually gives us some minimum of two observables - their product which will be smaller than any efficient.

Thus, the only question which is still unresolved concerns a definition of negation. In case of quantum propositions due to the two-valuedness of their spectrum the definition is obvious. The case of many-valued spectrum is much more obscure. Here the possible remedy will be to inspect algebraic properties of observables.

The matter of fact is that in case of a real vector space its elements regarding the operation + form an abelian group. But then for any A there is an inverse observable $-A$ which for values of observables in some state gives $\langle\varphi; -A\rangle = -\langle\varphi; A\rangle$ and beside this $- - A = A$. But it prompts an idea of the introduction on logical matrix a negation operation as $\neg(x) = -x$.

Early we have putted the question: which relation in **QLO** should replace an orthogonality? It appears to be that the acceptation of a such matrix negation preserves an orthogonality but make it partial. For by virtue of the uniqueness of the value of observable in some state one can consider an orthogonality in a model of quantum orthologic type only for those subsets on which an observable has the same value (and which obviously does not take place in all cases).

Concerning elements O and I one can quickly discover that if O could be constructed from elements of an algebra of observables or operators of logical matrix (e.g. as $A - A$ or $\neg(x \vee \neg x)$) then for I such constructing becomes problematic. The matter of fact is that relatively multiplication elements of an algebra $\mathcal{A}$ does not form a group which implies the lack of indispensable inverse elements. For logical matrix such a difficulty could be bypassed by means of introduction of a constant 1 using the definition $1(x) = 1$.

The only issue still unresolved is an issue of what would be accepted as the distinguished values of logical matrix. Indeed, the definition of negation operation immediately suggests that this role should be played by $\mathbf{R}^+$, that is, the real positive number line. The support for such an idea is that a class of positive elements of C^+-algebra to which belongs an algebra of observables (under condition of the further specification of its structure in order to satisfy the peculiarities of an algebraic approach) is the most important in general algebraic formalism.

To sum up the definitions introduced, a logical matrix of quantum logic of observables should be the following structure:

$$\mathcal{M}_{QLO} = \langle \mathbf{R}^+, \{\mathbf{R}^+\}, \neg, \vee, \wedge, J_\alpha, 1 \rangle$$

where for $x, y \in \mathbf{R}$ we have:

$$\neg(x) = -x; (x) \vee (y) = x + y; (x) \wedge (y) = xy; J_\alpha(x) = \alpha x \ (\alpha \in \mathbf{R}); 1(x) = 1.$$

Note, that a logical matrix of **QLO** allows to bypass the following difficulty. The matter of fact is that in every state φ for any observable A there are observables defined relatively to $+, \circ$ and multiplication on some real value. Hence,the closure under those operations is concerned. Then beside A in a state φ there will be an observable $-A$ which gives $\langle \varphi; -A \rangle = -\langle \varphi; A \rangle$. But what is the difference between this observable and the negation of A? A logical matrix allows to classify such an observable as the $\neg 1 \wedge (x)$ which coincides, in effect, with $\neg(x)$ but formally is different.

2 QLO: A Syntax

Notwithstanding we have in our disposal a logical matrix for **QLO** an attempt to yield a direct reconstruction of syntax yet fails without resort to additional analogies and hypotheses. A first proposal consists in considering **QLO** as a many-valued extension of R.Goldbaltt's quantum orthologic from [6]. Many-valuedness will means that propositional variables correspond the observables with many-valued spectrum instead of idempotent observables as it commonly accepted. As in case of quantum orthologic we shall write $A \vdash B$ if B is deducible from A.

Primitive terms of an objective language of **QLO** are:

(i) a denumerable set $\{p_i : i < \omega\}$ of propositional variables;
(ii) connectives $\neg, \vee, \wedge$;
(iii) J-function and a constant 1;
(iv) left and right parentheses (,).

A set of well-formed formulas (wff) Φ is constructed from those in a usual way. Letters $A, B, ...$ are used as metavariables over Φ. The end of proof will be denoted with $\Box$. $A \dashv\vdash B$ means that $A \vdash B$ and $B \vdash A$.

Schemes of axioms of **QLO**:

A1. $A \vdash A$
A2. $A \dashv\vdash \neg\neg A$
A3. $A \wedge (B \wedge C) \dashv\vdash (A \wedge B) \wedge C$
A4. $A \vee (B \vee C) \dashv\vdash (A \vee B) \vee C$
A5. $A \wedge (B \vee C) \dashv\vdash (A \wedge B) \vee (A \wedge C)$
A6. $\neg(A \vee \neg A) \vdash B \wedge B$
A7. $A \wedge \neg A \vdash \neg(B \vee \neg B)$
A8. $1 \wedge A \dashv\vdash A$
A9. $J_0 A \dashv\vdash \neg(A \vee \neg A)$
A10. $J_1 A \dashv\vdash A$
A11. $J_\alpha(A \wedge B) \dashv\vdash J_\alpha B \wedge A$
A12. $J_\alpha(A \vee B) \dashv\vdash J_\alpha B \vee J_\alpha A$
A13. $\neg J_\alpha A \dashv\vdash J_\alpha \neg A$
A14. $J_{\alpha+\beta} A \dashv\vdash J_\alpha A \vee J_\beta B$
A15. $J_{\alpha\beta} A \dashv\vdash J_\alpha J_\beta A$

Rules of **QLO**:

R1. $\dfrac{A \vdash B}{\neg B \vdash \neg A}$

R2. $\dfrac{A \vdash B}{J_{|\alpha|} A \vdash J_{|\alpha|} B}$

R3. $\dfrac{A \vdash B \quad B \vdash C}{A \vdash C}$

R4. $\dfrac{A \vdash B \quad C \vdash D}{A \vee C \vdash B \vee D}$

R5. $\dfrac{A \wedge A \vdash B \quad C \wedge C \vdash D}{(A \wedge A) \wedge (C \wedge C) \vdash B \wedge D}$

An uncertainty principle:

A0. $\neg((A \wedge B) \vee (B \wedge A))^2 \vdash (A \vee \neg A)^2 \vdash (B \vee \neg B)^2$ (where $A^2 = A \wedge A$).

This principle has a special meaning. The matter of fact is that the respective axiom concerning the case of real algebra of observables claims (see [5]):

An every pair of observables A and B in $\mathcal{A}$ relates to an observable $C \in \mathcal{A}$ which determines the lower limit of simultaneous observability of A and B in a sense that

$$\langle\varphi; (A - I\langle\varphi; A\rangle)^2\rangle\langle\varphi; (B - I\langle\varphi; B\rangle)^2\rangle \geq \langle\varphi; C\rangle^2 \ \ \forall\varphi \in \mathcal{S}$$

An attempt to include the counterpart of this axiom in **QLO** leads to quantification over propositions which is highly undesirable. In case of a complex algebra of observables an uncertainty principle is

$$\langle\varphi;(A - I\langle\varphi;A\rangle)^2\rangle\langle\varphi;(B - I\langle\varphi;B\rangle)^2\rangle \geq \langle\varphi;i[A,B]\rangle^2$$

The last formulation is more suitable because both on the left and on the right we have the same variables which allows to avoid quantification over propositions. But it is completely unclear how to proceed with the expression $i[A,B]$. Common usage purports $C = -i[A,B]$ which practically eliminates the difference between real and complex case. But since $[A,B] = AB - BA$ then squaring gives us $\langle\varphi;i[A,B]\rangle^2 = \langle\varphi;-[A,B]^2\rangle$. Hence, it is easy to obtain the required formulation of A0.

2.1. DEFINITION. Let **QLO** be a quantum logic of observables and Γ a non-empty set of wff. A wff A is-said to be QLO-*derivable* from Γ,$\Gamma \vdash A$, if there exist $B_1, ..., B_n \in \Gamma$ such that

(a) either $B_1 \vee ... \vee B_n \vdash A$;
(b) or $(B_1 \wedge ... \wedge B_n) \wedge (B_n \wedge ... \wedge B_1) \vdash A$;
(c) or $J_{|\alpha_i|}B_i \vdash A, i = 1, 2, ..., n$.

If A is QLO-derivable from $\neg(A \vee \neg A)$ then A is QLO-derivable or is an QLO-theorem which writes $\vdash_{QLO} A$. Γ is QLO-*consistent* if there is at least one wff not QLO-derivable from Γ, and QLO-*inconsistent* otherwise. (It can be shown that Γ is QLO-consistent iff for no A do we have both $\Gamma \vdash A$ and $\Gamma \vdash \neg A$.) Γ is QLO-*full* iff it is QLO-consistent and closed under $\neg, \vee, \wedge, J$ and QLO-derivability, i.e. iff

(1) for some wff, not $\Gamma \vdash A$;
(2) if $A \in \Gamma$ and $A \vdash B$ then $B \in \Gamma$;
(3) $A, B \in \Gamma$ implies $A \wedge B, A \vee B \in \Gamma$;
(4) $A \in \Gamma$ implies $J_{|\alpha|}A \in \Gamma$.

2.2. LEMMA. If $x \subseteq \Phi$ (where Φ is a set of wff) is QLO-full, then

(i) $x \vdash A$ iff $A \in x$;
(ii) $\neg(A \vee \neg A) \in x$, for all wffs A.

PROOF. (i) Since by A1 $A \vdash A$, sufficiency following from the definition of QLO-derivability. Necessity uses 2.1(2),(3),(4).

(ii) By definition x is non-empty, so there exists $B \in x$. But by 2.1(4) $J_0B \in x$, so the result follows by A3 and 2.1(2).□

QLO-full sets and QLO-derivability are linking with the following version of Lindenbaum's Lemma.

2.3. THEOREM. $\Gamma \vdash A$ iff A belongs to every QLO-full extension of Γ.

PROOF. If $\Gamma \vdash A$ then there exist $B_1, ..., B_n \in \Gamma$ such that either $(B_1 \wedge ... \wedge B_n) \wedge (B_n \wedge ... \wedge B_1) \vdash A$ or $B_1 \vee ... \vee B_n \vdash A$, or $J_{|\alpha_i|}B_i \vdash A$ $(i = 1, 2, ..., n)$. If x is QLO-full and $\Gamma \subseteq x$, we have $B_1, ..., B_n \in x$. Applying 2.1(3),(4) and then 2.1(2) we obtain $A \in x$.

For the converse, suppose A is not QLO-derivable from Γ. Let $x = \{B : \Gamma \vdash B\}$. From A1 we have $\Gamma \subseteq x$ and by hypothesis $A \notin x$. Our proof will therefore be completed if we can show that x is QLO-full. Now if $B \in x$ and $B \vdash C$ then there exists $B_1, ..., B_n \in \Gamma$ such that either $B_1 \vee ... \vee B_n \vdash B$ or $(B_1 \wedge ... \wedge B_n) \wedge (B_n \wedge ... \wedge B_1) \vdash B$, or $J_{|\alpha_i|}B_i \vdash B$ $(i = 1, 2, ..., n)$. Hence by R3, either $B_1 \vee ... \vee B_n \vdash C$ or $(B_1 \wedge ... \wedge B_n) \wedge (B_n \wedge ... \wedge B_1) \vdash C$, or $J_{|\alpha_i|}B_i \vdash C$ and so $\Gamma \vdash C$ i.e. $C \in x$.

If on the other hand $B, C \in x$ then there exist $B_1, ..., B_n, C_1, ..., C_m \in \Gamma$ such that $B_1 \vee ... \vee B_n \vdash B$ and $C_1 \vee ... \vee C_m \vdash C$. Then by R4 we obtain $B_1 \vee ... \vee B_n \vee C_1 \vee ... \vee C_m \vdash B \vee C$. So we have $\Gamma \vdash B \vee C$ and thus $B \vee C \in x$.

Furthermore, if $B \in x$ then there is $B' \in \Gamma$ such that $J_{|\alpha'|}B' \vdash B$. But by R2 $J_{|\beta|}J_{|\alpha'|}B' \vdash J_{|\beta|}B$ and by A15 we obtain $J_{|\alpha\beta|}B' \vdash J_{|\beta|}B$,and so $\Gamma \vdash J_{|\beta|}B$, $J_{|\beta|}B \in x$.

Again, let $B, C \in x$. Then there is $B_1, ..., B_n, C_1, ..., C_m \in \Gamma$ such that $(B_1 \wedge ... \wedge B_n) \wedge (B_n \wedge ... \wedge B_1) \vdash B$ and $(C_1 \wedge ... \wedge C_m) \wedge (C_m \wedge ... \wedge C_1) \vdash C$. Then by R5 we obtain $(B_1 \wedge ... \wedge B_n) \wedge (C_1 \wedge ... \wedge C_m) \wedge (B_n \wedge ... \wedge B_1) \wedge (C_m \wedge ... \wedge C_1) \vdash B \wedge C$. Hence, $\Gamma \vdash B \wedge C$, $B \wedge C \in x$.

This shows that x is closed under QLO-derivability, conjunction, disjunction and J-operator. Hence, since $A \notin x$, A is not QLO-derivable from x, and so x is QLO-consistent. Theorem 2.3. is established. □

2.4. THEOREM. If x is QLO-full and $\neg A \notin x$, then there exists an QLO-full set y such that $A \in y$, and for all B, either $\neg B \notin x$ or $B \notin y$.

PROOF. Let $y = \{B : A \vdash B\}$. By A1 $A \in y$.Now let $\neg B \in x$. Then $B \notin y$, or else $A \vdash B$, whence $\neg B \vdash \neg A$ by R1, and so in turn by 2.1(2), $\neg A \in x$, contrary to hypothesis. By 2.2 $\neg(A \vee \neg A) \in x$. According to what we just proved it follows that $A \vee \neg A \notin y$. Proceeding in a similar manner to 2.3 we can show that y is closed under $\neg, \vee, \wedge, J$ and QLO-derivability. Hence that $A \vee \neg A$ is not QLO-derivable from y i.e. y is QLO-consistent, and therefore y is QLO-full as required. □

It worth noting that QLO have some peculiarities featuring quantum orthologic. Both in QLO and quantum orthologic the proof of Lindenbaum's Lemma does not require such power tools as, for example, Zorn's Lemma, which was in case of orthologic counted as unprecedented for logical systems. As to the QLO-full sets, then from topological point of view they are proper filters and not the ultrafilters because of 2.1(4). This, in turns, leads that both for quantum orthologic and QLO there is no necessity in usage of some version of an axiom of choice which would be needed in order to prove an existence of ultrafilters. If we take into account that negation in case of QLO according to 2.4 have the same property as in case of quantum orthologic then one can expect that a dual space for QLO will be a compact space analogously to Stone space of orthostructure (cf. [7]).

Besides, a definition of QLO-full sets gives rise to an interest situation. 9th axiom of algebraic approach claims that:

an algebra $\mathcal{A}$ should be identified with the set of all self-adjoint elements of real or complex associative algebra $\mathfrak{R}$ with involution which satisfies two following conditions:

1) for any element $R \in \mathfrak{R}$, there is the element $A \in \mathcal{A}$ such that $R^*R = A^2$;
2) if $R^*R = 0$ then $R = 0$.

But it appears that requirement of $\mathcal{A}$ be identified with the set of all self-adjoint operators from the mathematical point of view should be replaced with the requirement of invertibility of an algebra $\mathcal{A}$ to $\mathfrak{R}$. It means that $\mathcal{A}$ allows the following generalization of symmetrical product: for any finite sequence (A_i) of elements of an algebra $\mathcal{A}$ the element $\prod_{i=1}^{n} A_i \times \prod_{i=n}^{1} A_i$ belongs to $\mathcal{A}$ too. But if we consider a definition of QLO-full sets then it is clear that this role plays 2.1(b). Hence, one can expect that Lindenbaum's algebra of QLO behaves as an algebra $\mathcal{A}$ in 9th axiom which allows to reduce complex and real cases of algebra of observables to real subalgebra while matrix of QLO is real by definition.

3 QLO: Semantics

First of all we need to check whether an algebra corresponding to QLO would be an algebra of observables satisfying axioms of algebraic approach which would be found in [5].

3.1. Definition. Wffs A and B are equivalent, $A \approx B$, if $\vdash_{QLO} A \dashv\vdash B$.

Actually, a relation $\approx$ is reflexive by A1, transitive by R3 and explicitly symmetric. Let now $P/_{\approx} = \{A : P \approx A\}$ and let $F = \{P/_{\approx} : P \text{ is a formula}\}$. Denoting a set $A/_{\approx}$ as $[A]$ we define

$[A] + [B] = [A \vee B]$;
$[A] \circ [B] = [A \wedge B]$;
$-[A] = [\neg A]$;
$[1] = 1$;
$\alpha[A] = [J_\alpha A]$.

3.2. Lemma. A structure $\mathfrak{F} = \langle F, +, \circ, -, \alpha, 0, 1\rangle$ (where $\alpha \in \mathbf{R}$) is an algebra (of observables) while $\langle F, +, -, \alpha, 0\rangle$ be a vector (linear) space. 0 is a unit relatively to $+$, and 1 is a unit relatively to $\circ$.

Proof. Commutativity of $+$ we obtain as follows. By A1 we have $\vdash_{QLO} A \dashv\vdash J_1 A$ and $\vdash_{QLO} B \dashv\vdash J_1 B$ and then by R4 we get $\vdash_{QLO} A \vee B \dashv\vdash J_1 A \vee J_1 B$. But by A12 we have $J_1 A \vee J_1 B \dashv\vdash J_1(A \vee B)$ and under R4, A10 we arrive at $\vdash_{QLO} A \vee B \dashv\vdash B \vee A$. Hence, $[A] + [B] = [B] + [A]$. Associativity of $+$ is due to A4.

In case of $[A] + 0$ by using A14,A9,A10 we get $[A] + 0 = [A]$. Axioms for multiplication on a real number will obviously result from axioms for J-operators. Thus, it follows that $\langle F, +, -, \alpha, 0\rangle$ be a vector space. Elements $(-1)(-[A])$ and $-((-1)[A])$ coincides because of an axiom A13.

In order to prove commutativity of $\circ$ we obtain by A11 and A10, $[J_1(A \wedge B)] = [J_1 B \wedge A] = [J_1 B] \circ [A] = [B] \circ [A]$ and $[J_1(A \wedge B)] = [A \wedge B] = [A] \circ [B]$. Hence, $[A] \circ [B] = [B] \circ [A]$. Associativity of $\circ$ be a consequence of A3. Distributivity relatively to $+$ and $\circ$ will be determined by A5. Furthermore, $1 \circ 0 = 0$ according to A10, and $1 + 0 = 2$ by virtue of a property proved above. Thus, **F** be an algebra (of observables). It is easy to see that operations introduced do not depend on concrete representatives, and hence are well-defined (i.e. their implementation do not lead to elements outside the scope of **F**).

As to the ordering on **F** then $[A] \leq [B]$ is defined as truthfulness of $C \vdash D$ for any $C \in [A], D \in [B]$. Clearly, an order so defined is correlated with the second axiom of algebraic approach in [5].□

Now we start to define frames and models for QLO. One is quickly observed that an orthogonality relation below differs nothing from that in case of quantum orthologic.

3.3. Definition. If X is a non-empty set, $\perp$ is an orthogonality relation on X if $\perp \subseteq X \times X$ is irreflexive and symmetric. We write $x \perp Y$ (where $x \in X, y \subseteq Y$) if for any $y \in Y$ we have $x \perp y$ and denote Y^* a set $\{x : x \perp y\}$. $Y \subseteq X$ is said to be $\perp$-closed if $Y^{**} = Y$. Besides this, $\emptyset \perp X$.

3.4. Definition. $\Im = \langle X, \perp, \xi \rangle$ is a QLO-frame iff:

(1) X is non-empty set (carrier of X);
(2) $\perp$ is an orthogonality relation on X;
(3) ξ is a non-empty collection of $\perp$-closed subsets of X closed under set intersection and the operation *.

3.5. Definition. $\mathcal{M} = \langle X, \perp, \xi, v \rangle$ is a QLO-model if:

(i) $\langle X, \perp, \xi \rangle$ is a QLO-frame;
(ii) v is a function assigning to each propositional variable and formula of QLO recursively in every point (every element) of X a real number, i.e. $v : S \cup \Phi \times X \to \mathbf{R}$ where S is a set of propositional variables and Φ is a set of wff.

Denoting the set $\{x \in X : v(A, x) = a\}$ by $\|A\|_a$ we define recursively the value of a wff in QLO-model as follows:

(1) $\|p_i\|_a = \{x \in X : v(p_i, x) = a\} \in \xi$;
(2) $\|A \vee B\|_a = \{x \in X : x \in \|A\|_b \cap \|B\|_c \& a = b + c\}$;
(3) $\|A \wedge B\|_a = \{x \in X : x \in \|A\|_b \cap \|B\|_c \& a = bc\}$;
(4) $\|\neg A\|_a = \{x \in X : x \in \|A\|_{-a} \& v(\neg A, x) = a\}$;
(5) $\|J_\alpha A\|_a = \{x \in X : x \in \|A\|_b \& a = \alpha b\}$;
(6) $\|1\|_1 = X$, i.e. $v(1, x) = 1$ for all $x \in X$.

If Γ is a non-empty set of wff then we say that Γ *implies* A at x in $\mathcal{M}$, $\mathcal{M} : \Gamma \models_x A$, iff for any $B \in \Gamma$, $v(B,x) \leq v(A,x)$, Γ $\mathcal{M}$-*implies* A in $\mathcal{M}$, $\mathcal{M} : \Gamma \models A$, iff either there is $B \in \Gamma$ such that $x \notin \|B\|_{(-)}$, i.e. when B is not verified in x (verification but not truthfulness, because we deal with many-valued logical matrix), or else Γ implies A at all x in $\mathcal{M}$. If $\Im$ is a QLO-frame then Γ $\Im$-*implies* A, $\Im : \Gamma \models A$ iff $\mathcal{M} : \Gamma \models A$, for all QLO-models $\mathcal{M}$ on $\Im$. If $\wp$ is a class of QLO-frames, Γ $\wp$-*implies* A, $\wp : \Gamma \models A$, iff $\Im : \Gamma \models A$ for all $\Im \in \wp$. A class $\wp$ is said to *determine* QLO iff for all $A, B \in \Phi$, $A \vdash B$ iff $\wp : A \models B$. $\wp$ strongly determines QLO iff for all Γ, A, $\Gamma \vdash A$ iff $\wp : \Gamma \models A$.

If we define a range of values of a formula A as $\|A\| = \bigcup_{a\in\mathbf{R}} \|A\|_a$ then extending this definition on 3.5(1)-(5) hereafter we denote as $\|p_i\|$, $\|A \vee B\|$, $\|A \wedge B\|$, $\|\neg A\|$, $\|J_\alpha A\|$, $\|1\|$ the ranges of values of respective formulas while $\|A\|_{(-)}$ will means an arbitrary value of respective formula.

3.6. LEMMA. For any QLO-model $\mathcal{M}$ and any $A \in \Phi$, $\|A\|_{(-)} \in \xi$.

PROOF. By induction on the length of A, using 3.4(2) and 3.5(2)-(5). □

3.7. THEOREM (SOUNDNESS OF QLO). $\Gamma \vdash A$ if $\Theta : \Gamma \models A$, where Θ is the class of all QLO-frames.

PROOF. The proof, by induction on QLO-derivability, proceeds by showing that the result holds for A0-A15 and is preserved by application of R1-R5. We consider only the less obvious cases.

A2. Let $x \in \|A\|_a$. Then if $y \in \|\neg A\|_{-a}$, by 3.5(4) $y \perp x$ and hence (symmetry) $x \perp y$. 3.5(4) again gives $x \in \|\neg\neg A\|_a$.

Now let $x \in \|\neg\neg A\|_a$. Then $y \in \|\neg A\|_{-a}$ only if $x \perp y$, i.e. $y \perp \|A\|_a$ only if $x \perp y$. But $\|A\|_a$ is $\perp$-closed (by 3.6) and, thus, $x \in \|A\|_a$.

A6. It is easy to make sure that $v(\neg(A \vee \neg A), x) = 0$ at any point $x \in X$ and alike $v(A \vee \neg A, x) = 0$. But if $x \in \|\neg(A \vee \neg A)\|_0$, then $y \in \|A \vee \neg A\|_0$ only if $x \perp y$. By 3.5(2) $y \in \|A\|_b \cap \|\neg A\|_c$ and $b+c = 0$, i.e. $c = -b$. But then by 3.5(4) $y \perp y$ contrary to the irreflexivity of $\perp$. Hence, there is no y in any $\mathcal{M}$ for which we have $y \perp \|A \vee \neg A\|_0$ and whence by 3.3 $x \in \|\neg(A \vee \neg A)\|_0$ for any x. Besides, for all B, by 3.5(3), $v(B \wedge B, x) \geq 0$.

A7. Let $x \in \|1 \wedge A\|_a$. Then $x \in \|1\|_1 \cap \|A\|_a$ by 3.5(3) and $v(1 \wedge A, x) = v(1,x) \circ v(A,x)$. But $v(1,x) = 1$ at any x by virtue of the definition of a constant 1 (see 3.5(6)). So $v(1 \wedge A, x) = v(A,x)$ and thus $\mathcal{M} : 1 \wedge A \models A$ and $\mathcal{M} : A \models 1 \wedge A$ for any A.

A11. Let $x \in \|J_\alpha(A \wedge B)\|_a$. Then by 3.5(5) $x \in \|A \wedge B\|_b$ and $a = \alpha b$. By 3.5(3) $x \in \|A\|_c \cap \|B\|_d$ and $b = cd$. Hence, $a = \alpha cd$. But by 3.5(5) $x \in \|J_\alpha A\|_{\alpha c} \cap \|B\|_d$ and by 3.5(3) $x \in \|J_\alpha A \wedge B\|_{acd=a}$.

A13. Let $x \in \|A\|_a$. Then by 3.5(5) $x \in \|J_\alpha A\|_{\alpha a}$ and by 5.5(4) $y \in \|\neg J_\alpha A\|_{-\alpha a}$ only if $x \perp y$. But then $y \in \|\neg A\|_{-a}$ because of $x \perp y$ and by 3.5(5) $y \in \|J_\alpha \neg A\|_{-\alpha a}$.

R1. Suppose $\mathcal{M} : A \models B$, and let $x \in \|\neg B\|_{-a}$. Then $y \in \|A\|_b$ only if $y \in \|B\|_a$ (by inductive hypothesis), only if $x \perp y$. This shows that $x \in \|\neg A\|_{-b}$.

The rest is obvious. □

3.8. DEFINITION. Let L be a quantum logic of observables. The *canonical QLO-model for L* is the structure $\mathcal{M}_L = \langle X_L, \perp_L, \xi_L, v_L \rangle$ where:

(1) $X_L = \{x \subseteq \Phi : x \text{ is a QLO-full set}\}$;
(2) $x \perp_L y$ iff there is a wff A such that $\neg A \in x, A \in y$;
(3) $\xi_L = \{\|A\|^L : A \in \Phi\}$, where $\|A\|^L$ is $\{x \in X_L : A \in x\}$;
(4) $v_L : \Phi \cup S \times X_L \to \mathbf{R}$.

Denoting $\{x \in X_L : v_L(A, x) = a\}$ as $\|A\|^L_\alpha$ we come to the definition of the value of formula and ranges of valuation in canonical model $\mathcal{M}_L$ analogously to 3.5(1)-(6).

3.9. LEMMA. $\Im_L = \langle X_L, \perp_L, \xi_L \rangle$ is a QLO-frame.

PROOF. Let $x \in X_L$. Then for any A neither $\neg A, A \in x$ nor x is QLO-compatible (by A7). Hence, $x \perp x$ does not take place. If $x \perp_L y$ then for some wff A we have $\neg A \in x$, $A \in y$. By means of A2 we come to the conclusion that $\neg B \in y$, $B \in x$, where $B = \neg A$. Thus, $x \perp_L y$ and $\perp_L$ is an orthogonality relation. To check whether $\|A\|^L$ be $\perp_L$-closed, suppose that $x \notin \|A\|^L$, i.e. $A \notin x$. By A2 $\neg\neg A \notin x$ and so by 2.4 there is $y \in X_L$ such that $x \perp_L y$ fails and $\neg A \in y$. Meanwhile if $z \in \|A\|^L$ then $A \in z$ and, hence, $y \perp_L z$. Thus, $y \perp_L \|A\|^L$ as it was required. Clearly, ξ_L will be closed under intersections (by virtue of properties of QLO-derivability and QLO-fullness).□

3.10. THEOREM (FUNDAMENTAL THEOREM FOR QLO). For all wff A and all $x \in X_L$, $x \in \|A\|^L$ iff $A \in x$.

PROOF. By induction on the length of A. In case of $A = B \wedge C$, $A = B \vee C$, $A = J_\alpha B$ it is easy to see that $B, C \in x$ follows form the condition $B \wedge C, B \vee C, J_\alpha B \in x$. Suffice it to use 3.5(2),(3),(5). Conversion follows from 2.1(3),(4).

Suppose that $A = \neg B$ and that for B the theorem is true. Let $\neg B \in x$. If $y \in \|B\|^L_{(-)}$ then by inductive hypothesis $B \in y$ and hence $x \perp_L y$. By 3.5(4) it follows that $x \in \|\neg B\|^L_{(-)}$. Again, if $\neg B \notin x$ then according to 2.4 there is $y \in X_L$ such that $B \in y$ and thus by inductive hypothesis $y \in \|B\|^L_{(-)}$ but $x \perp_L y$ fails. By 3.5(4) we come to the conclusion that $y \in \|B\|^L_{(-)}$. □

3.11. COROLLARY. $\Gamma \vdash A$ iff $\mathcal{M}_L : \Gamma \models A$.

PROOF. If $\Gamma \vdash A$ then there are $B_1, ..., B_n \in \Gamma$ such that either $B_1 \vee ... \vee B_n \vdash A$ or $(B_1 \wedge ... \wedge B_n) \wedge (B_n \wedge ... \wedge B_1) \vdash A$, or $J_{|\alpha_i|} B_i \vdash A$ $(1 \leq i \leq n)$. If $x \in \|B\|^L_{(-)}$ for all $B \in \Gamma$ then by 3.10 $B_1, ..., B_n \in x$. By 2.1(2)-(4) it follows that $A \in x$ and thus $x \in \|A\|^L_{(-)}$.

The other way round, if A is not QLO-derivable from Γ then by 2.3 there is $x \in X_L$ such that $\Gamma \subseteq x$ and $A \notin x$. Then by 3.10 $x \in \|B\|^L_{(-)}$ for all $B \in \Gamma$ but $x \notin \|A\|^L_{(-)}$. □

3.12. THEOREM. $\Gamma \vdash A$ iff $\Im_L : \Gamma \models A$.

PROOF. Let $\mathcal{M}$ be an arbitrary QLO-model on $\Im_L$. For every $i < \omega$, $\|p_i\|^{\mathcal{M}} \in \xi_L$ there is B_i such that $\|p_i\|^{\mathcal{M}}_{(-)} = \|B_i\|^{L}_{(-)}$ ($\|B_i\|^{L}_{(-)}$ is defined as in 3.8(3)) and $\|B_i\|^{L}_{(-)} = \|B_i\|^{\mathcal{M}_L}_{(-)}$. For any wff C, let C' be the result of uniformly replacing each p_i occurring in C by B_i. Clearly, there are in Γ such $A_1, ..., A_n$ that either $A_1 \vee ... \vee A_n \vdash A$ or $(A_1 \wedge ... \wedge A_n) \wedge (A_n \wedge ... \wedge A_1) \vdash A$, or $J_{|\alpha_i|}A_i \vdash A$ $(i = 1, 2, ..., n)$ and so we have $A'_1 \vee ... \vee A'_n \vdash A'$ etc. Then by 3.11 either $\mathcal{M}_L : A'_1 \vee ... \vee A'_n \models A'$ or $\mathcal{M}_L : (A'_1 \wedge ... \wedge A'_n) \wedge (A'_n \wedge ... \wedge A'_1) \models A'$, or $\mathcal{M}_L : J_{|\alpha_i|}A'_i \models\vdash A'$. But a simple induction shows that $\|C\|^{\mathcal{M}} = \|C'\|^{\mathcal{M}_L}$ and so either $\mathcal{M} : A_1 \vee ... \vee A_n \models A$ or $\mathcal{M} : (A_1 \wedge ... \wedge A_n) \wedge (A_n \wedge ... \wedge A_1) \models A$, or $\mathcal{M} : J_{|\alpha_i|}A_i \models\vdash A$ $(i = 1, 2, ..., n)$ whence $\mathcal{M} : \Gamma \models A$. Since this holds for all models $\mathcal{M}$ on $\Im_L$, we conclude $\Im_L : \Gamma \models A$.□

3.13. COROLLARY (STRONG COMPLETENESS FOR QLO). $\Theta : \Gamma \models A$ only if $\Gamma \vdash A$.

PROOF. Since by 3.9 $\Im_L$ be a QLO-frame then Θ contains $\Im_L$ as its element. The rest is obvious.□

Thus corollary 3.13 shows that QLO is strongly determined by the class of all QLO-frames.

Let us return now to quantum propositions, i.e. elements of an algebra of observables satisfying the condition $P^2 = P$. It is easy to show that for propositions QLO-models would be transformed to Goldblatt's models for orthologic. Recall in this connection definitions of those [6, p.32-33].

3.14. DEFINITION. $\mathfrak{R} = \langle X, \perp \rangle$ is an orthoframe iff X is non-empty set and $\perp$ is an orthogonality relation on X, i.e. $\perp \subseteq X \times X$ is irreflexive and symmetric. $\Im = \langle X, \perp, \xi \rangle$ is a quantum frame if $\langle X, \perp \rangle$ is an orthoframe and ξ is a non-empty collection of $\perp$-closed subsets of X such that:

(i) ξ is closed under set intersection, and the operation * defined by $Y^* = \{x : x \perp Y\}$;
(ii) if $Y, Z \in \xi$, then $Y \subseteq Z$ only if Y is $\perp$-closed in Z.

$\mathcal{M} = \langle X, \perp, \xi, v \rangle$ is a quantum model iff v is a function assigning to each p_i a member of ξ. The truth of a wff A at x in $\mathcal{M}$ is defined recursively as follows:

(1) $\|p_i\|^{\mathcal{M}} = v(p_i)$;
(2) $\|A \wedge B\|^{\mathcal{M}} = \|A\|^{\mathcal{M}} \cap \|B\|^{\mathcal{M}}$;
(3) $\|\neg A\|^{\mathcal{M}} = \{x : X \perp \|A\|^{\mathcal{M}}\}$ (here $\|A\|^{\mathcal{M}} = \{x \in X : \mathcal{M} \models_x A\}$).

3.15. THEOREM. For quantum propositions QLO-models is transformed to quantum orthomodels.

PROOF. As to the orthogonality relation then in QLO-model it coincides with this for quantum orthomodel. However, if for quantum orthoframe we need a coherence of set-theoretical inclusion and orthogonality (by 3.14(ii)) then for QLO-frame such a condition is dispensable. Meanwhile, such condition is fulfilled for QLO-models in case of two-valued spectrum since in this case the negation assigns to an every wff A a subset $Y \subseteq X$ at points of which a value of A is 1. Since we have now $\{0,1\}-$valuation then it leads to disappearing of J-operators and for the conjunction the truth value is determined by intersection while the disjunction is disappeared too. Thus, we are forced to come to the definition of truth values for wff in quantum orthomodels.□

4 A Normed Quantum Logic of Observables

Our formulation of QLO left aside aspects of an algebraic approach to the theory of quantum system concerning the topological structure of algebra of observables. It is known that in order to obtain a natural topology on algebra of observables we need that a set of observables would be endowed with the structure of a real Banach space relatively to the natural norm induced on it [5]. A norm introduced in this case for any observable $A \in \mathcal{A}$ coincides with

$$|A| = \sup_{\varphi \in \mathcal{S}} |\langle \varphi; A \rangle| \,.$$

From the phenomenological point of view it is usually remarked that actually in the laboratory one never deals with the observables for which a matrix element $\langle \varphi; A \rangle$ would be infinite.

In the framework of QLO this fact bespeaks that in matrix $\mathcal{M}_{QLO}$ the truth value of observable whatsoever should be finite if we shall endeavour to extend our approach on case of a quantum logic of observables with the norm. Besides, as a norm of any observable in $\mathcal{M}_{QLO}$ one can adopt some *sup* on $X \subseteq \mathbf{R}^+$ where X be the set of truth values for this observable.

Similar situation arises in first-order many-valued logics. In particular, for logic with positive and negative truth values C.C. Chang [9] defines in logical matrix the following operation:

$$\text{for } x, y \in [-1,+1] \text{ and } X \subseteq [-1,+1],\ \exists(X) = \sup X.$$

One can easy discover such a tendency of matrix *sup*-interpretation of existential quantifier in continuous logics [4].

On this basis we define in $\mathcal{M}_{QLO}$ an analogue of the norm as

$$\exists(X) = \sup |X|$$

which is, clearly, determined by the definition of the norm of observable above. Denoting the matrix modified (i.e. endowed with the norm operation) as $\mathcal{M}_{NQLO}$ we shall below attempt to reconstruct a syntax of logic of normed observables. The system of this logic will be called a normed quantum logic of observables and will be denoted NQLO.

From the outset we must take into account that in NQLO a quantification concerns all variables in wff since according to the definition of matrix operation the variables quantified does not appear explicitly but are just purported. Thus, the quantifier binds all variables in wff and always will be written as $\exists u$, i.e., in fact, we purport that all observables are the predicates with one and the same quantified variable.

So, a normed quantum logic of observables is obtained by adding the following axiom schemes to QLO:

A16. $A \vdash \exists u A$
A17. $\neg \exists u A \vdash \exists u \neg A$
A18. $\exists u J_\alpha A \dashv\vdash J_\alpha \exists u A$
A19. $\exists u (A \vee B) \dashv\vdash \exists u A \vee \exists u B$
A20. $\exists u (A \wedge B) \dashv\vdash \exists u A \wedge \exists u B$

and the rule

R6. $\dfrac{A \dashv\vdash B}{\exists u A \dashv\vdash \exists u B}$

It is worth-while to remind that according to the 7th axiom of an algebraic approach the norm of any element of algebra of observables is finite, hence, $\sup |X| < \omega$, i.e. sup always exists and be finite.

4.1. Definition. A non-empty set wff Γ is a NQLO-full iff:

(i) Γ is QLO-full;
(ii) $A \in \Gamma$ implies $\exists u A \in \Gamma$.

A notion of NQLO-derivability coincides with this for QLO-derivability. Lindenbaum's Lemma for NQLO is as follows:

4.2. Theorem. $\Gamma \vdash A$ iff A belongs to every NQLO-full extension of Γ.

Proof. We need to check only the fulfillment of 4.1(ii). If $\Gamma \vdash A$ then $A \in \Gamma$. But by A16 and R3 we have $\exists u A \in \Gamma$.

If A is not QLO-derivable from Γ then let $x = \{B : \Gamma \vdash B\}$. By A1 $\Gamma \subseteq x$ and by inductive hypothesis $A \in x$. If $B \in x$ then according to aforesaid $\exists u B \in x$ and x is a QLO-full and closed relatively to $\exists u$-quantification. Since $A \notin x$ then A is not QLO-derivable from x and therefore x is QLO-consistent and x is NQLO-full.□

4.3. Lemma. A structure $\mathcal{NF} = \langle F, +, \circ, -, \alpha, |...|, 0, 1 \rangle$ be a Banach space.

Proof. According to 3.2 $\mathcal{F} = \langle F, +, \circ, -, \alpha, 0, 1 \rangle$ is an algebra. An operation $|...|$ would be defined as $|[A]| = [\exists u A]$. Its well-definiteness is obvious. For normed algebra $\mathcal{NF}$ we need the following conditions will be satisfied: $|\lambda A| = |\lambda| \cdot |A|$; $|A + B| = |A| + |B|$; $|A \circ B| = |A| \circ |B|$; $|A| = 0$ only if $A = 0$. It is easy to see that first three conditions are satisfied by A18-A20. Fourth condition is satisfied by A19, A17. □

A QLO-model becomes NQLO-model with the help of 3.5 and the following definition:

4.4. Definition. $\|\exists uA\|_a = \{x \in X : x \in \|A\|$ and $\forall y \in \|A\| (a = \sup |v(A, y)|\}$.

Clearly, $\|\exists uA\|_a = \|\exists uA\|$. Beside this it is obvious that the lemma 3.6 will also take place.

4.5. Theorem (soundness of NQLO). $\Gamma \vdash A$ if $\Theta : \Gamma \models A$, where Θ is the class of all NQLO-frames.

Proof. We consider only the less obvious cases.

A17. Let $x \in \|\exists uA\|_a$. Then by 3.5(4) $y \in \|\neg\exists uA\|_{-a}$ only if $y \perp x$. But by 4.4 $x \in \|A\|_a$ and by virtue $y \perp x$ we obtain $y \in \|\neg A\|_{-a}$. Whence by 4.4 $y \in \|\exists u\neg A\|_a$. Then $v(\neg\exists uA, y) \leq v(\exists u\neg A, y)$ and, finally, $\mathcal{M} : \neg\exists uA \models \exists u\neg A$.□

Concerning the canonical NQLO-model we can quickly observe that it obtaining from canonical QLO-model by reinforcement of the range of $\|\exists uA\|^{NL}$ with the help of the respective definition. Lemma 3.9 remains true under this procedure.

4.6. Theorem (fundamental theorem for NQLO). For all wff A and all $x \in X_{NL}$, $x \in \|A\|^{NL}$ iff $A \in x$.

Proof. By induction on the length of A. In case of $A = \exists uB$ if $B \in x$ then by A16 by virtue of properties of QLO-derivability $\exists uB \in x$. Conversion follows from the definition 4.4 since $\|\exists uA\|$ and $\|A\|$ coincide. □

4.7. Corollary. $\Gamma \vdash A$ iff $\mathcal{M}_{NL} : \Gamma \models A$.

Proof. As in case of 3.11.□

4.8. Theorem. $\Gamma \vdash A$ iff $\Im_{NL} : \Gamma \models A$.

Proof. As in case of 3.12. □

4.9. Corollary (strong completeness for NQLO). $\Theta : \Gamma \models A$ only if $\Gamma \vdash A$.

Proof. Since $\Im_{NL}$ is NQLO-full then analogously to the proof of 3.15 Θ contains $\Im_{NL}$ as its element.□

5 A Temporal Quantum Logic of Observables

An account of dynamic aspects *per se* indeed redounds to improve the descriptions of quantum systems by means of QLO but immediately the row of questions arises concerning the nature of time and forcing us to put hypotheses on properties and structure of time. An attempt of such a kind was made way back in the sixties by G. Mackey when he added to his system of axioms a new one describing a presence of some semigroup of transformations on the set of states [8].

For a system of quantum orthologic similar extension was proposed by the author in [9] (it was done for H.Nishimura's sequential formulation but the equivalence of both formulations was proved). If enlarge such a system with an axiom of reversibility, i.e. by introducing the reverse von Wright's operator, then one can treat it as a quantum logical description of the symmetry of quantum systems. As it turned out, such an extension would be made in case of QLO too.

So, if we use von Wright's temporal operator $\bigcirc$ which intuitively means "next moment", "tomorrow", then a system of temporal quantum logic of observables describing the dynamics of quantum system would be obtained by enlarging QLO with the following axiom schemes:

A21. $\bigcirc\neg A \dashv\vdash \neg \bigcirc A$
A22. $\bigcirc(A \vee B) \dashv\vdash \bigcirc A \vee \bigcirc B$
A23. $\bigcirc(A \wedge B) \dashv\vdash \bigcirc A \wedge \bigcirc B$
A24. $\bigcirc J_\alpha A \dashv\vdash J_\alpha \bigcirc A$
A25. $\bigcirc^{-1} \bigcirc A \dashv\vdash A$

and $\bigcirc$-rule

R7. $\dfrac{A}{\bigcirc A}$

A temporal extension of QLO will be denoted as TQLO.

5.1. Definition. A non-empty set wff Γ is a TQLO-full iff:

(i) Γ is QLO-full;
(ii) $A \in \Gamma$ implies $\bigcirc A \in \Gamma$.

A notion of TQLO-derivability coincides with this for QLO-derivability. Lindenbaum's Lemma for TQLO is as follows:

5.2. Theorem. $\Gamma \vdash A$ iff A belongs to every TQLO-full extension of Γ.

Proof. We need to check only the fulfillment of 5.1(ii). If $\Gamma \vdash A$ then $A \in \Gamma$. But by R7 we have $\bigcirc A \in \Gamma$.

If A is not QLO-derivable from Γ then let $x = \{B : \Gamma \vdash B\}$. By A1 $\Gamma \subseteq x$ and by inductive hypothesis $A \in x$. Since by R7 for any $B \in x$ we have $\bigcirc B \in x$ then x is a QLO-full and closed relatively to $\bigcirc$-quantification. Since $A \notin x$ then A is not QLO-derivable from x and therefore x is QLO-consistent and x is TQLO-full.□

The relational semantics for TQLO is obtained by modifying semantics for QLO.

5.3. Definition. $\Im = \langle X, \perp, \xi, R\rangle$ is a TQLO-frame iff:

(1) $\langle X, \perp, \xi\rangle$ is a QLO-frame;

(2) R is a binary "successor" relation on X, irreflexive and antisymmetric.

Considering TQLO-frame as functional (i.e. $\forall\, x, y, z(xRy \& xUz \Longrightarrow y = z)$) and serial (i.e. $\forall x \exists y(xRy)$) we introduce on the set X a partially defined function f and given $Y \subseteq X$ we denote as X^* the set $\{x : \forall y \in Y(yRx)\}$ or $\{x : \forall y \in Y(f(y) = x)\}$. Besides, hereafter we put $x^+ = f(x)$.

Informally it means that the conception of time upon which we focus has two important features. For one thing, moment follows upon moment: there is no last moment. For another thing, every following moment is unique: no moment has two next moments. The last property force us in view of the presence of reverse von Wright's operator determine that there is no branching in the past and thus we need to introduce the reverse function f^{-1}.

5.4. Definition. The succesor relation and orthogonality relation are correlated on X, i.e. if $x \perp y$ then $x^+ \perp y^+$.

5.5. Definition. $\mathcal{M} = \langle X, \perp, \xi, R, v\rangle$ is a TQLO-model if:

(i) $\langle X, \perp, \xi, R\rangle$ is a TQLO-frame;
(ii) v coincides with that for QLO-model. The new additional point is:
(1) $\|\bigcirc A\|_a = \{x \in \|A\|^+ : a = v(A, x)\}$.

5.6. Lemma. Given TQLO-frame $\Im = \langle X, \perp, \xi, R\rangle$

(1) if $Y^+, Z^+ \subseteq X$ are $\perp$-closed then their intersection $Y^+ \cap Z^+$ is closed too;
(2) for any $Y \subseteq X$, Y^+ is $\perp$-closed;
(3) for any wff A, $\|A\|^+ \subseteq X$ is $\perp$-closed.

Proof. All assertions are consequences of the functionality of TQLO-frame.□

5.7. Lemma. Let $\Im = \langle X, \perp, \xi, R\rangle$ be a TQLO-frame. Then ξ will be closed relatively to $^+$.

Proof. Follows from 5.6. □

5.8. Lemma. For any TQLO-model $\mathcal{M}$ and any $A \in \Phi$, $\|A\|_{(-)} \in \xi$.

Proof. For any QLO-wff it follows from 3.6. In case of $A = \bigcirc B$ it follows by 5.6.□

5.9. Theorem (soundness of TQLO). $\Gamma \vdash A$ if $\Theta : \Gamma \models A$, where O is the class of all TQLO-frames.

Proof. We consider only less obvious new cases.

A21. Let $x \in \|A\|_a$. Then if $y \in \|\neg A\|_{-a}$ we have $x \perp y$ by 3.5(4). Furthermore, by 5.5(1) we obtain $y^+ \in \|\bigcirc A\|_{v(\neg A, y^+)}$.

Now let $x^+ \in \|\bigcirc A\|_{v(A,x^+)}$. By 5.3 if $x \perp y$ then $x^+ \perp y^+$ and, hence, $y^+ \in \|\neg \bigcirc A\|_{-v(A,x^+)}$. But by 3.5(4) $v(\neg A, y^+) = -v(A, x^+)$. Thus, $\|\neg \bigcirc A\|$ and $\|\bigcirc \neg A\|$ coincides.

A25. Here we need to take into account the reverse von Wright's operator. Recall in this connection that early it was said that there is any branching into past. Denote $f^{-1}(x)$ as x^{-} and let $x \in \|A\|_a$. Then $x^{+} \in \|\bigcirc A\|_{v(A,x^{+})}$ and $(x^{+})^{-} = x \in \|\bigcirc^{-1} \bigcirc A\|_{v(A,(x^{+})^{-})}$. But $v(\bigcirc A,(x^{+})^{-}) = v(\bigcirc A, x) = v(A,x)$ by virtue of the composition of f and f^{-} and 5.5(1).

R7. Follows from 5.5(1), since $v(\bigcirc A, x) = v(A,x)$.□

5.10. Definition. Let T be a temporal quantum logic of observables. The *canonical TQLO-model for* T is the structure $\mathcal{M}_T = \langle X_T, \bot_T, \xi_T, R_T, v_T\rangle$ where:

(1) $X_T = \{x \subseteq \Phi : x$ is a TQLO-full set$\}$;
(2) $x \bot_T y$ iff there is a wff A such that $\neg A \in x, A \in y$;
(3) $\xi_T = \{\|A\|^T : A \in \Phi\}$, where $\|A\|^T$ is $\{x \in X_T : A \in x\}$;
(4) xR_Ty iff when there is wff A such that $\bigcirc A \in y, A \in x$ (but not the other way round);
(5) $v_T : \Phi \cup S \times X_T \to \mathbf{R}$.

Denoting $\{x \in X_T : v_T(A,x) = a\}$ as $\|A\|^T_\alpha$ we come to the definition of the value of formula and ranges of valuation in canonical model $\mathcal{M}_T$ analogously to 3.5(1)-(6), 5.5(1).

5.11. Lemma. $\Im_T = \langle X_T, \bot_T, \xi_T, R_T\rangle$ is a TQLO-frame.

Proof. According to 3.9 $\langle X_T, \bot_T, \xi_T\rangle$ is a TQLO-frame. We need to check whether R_T will be the successor relation. Let for some $x \in X_T$, $A, \bigcirc A \in x$. Then $\neg A \in y$, where $y \perp x$, and $\bigcirc\neg A \in y^{+}$. Furthermore, since $\bigcirc A \in x$ then $\neg \bigcirc A \in y$. But then $\bigcirc\neg A \in y^{+}$ and $\neg \bigcirc A \in y$, hence A21 is fulfilled only if $y^{+} = y$ and the relation R fails. Thus, defined in this way R_T is irreflexive. Supposing that xR_Ty and yR_Tx we obtain the violation of the condition 5.10(4). Hence, R_T is antisymmetric. Correlativity of R_T and $\bot_T$ follows from A21.□

5.12. Theorem (fundamental theorem for TQLO). For all wff A and all $x \in X_T$, $x \in \|A\|^T$ iff $A \in x$.

Proof. By induction on the length of A. In case of $A = \bigcirc B$ let for B the theorem take place and let $\bigcirc B \in x$. Then if $y \in \|B\|^T_{(-)}$, by inductive hypothesis $B \in y$ and thus yR_Tx. By 5.5(1) we obtain $x \in \|B\|^{T+}_{(-)}$. Again, if $\bigcirc B \notin x$ then $\bigcirc B \in y$ and xR_Ty (according to 5.10(4)). But then by inductive hypothesis $y \in \|\bigcirc B\|^T_{(-)}$ and by 5.5(1) $y \in \|B\|^{T+}_{(-)}$.□

5.13. Corollary. $\Gamma \vdash A$ iff $\mathcal{M}_T : \Gamma \models A$.

Proof. Is proved because of 3.11 since TQLO-derivability coincides with QLO-derivability and $\mathcal{M}_T$ turns out to be an extension of $\mathcal{M}_{QLO}$ correlative with $\bot$.□

5.14. Theorem. $\Gamma \vdash A$ iff $\Im_T : \Gamma \models A$.

PROOF. As in case of 5.12. □

5.15. COROLLARY (STRONG COMPLETENESS FOR TQLO). $\Theta : \Gamma \models A$ only if $\Gamma \vdash A$.

PROOF. Since $\Im_T$ by 5.11 is a TQLO-frame then Θ contains $\Im_T$ as its element. The rest is obvious.□

Should we "temporize" with the help of von Wright's operator a normed quantum logic of observables? The evident proposal consists in addition to axiom schemata A0-A15 and rules R1-R7 the following new axiom scheme:

A26. $\bigcirc \exists u A \vdash \exists u \bigcirc A$

Let us denote the system obtained as TNQLO. Clearly, TNQLO-derivability coincides with QLO-derivability while the definition of TNQLO-full sets is obtained joining definitions 4.1 and 5.1, i.e. definitions of NQLO- and TQLO-full sets. It is easy to check that Lindenbaum's Lemma will be satisfied too for TNQLO since, in fact, its proof consists in joining respective proofs for NQLO and TQLO. In such a way the notions of TNQLO-frame and TNQLO-models are obtained.

5.16. THEOREM (SOUNDNESS OF TNQLO). $\Gamma \vdash A$ if $\Theta : \Gamma \models A$, where Θ is the class of all TNQLO-frames.

PROOF. In case of A26 let $x \in \|A\|_a$. Then by 4.4 $x \in \|\exists u A\|_b$, where $b = \sup|v(A,(-))|$. By 5.5(1) we obtain $x^+ \in \|\exists u \bigcirc A\|_c$ where $c = v(\exists u a A, x^+) = b$.
Again, by 5.5(1) $x^+ \in \|\bigcirc A\|_d$ where $d = v(A, x^+)$. By 4.4. $x^+ \in \|\exists u \bigcirc A\|_e$ where $c = \sup|v(\bigcirc A,(-))|$. But since $\sup|v(\bigcirc A,(-))| \leq \sup|v(A,(-))|$ then $e \leq b$ (such situation would arise by virtue of that we could not exclude the case of $v(A,x) \geq v(\bigcirc A, x^+)$).□

As to the rest results (fundamental theorem, strong completeness) then it is easy to make sure that they would be achieved joining the respective proofs for TQLO and NQLO.

A few words would be said about the role of von Wright's $\bigcirc$-operator in QLO. From matrix point of view it could be described as an endomorphism on $\mathfrak{M}_{QLO}$. The introduction of the reverse von Wright's operator in A25 was stimulated by the case of reciprocal quantum systems since the request for reciprocity always plays physically important role in considerations concerning the time translations in the framework of algebraic approach to quantum system. In fact, the introduction of the reverse von Wright's operator led to the introduction on QLO-frame not only a successor relation but also a "predecessor" relation R^{-1} for which $xRyR^{-1}x$. But since changings caused by this extension of QLO-framework, in effect, are not too substantial and evident enough, then explicit introduction of R^{-1} was not made.

6 Final Remarks

It would not be suspected that a quantum logic of observables, as it was formulated here, is just the same algebra of observables logically "disguised" Even if one is convinced that the formalism of algebra of observables and Hilbert spaces is the only language of quantum theory, then a quantum logic of observables should be regarded as a translation into another language which principles (just as in a case of natural language) are not so apparent. As a consequence, this leads to the blatant equivocation of translation apart from that the translation into another language always works towards engaging new readers (logicians, in our case).

Since even Birkhoff and von Neumann reconstructed their calculus of quantum propositions as a kind of formal system for which one can construct many models, then the same concerns a quantum logic of observables. As such, a strong determination of QLO with a class of all QLO-frames might be considered absolutely independently from the properties of quantum systems, required by algebraic approach, and serves as a genuine feature of such systems distinguishing them within a class of physical systems whatsoever.

In this connection it worth denoting that the peculiarity of QLO (common for all logical systems) consisting in its interpretability for all valuations in all frames of some sort (on which the proof of soundness and completeness is based) is strikingly similar to the requirement of Gelfand-Neimark-Segal construction to the contemporary physical theory which is completely beyond the scope of traditional formulation of quantum theory (developed, for instance, in Dirac's way). According to this requirement for a Hilbert space, in which observables act as operators, is not obligatory to be inherent for neither a theory nor a model investigated and must depends on the state or just the way of obtaining an exploring system. Thus, the essential feature of a theory would be its multi-space character. An analogy with the strong determination of QLO appears to be a very transparent.

By and large we can think that formulations of QLO and NQLO take into account and logically interpret all 10 axioms of a general justification of algebraic approach in [8]. Moreover, systems of TQLO and TNQLO describe some cases related to issues of the global theories of quantum systems. Meanwhile, the following questions arise: how describe in the framework of QLO the fact that a fundamental space of quantum mechanics is an infinite-dimensional Hilbert space under the field of complex numbers? How QLO is connected with an axiomatic probability theory, i.e. how to introduce the notion of probability in QLO?

The first question was asked way back in sixtieth by J.M. Jauch but he mainly was interested the nature of this phenomenon. Curiously enough in the framework of QLO such question arises just during the formulation the notion of QLO-full sets but it does not act on the axiomatics. But if we try directly pass to the case of complex algebra of observables, then in QLO it will result on the formulation of the valuation at QLO-frame or $\mathcal{M}_{QLO}$-

matrix. It may be shown that if we introduce one more operator of negation (i.e. if we shall distinguish kinds of negation analogously, for example, the distinction between the contrary and contradictory negation in logics), then an axiomatics of QLO will be modified very insignificantly But since algebraically to this operator will correspond an involution, then $\mathcal{M}_{QLO}$-matrix becomes a complex one. Hence, the question arises: what is a complex truth-value from a logical point of view?

It turns out that in this case a logical explication should be done (e.g. to interpret its complex part as contextual truthfulness). Moreover, we can yield a complex-valued extensions of some known logical systems. The result in this direction should be found in [10].

The second question would be answered resorting to the following reasoning. An algebraic approach to the global theory of quantum systems attends to a notion of the state as a positive linear functional with the unit norm. If one interpret such state as an operator of quantum modality (e.g. "possibility") then at $\mathcal{M}_{QLO}$-matrix it leads to the formulation $\triangle(x) \in [0,1]$, i.e. $\triangle A$ should characterize an opportunity of measuring (existence) of the observable A in some state. Then the value $\rho(\varphi, A, M)$, which in Mackey's formalism describes the probability for A in a state φ to have during measuring a value belonging to the set $M \in \mathcal{M}_{QLO}$, should be counted as a matrix definition of the operator of quantum necessity.

Finally, regarding the opportunity of yielding a relativistic version of QLO we can expect, that the passage to a system with more sophisticated temporal operator and next with the spatio-temporal one will meet the needs of a relativistic theory of quantum systems.

References

1. N.D. Belnap, T.B. Steel: *The Logic of Questions and Answers* (Yale University Press, New Haven, London 1976)
2. G. Birkhoff, J. v. Neumann: 'The Logic of Quantum Mechanics'. Ann. of Math. **37** (1936) pp. 823–843
3. C.C. Chang: 'Logic With Positive and Negative Truth Values'. Acta Phil. Fennica Fasc. **XVI** (1963) pp. 19–39
4. C.C. Chang, H.J. Keisler: *Continuous Model Theory* (Princeton University Press, Princeton 1966)
5. G.G. Emch: *Algebraic Methods in Statistical Mechanics and Quantum Field Theory* (Wiley-Interscience, New York 1972)
6. R.I. Goldblatt: 'Semantic Analysis of Orthologic'. Journal of Philosophical Logic **3** 1–2 (1974) pp. 19–35
7. R.I. Goldblatt: 'The Stone Space of an Ortholattice'. Bull. London. Math. Soc. **7** 1 (1975) pp.45–48
8. G.G. Mackey: *The Mathematical Foundations of Quantum Mechanics* (W.A. Benjamin, New York 1963)

9. V.L. Vasyukov: 'Quantum Logic of Time'. In: *Logical Investigations* (Proceedings of the Scientific-Research Seminar on Logic of the Institute of Philosophy USSR Academy of Sciences, Moscow 1983) pp. 93–102 (in Russian)
10. V.L. Vasyukov: 'Complex-valued Logics or How to Regard Context in Logical Systems'. In: *Non-Standard Semantics for Non-Classical Logics* (Proceedings of the Scientific-Research Seminar on Logic of the Institute of Philosophy USSR Academy of Sciences, Moscow 1987) pp. 15–38 (in Russian)
11. V.L. Vasyukov: 'A Quantum Logic of Observables: Reconstruction and Semantic Analysis'. In: *A Syntactic and Semantic Investigations of Non-Extensional Logics* ed. by V.A. Smirnov (Nauka, Moscow 1989) pp. 120–169 (in Russian)
12. V.L. Vasyukov: *The Completeness of the Factor Semantics for Łukasiewicz's Infinite-valued Logics.* Studia Logica **52** 1 (1993) pp. 143–167

An Unsharp Quantum Logic from Quantum Computation

Gianpiero Cattaneo, Maria Luisa Dalla Chiara, and Roberto Giuntini

1 Introduction

One of the most interesting logical suggestions that arise from quantum-computation theory is to use the quantum-theoretical formalism in order to represent parallel reasoning ([3], [4]). As is well known, the unit of measurement in classical information theory is the *bit*: one *bit* measures the information quantity that can be either transmitted or received whenever one chooses one element from a set consisting of two elements (say, from the set $\{0, 1\}$). From the intuitive point of view, both the objects 0 and 1 can be imagined as a well-determined *state* of a classical physical system (for instance, the state of a tape cell in a given machine).

Let us now refer to a quantum-physical system. Following von Neumann's axiomatization, the *pure states* of our system are mathematically represented by unitary vectors in a convenient abstract space: a *Hilbert space* $\mathcal{H}$. From the intuitive point of view, a pure state represents *maximal information* about the physical system under investigation: information that cannot be consistently extended to richer information in the framework of the theory. Let us suppose the simplest situation, where our Hilbert space $\mathcal{H}$ has dimension 2. In such a case $\mathcal{H}$ will have a *basis* consisting of two unitary elements: thus any vector of the space will be representable as a *superposition* (or *linear combination*) of the basis-elements. Following Dirac's notation, we will indicate the vectors of $\mathcal{H}$ by $|\psi\rangle$, $|\varphi\rangle$,...; while the basis-elements will be indicated by $|0\rangle$, $|1\rangle$. As a consequence, for any unitary $|\psi\rangle$ we will have:

$$|\psi\rangle = a_0|0\rangle + a_1|1\rangle,$$

where the coefficients a_0, a_1 are complex numbers such that their modules $|a_0|$, $|a_1|$ satisfy the following relation:

$$|a_0|^2 + |a_1|^2 = 1.$$

According to the standard interpretation of the quantum formalism, pure states that are superpositions of other states (say, of basis-elements) codify a characteristic quantum indeterminacy. In contrast to classical mechanics, a quantum pure state generally assigns to a given physical property only a *probability-value*, and not a classical *truth-value* (either *True* or *False*). Any state $|\psi\rangle$ has some *certain* properties (with probability-value 1), some *impossible* properties (with probability-value 0) and many *indeterminate* properties (with probability-value different both from 1 and from 0). When $|\psi\rangle$ has the form $a_0|0\rangle + a_1|1\rangle$, this means that the physical system in state $|\psi\rangle$ *might* satisfy with probability $|a_0|^2$, those properties that are certain for state $|0\rangle$, and might satisfy with probability $|a_1|^2$, those properties that are certain for state $|1\rangle$.

The axiom of quantum theory that determines such probability-values is the so-called "Born rule". Since $|\psi\rangle$ is a unitary vector, one can easily show that $|a_0|^2, |a_1|^2 \in [0,1]$.

Let us now try and apply such a formalism to a quantum-information theory. The basic idea is to generalize the concept of *bit*, by introducing the notion of *qubit* or *quantum bit*. A *qubit* is any unitary vector in the Hilbert space $\mathcal{H}$ (of dimension 2) based on the set of all ordered pairs of complex numbers. Thus any *qubit* will have the form:

$$|\psi\rangle = a_0|0\rangle + a_1|1\rangle.$$

The interpretation is determined by the Born rule. Suppose that (similarly to the classical case) the pure states $|0\rangle$ and $|1\rangle$ represent two maximal and precise pieces of information. Then the superposition-state

$$|\psi\rangle = a_0|0\rangle + a_1|1\rangle$$

will represent information that involve a certain degree of *uncertainty*. In particular, the number $|a_0|^2$ will correspond to the probability-value of the information described by the basic state $|0\rangle$; while $|a_1|^2$ will correspond to the probability-value of the information described by the basic state $|1\rangle$.

In this context, it makes sense to imagine $|\psi\rangle$ as an *epistemic state* that stocks two precise pieces of information *in parallel*: the information $|0\rangle$ and the information $|1\rangle$. As a consequence, differently from the classical *bit*, the *qubit* permits us to represent two atomic pieces of information, in parallel.

Let us now consider a situation characterized by many *bits* or *qubits*. In classical information theory, a system consisting of n *bits* is naturally represented by a sequence of n elements belonging to the set $\{0,1\}$ (i.e., as an element of the set $\{0,1\}^n$). In the framework of quantum computation, it is convenient to adopt the tensor product formalism, which is used in quantum theory in order to represent compound physical systems. Assume a two-particle quantum system:

$$S = S_1 + S_2.$$

For instance, S_1 and S_2 might correspond, respectively, to the two electrons in a given helium atom. In such a case, the Hilbert space $\mathcal{H}^S$ associated to S will be the tensor product $\mathcal{H}^{S_1} \otimes \mathcal{H}^{S_2}$ of the two Hilbert spaces $\mathcal{H}^{S_1}$ e $\mathcal{H}^{S_2}$, which are associated to S_1 and S_2, respectively. Thus any pure state of S will be a unitary vector in the space $\mathcal{H}^S$.

A particularly interesting case is represented by those vectors $|\psi\rangle$ of $\mathcal{H}^S$ that can be expressed as tensor product of two vectors $|\psi_1\rangle$ e $|\psi_2\rangle$, belonging to $\mathcal{H}^{S_1}$ and to $\mathcal{H}^{S_2}$, respectively. In other words:

$$|\psi\rangle = |\psi_1\rangle \otimes |\psi_2\rangle.$$

In such cases, one usually speak of *factorized states.* It is worthwhile noticing that not all vectors of $\mathcal{H}^S$ can be expressed in such a simple form.

How to represent, in this framework, a system consisting of n *qubits*? It seems quite natural to describe our system as the pure state of a compound physical system consisting of n quantum objects. On this basis our n-*qubit* system can be identified with a unitary vector of the product space

$$\underbrace{\mathcal{H} \otimes \ldots \otimes \mathcal{H}}_{n-times}.$$

Instead of $\underbrace{\mathcal{H} \otimes \ldots \otimes \mathcal{H}}_{n-times}$ we will also write: $\otimes^n\mathcal{H}$. Particularly interesting examples will be represented by the vectors of $\otimes^n\mathcal{H}$ whose form is:

$$|x_1\rangle \otimes \ldots \otimes |x_n\rangle,$$

where each $|x_i\rangle$ is an element of the basis of $\mathcal{H}$ (i.e., $|x_i\rangle = |0\rangle$ or $|x_i\rangle = |1\rangle$). One can prove that the set of all vectors having this form represents a basis for the product space $\otimes^n\mathcal{H}$.

How can we deal with the concept of *quantum computation*, in this framework? The basic idea is to describe a computation by means of that kind of process that corresponds to the dynamic evolution of a quantum system. Suppose a physical system S, whose pure state at time t_0 is the vector $|\psi(t_0)\rangle$ (where $|\psi(t_0)\rangle$ belongs to the Hilbert space $\mathcal{H}^S$ associated to S). Owing to Schrödinger's equation, for any time t (where either $t \leq t_0$ or $t_0 \leq t$), there exists an operator U^t that determines the state of the system at time t as a function of the state of the system at time t_0. In other words:

$$|\psi(t)\rangle = U^t|\psi(t_0\rangle.$$

Any operator U^t of this kind is *unitary.* This means that our operator preserves the length of the vectors and the orthogonality relation. Furthermore, it is *reversible*: as a consequence one can go from $|\psi(t_0)\rangle$ to $|\psi(t)\rangle$ and *vice versa*, without any dissipation of information.

Thus, it makes sense to represent a quantum computation by means of convenient unitary operators assuming arguments and values in particular sets of *qubit* systems. As we have seen, *qubits* are generally superposition-states; as a consequence we will obtain some typical *parallel configurations.* Let us describe a particular example. Consider the basis $\{|0\rangle, |1\rangle\}$ of $\mathcal{H}$. The quantum-theoretical formalism permits us to transform each element of the basis (which represents precise information) into a *qubit*, which represents a "maximally uncertain information". It is sufficient to apply the following transformations:

$$|0\rangle \mapsto \tfrac{1}{\sqrt{2}}|0\rangle + \tfrac{1}{\sqrt{2}}|1\rangle \qquad \text{H1}$$
$$|1\rangle \mapsto \tfrac{1}{\sqrt{2}}|0\rangle - \tfrac{1}{\sqrt{2}}|1\rangle \qquad \text{H2.}$$

In both cases, the probability that the transformed *qubit* satisfies those properties that are certain for $|0\rangle$ (respectively, for $|1\rangle$) is $\frac{1}{2}$. In contrast to from classical theory, in the framework of quantum computation, one can transform a certain information into an uncertain information in a *reversible* way! What seems to be *prima facie* a disadvantage, turns out to be instead one of the most powerful resources of quantum computation. As an example, suppose we want to represent classically (in binary form) all the natural numbers between 0 and 7. We will need 8 strings of 3 *bits*: the number 0 will be represented by the string 000, the number 1 by the string 001,..., the number 7 by the string 111. Hence, a 3-*bit* system can only describe one of the numbers between 0 and 7. Let us now turn to the quantum case. Here, the 8 classical strings may be associated to the vectors of the basis of $\otimes^3\mathcal{H}$:

$$\begin{aligned} 0 &\mapsto |000\rangle \\ 1 &\mapsto |000\rangle \\ &\vdots \\ 7 &\mapsto |111\rangle . \end{aligned}$$

However, the crucial point is that, in such a case, any vector $|\psi\rangle$ of $\otimes^3\mathcal{H}$ is able to "stock" the 8 precise pieces of information ($|000\rangle, |001\rangle, \ldots, |111\rangle$) in a parallel way. It is sufficient to consider, for instance, the state

$$|\psi\rangle := \frac{1}{\sqrt{8}}(|000\rangle + |001\rangle + |010\rangle + |011\rangle + |100\rangle + |101\rangle + |110\rangle + |111\rangle).$$

How can we obtain the state $|\psi\rangle$? By applying the transformation H1 (which transforms a certain information into an uncertain one) to each *qubit* $|0\rangle$ of state $|000\rangle$. At the same time, we also know how to render uncertain the information described by $|1\rangle$ (by applying the transformation H2). As a consequence, we will be able to transform each vector of the basis of $\otimes^3\mathcal{H}$. Let us call H such a transformation (where "H" stands for *Hadamard*). We will have, for instance:

$$|101\rangle \mapsto_H \frac{1}{\sqrt{8}}(|000\rangle - |001\rangle + |010\rangle - |011\rangle - |100\rangle + |101\rangle - |110\rangle + |111\rangle).$$

Our H represents an important example of a unitary operator.

It is interesting to isolate some important unitary operators that turn out to have a special role in quantum computation. Such operators have been called *logical gates*.[1] The name "logical gate" is due to the following reason: when restricted to the elements of the basis ($|0\rangle$, $|1\rangle$), our operators turn out to behave as classical truth-functions.

All this suggests, in a natural way, a new form of quantum-logical semantics, whose main features are quite different from orthodox *quantum logic* (introduced by Birkhoff and von Neumann)([1], [2]). In the next sections, we will describe the general ideas of this semantics. Technical details and further developments will be given in other papers.

2 Quantum-Logical Gates

Consider the two-dimensional Hilbert space $\mathcal{H}$, based on dC^2 (the second cartesian power of the set of all complex numbers dC). Any vector $|\psi\rangle$ will be represented as a pair of complex numbers. Let $B = \{|0\rangle, |1\rangle\}$ be an *orthonormal basis* for $\mathcal{H}$. Thus the elements of B are two particular unitary vectors that are mutually *orthogonal* (i.e. their inner product is 0).

Definition 1. (*Qubit*)
A qubit is any unitary vector $|\psi\rangle$ of the space $\mathcal{H}$.

Hence, any *qubit* will generally have the following form:

$$|\psi\rangle = a_0|0\rangle + a_1|1\rangle,$$

where $a_0, a_1 \in$ dC and $|a_0|^2 + |a_1|^2 = 1$.

Let us first introduce some notation. We will use $x, y, \ldots$ as variables ranging over the set $\{0, 1\}$. At the same time, $|x\rangle, |y\rangle, \ldots$ will range over the vectors $\{|0\rangle, |1\rangle\}$.
Furthermore, we will use the following abbreviations:
$\otimes^2\mathcal{H} := \mathcal{H} \otimes \mathcal{H}$
$\otimes^n\mathcal{H} := \underbrace{\mathcal{H} \otimes \ldots \otimes \mathcal{H}}_{n-times}$
$|x_i\rangle^n := |x_1\rangle \otimes \ldots \otimes |x_n\rangle \in \otimes^n\mathcal{H}$.
We will also write $|x_1, \ldots, x_n\rangle$ instead of $|x_1\rangle \otimes \ldots \otimes |x_n\rangle$.
By $|\psi\rangle^n$ we will indicate any vector of the product space $\otimes^n\mathcal{H}$. One can prove that the dimension of such a space is 2^n. Furthermore, the set of all vectors having the form $|x_i\rangle^n$ represents an orthonormal basis for the space $\otimes^n\mathcal{H}$.

Definition 2. (*n-qubit system or n-register*)
An n-qubit system or n-register is any unitary vector $|\psi\rangle$ in the product space $\otimes^n\mathcal{H}$.

[1] Notice that the term "logical gate" is used somewhat ambiguously in the quantum-computation literature. Sometimes "logical gates" are defined as unitary operators having a special form. Sometimes, instead, "logical gate" is simply used as a synonymous for unitary operator.

Now we want to introduce some examples of *quantum-logical gates*. Generally, a quantum-logical gate can be described as a special unitary operator, assuming arguments and values in a product Hilbert space $\otimes^n \mathcal{H}$. First, we will study the so-called *Toffoli logical gate*, which we will simply call *Toffoli gate* ([6]). It will be expedient to start by analysing the simplest case, when our Hilbert space has the form:

$$\otimes^3 \mathcal{H} = \mathcal{H} \otimes \mathcal{H} \otimes \mathcal{H}.$$

In such a case, we will have an example of a Toffoli gate, that transforms the vectors of $\otimes^3 \mathcal{H}$ into vectors of $\otimes^3 \mathcal{H}$. In order to stress that our operator is defined on the product space $\otimes^3 \mathcal{H}$, we will indicate it by $T^{(1,1,1)}$. Since we want to define a unitary operator, it will be sufficient to determine its behaviour for the elements of the basis. Let us recall that unitary operators are *linear*. Hence, they preserve superpositions: in other words, any linear combination of a given set of vectors is transformed by our operator into the same linear combination of the transformed vectors. As we already know, the elements of the basis of $\otimes^3 \mathcal{H}$ have the form $|x\rangle \otimes |y\rangle \otimes |z\rangle$ (where $x, y, z \in \{0, 1\}$).

Definition 3. (*The Toffoli gate* $T^{(1,1,1)}$)
TheToffoli gate $T^{(1,1,1)}$ *is the linear operator* $T^{(1,1,1)} : \otimes^3 \mathcal{H} \to \otimes^3 \mathcal{H}$ *that is defined for any element* $|x\rangle \otimes |y\rangle \otimes |z\rangle$ *of the basis as follows:*

$$T^{(1,1,1)}(|x\rangle \otimes |y\rangle \otimes |z\rangle) = |x\rangle \otimes |y\rangle \otimes |min(x, y) \oplus z\rangle,$$

where $\oplus$ *represents the sum modulo* 2.

Clearly, $T^{(1,1,1,)}$ transforms any product vector $|x\rangle \otimes |y\rangle \otimes |z\rangle$ into the product that is obtained by leaving unchanged the first two factors ($|x\rangle$ and $|y\rangle$) and by transforming the third factor $|z\rangle$ as follows::

$$|z\rangle \mapsto |\min(x, y) \oplus z\rangle.$$

From the intuitive point of view, it seems quite natural to "see" the gate $T^{(1,1,1)}$ as a kind of "truth-table" that transforms triplets of *zeros* and of *ones* into triplets of *zeros* and of *ones*. The "table" we will obtain is the following:

$$\begin{array}{lcl}
|0,0,0\rangle & \mapsto & |0,0,0\rangle \\
|0,0,1\rangle & \mapsto & |0,0,1\rangle \\
|0,1,0\rangle & \mapsto & |0,1,0\rangle \\
|0,1,1\rangle & \mapsto & |0,1,1\rangle \\
|1,0,0\rangle & \mapsto & |1,0,0\rangle \\
|1,0,1\rangle & \mapsto & |1,0,1\rangle \\
|1,1,0\rangle & \mapsto & |1,1,1\rangle \\
|1,1,1\rangle & \mapsto & |1,1,0\rangle.
\end{array}$$

In the first six cases, $T^{(1,1,1)}$ behaves like the identity operator; in the last two cases, instead, our gate transforms the last element of the triplet into the opposite element (0 is transformed into 1 and 1 transformed into 0).

One can easily show that $T^{(1,1,1)}$ has been well defined for our aims: one is dealing with an operator that is not only linear but also unitary. The matrix representation of $T^{(1,1,1)}$ is the following:

$$\begin{pmatrix} 1 & 0 & 0 & 0 & 0 & 0 & 0 & 0 \\ 0 & 1 & 0 & 0 & 0 & 0 & 0 & 0 \\ 0 & 0 & 1 & 0 & 0 & 0 & 0 & 0 \\ 0 & 0 & 0 & 1 & 0 & 0 & 0 & 0 \\ 0 & 0 & 0 & 0 & 1 & 0 & 0 & 0 \\ 0 & 0 & 0 & 0 & 0 & 1 & 0 & 0 \\ 0 & 0 & 0 & 0 & 0 & 0 & 0 & 1 \\ 0 & 0 & 0 & 0 & 0 & 0 & 1 & 0 \end{pmatrix}.$$

By using $T^{(1,1,1)}$, we can introduce a convenient notion of *conjunction*. Our conjunction, which will be indicated by AND, is characterized as a function whose arguments are pairs of vectors in $\mathcal{H}$ and whose values are vectors of the product space $\otimes^3\mathcal{H}$.

Definition 4. (AND)
For any $|\varphi\rangle \in \mathcal{H}$ *and any* $|\psi\rangle \in \mathcal{H}$*:*

$$\mathtt{AND}\,(|\varphi\rangle, |\psi\rangle) := T^{(1,1,1)}\,(|\varphi\rangle \otimes |\psi\rangle \otimes |0\rangle)\,.$$

Let us immediately check that AND represents a good generalization of the corresponding classical truth-function. For the arguments $|0\rangle$ and $|1\rangle$ we will obtain the following "truth-table":

$$\begin{array}{lll} (|0\rangle, |0\rangle) & \mapsto & T^{(1,1,1)}(|0\rangle \otimes |0\rangle \otimes |0\rangle) = |0\rangle \otimes |0\rangle \otimes |0\rangle \\ (|0\rangle, |1\rangle) & \mapsto & T^{(1,1,1)}(|0\rangle \otimes |1\rangle \otimes |0\rangle) = |0\rangle \otimes |1\rangle \otimes |0\rangle \\ (|1\rangle, |0\rangle) & \mapsto & T^{(1,1,1)}(|1\rangle \otimes |0\rangle \otimes |0\rangle) = |1\rangle \otimes |0\rangle \otimes |0\rangle \\ (|1\rangle, |1\rangle) & \mapsto & T^{(1,1,1)}(|1\rangle \otimes |1\rangle \otimes |0\rangle) = |1\rangle \otimes |1\rangle \otimes |1\rangle. \end{array}$$

One immediately realizes the difference with respect to the classical case. The classical truth-table represents a typical irreversible transformation:

$$\begin{array}{lll} (0,0) & \mapsto & 0 \\ (0,1) & \mapsto & 0 \\ (1,0) & \mapsto & 0 \\ (1,1) & \mapsto & 1. \end{array}$$

The arguments of the function determine the value, but not the other way around. As is well known, irreversibility generally brings about dissipation of information. Mathematically, however, any Boolean function

$f : \{0,1\}^n \to \{0,1\}^m$ can be transformed into a *reversible* function $\hat{f} : \{0,1\}^n \times \{0,1\}^m \to \{0,1\}^n \times \{0,1\}^m$ in the following way:

$$\forall u \in \{0,1\}^n \; \forall v \in \{0,1\}^m : \hat{f}((u,v)) = (u, v \oplus f(u)),$$

where $\oplus$ is the sum modulo 2 pointwise defined. As one can easily realize the function that is obtained by making reversible the irreversible classical "and" is precisely the Toffoli gate. The classical "and" is then recovered by fixing the third input bit to 0. Accordingly, the three arguments $(0,0)$, $(0,1)$, $(1,0)$ turn out to correspond to three distinct values, represented by the triplets $(0,0,0)$, $(0,1,0)$, $(1,0,0)$. The price we have paid in order to obtain a reversible situation is increasing complexity of our Hilbert space. The function AND associates to pairs of arguments, belonging to the two-dimensional space $\mathcal{H}$, values belonging to the space $\otimes^3\mathcal{H}$ (whose dimension is 2^3).

All this happens in the simplest situation, when one is only dealing with elements of the basis (in other words, with precise pieces of information). Let us examine the case where the function AND is applied to arguments that are superpositions of the basis elements in the space $\mathcal{H}$. Consider the following *qubit* pair:

$$|\psi\rangle = a_0|0\rangle + a_1|1\rangle, \; |\varphi\rangle = b_0|0\rangle + b_1|1\rangle.$$

By applying the definitions of AND and of $T^{(1,1,1)}$, we obtain:

$$\mathtt{AND}(|\psi\rangle, |\varphi\rangle) = a_1 b_1 |1,1,1\rangle + a_1 b_0 |1,0,0\rangle + a_0 b_1 |0,1,0\rangle + a_0 b_0 |0,0,0\rangle.$$

This result suggests a quite natural logical interpretation. The four basis-elements that occur in our superposition-vector correspond to the four cases of the truth-table for the classical conjunction:

$$(1,1,1), (1,0,0), (0,1,0), (0,0,0).$$

However, here, in contrast to the classical situation, each case is accompanied by a complex number, which represents a characteristic quantum *amplitude*. By applying the "Born rule" we will obtain the following interpretation: $|a_1 b_1|^2$ represents the probability-value that both the *qubit*-arguments are equal to $|1\rangle$, and consequently their conjunction is $|1\rangle$. Similarly in the other three cases.

How to deal in this context with the concept of negation? A characteristic of quantum computation is the possibility of defining a plurality of negation operations, all representing good generalizations of the classical negation. For our aims, we will chose a function NOT that can be easily defined in terms of the Toffoli gate.

Consider first the simplest case, concerning the negation of a single *qubit*. In such a case, the function NOT will be a unary function assuming arguments in the space $\mathcal{H}$ and values in the space $\otimes^3\mathcal{H}$.

Definition 5. (NOT)
For any $|\varphi\rangle \in \mathcal{H}$:

$$\texttt{NOT}\,(|\varphi\rangle) := T^{(1,1,1)}\,(|\varphi\rangle \otimes |1\rangle \otimes |1\rangle)\,.$$

One can immediately check that NOT represents a good generalization of the classical truth-table. Consider the basis-elements $|0\rangle$ and $|1\rangle$. In such a case we will obtain:

$$\texttt{NOT}(|1\rangle) = T^{(1,1,1)}(|1\rangle \otimes |1\rangle \otimes |1\rangle) = |1\rangle \otimes |1\rangle \otimes |0\rangle;$$

$$\texttt{NOT}(|0\rangle) = T^{(1,1,1)}(|0\rangle \otimes |1\rangle \otimes |1\rangle) = |0\rangle \otimes |1\rangle \otimes |1\rangle.$$

Let us examine the two values of the function NOT, that means the two products $|1\rangle \otimes |1\rangle \otimes |0\rangle$ and $|0\rangle \otimes |1\rangle \otimes |1\rangle$. Clearly, in both cases, the first factor corresponds to the classical argument, while the third factor corresponds to the classical value for that argument ($|1\rangle$ is transformed into $|0\rangle$ and viceversa). As to the second factor (which is always $|1\rangle$), we can interpret it as a kind of "open control channel".

Finally, how to introduce a reasonable disjunction? To this aim, it will be first expedient to define a new logical gate that will be called a *dual Toffoli gate* and will be indicated by $Q^{(1,1,1)}$. Similarly to $T^{(1,1,1)}$, the operator $Q^{(1,1,1)}$ also assumes arguments in the space $\mathcal{H}$ and values in the space $\otimes^3\mathcal{H}$. The dual Toffoli gate is obtained by making reversible the classical "or".

Definition 6. (*The dual Toffoli gate* $Q^{(1,1,1)}$)
The dual Toffoli gate $Q^{(1,1,1)}$ *is the linear operator* $Q^{(1,1,1)} : \otimes^3\mathcal{H} \to \otimes^3\mathcal{H}$ *that is defined for any element* $|x\rangle \otimes |y\rangle \otimes |z\rangle$ *of the basis in the following way:* $Q^{(1,1,1)}(|x\rangle \otimes |y\rangle \otimes |z\rangle) = |x\rangle \otimes |y\rangle \otimes |max(x,y) \oplus z\rangle$, *where* $\oplus$ *represents the sum modulo* 2.

Clearly $Q^{(1,1,1)}$ has a dual behaviour with respect to $T^{(1,1,1)}$.
The "truth-table" corresponding to $Q^{(1,1,1)}$ will be the following:

$$\begin{array}{lcl}
|0,0,0\rangle & \mapsto & |0,0,0\rangle \\
|0,0,1\rangle & \mapsto & |0,0,1\rangle \\
|0,1,0\rangle & \mapsto & |0,1,1\rangle \\
|0,1,1\rangle & \mapsto & |0,1,0\rangle \\
|1,0,0\rangle & \mapsto & |1,0,1\rangle \\
|1,0,1\rangle & \mapsto & |1,0,0\rangle \\
|1,1,0\rangle & \mapsto & |1,1,1\rangle \\
|1,1,1\rangle & \mapsto & |1,1,0\rangle.
\end{array}$$

The matrix representation of $Q^{(1,1,1)}$ is the following:

$$\begin{pmatrix} 1 & 0 & 0 & 0 & 0 & 0 & 0 & 0 \\ 0 & 1 & 0 & 0 & 0 & 0 & 0 & 0 \\ 0 & 0 & 0 & 1 & 0 & 0 & 0 & 0 \\ 0 & 0 & 1 & 0 & 0 & 0 & 0 & 0 \\ 0 & 0 & 0 & 0 & 0 & 1 & 0 & 0 \\ 0 & 0 & 0 & 0 & 1 & 0 & 0 & 0 \\ 0 & 0 & 0 & 0 & 0 & 0 & 0 & 1 \\ 0 & 0 & 0 & 0 & 0 & 0 & 1 & 0 \end{pmatrix}.$$

We can now introduce a disjunction function (OR), assuming as arguments pairs of vectors of $\mathcal{H}$ and as values vectors of $\otimes^3\mathcal{H}$.

Definition 7. (OR)
For any $|\varphi\rangle \in \mathcal{H}$ *and any* $|\psi\rangle \in \mathcal{H}$:

$$\mathtt{OR}\,(|\varphi\rangle, |\psi\rangle) := Q^{(1,1,1)}\,(|\varphi\rangle \otimes |\psi\rangle \otimes |0\rangle)\,.$$

The function OR also represents a good generalization of the corresponding classical truth-function. For the arguments $|0\rangle$ and $|1\rangle$ we will obtain the following "truth-function":

$$\begin{array}{lcl} (|0\rangle, |0\rangle) & \mapsto & Q^{(1,1,1)}(|0\rangle \otimes |0\rangle \otimes |0\rangle) = |0\rangle \otimes |0\rangle \otimes |0\rangle \\ (|0\rangle, |1\rangle) & \mapsto & Q^{(1,1,1)}(|0\rangle \otimes |1\rangle \otimes |0\rangle) = |0\rangle \otimes |1\rangle \otimes |1\rangle \\ (|1\rangle, |0\rangle) & \mapsto & Q^{(1,1,1)}(|1\rangle \otimes |0\rangle \otimes |0\rangle) = |1\rangle \otimes |0\rangle \otimes |1\rangle \\ (|1\rangle, |1\rangle) & \mapsto & Q^{(1,1,1)}(|1\rangle \otimes |1\rangle \otimes |0\rangle) = |1\rangle \otimes |1\rangle \otimes |1\rangle. \end{array}$$

The logical gates we have considered so far refer to a very special situation, characterized by a Hilbert space having the form $\otimes^3\mathcal{H}$. However, our procedure can be easily generalized. Both the Toffoli gate and its dual can be defined in any Hilbert space having the form:

$$\otimes^n\mathcal{H} \otimes \otimes^m\mathcal{H} \otimes \mathcal{H} \; (= \otimes^{n+m+1}\mathcal{H}).$$

Definition 8. (*The Toffoli gate* $T^{(n,m,1)}$)
The Toffoli gate $T^{(n,m,1)}$ *is the linear operator*

$$T^{(n,m,1)} : \otimes^n\mathcal{H} \otimes \otimes^m\mathcal{H} \otimes \mathcal{H} \;\rightarrow\; \otimes^n\mathcal{H} \otimes \otimes^m\mathcal{H} \otimes \mathcal{H}$$

that is defined for any basis-element $|x_1, \ldots, x_n\rangle \otimes |y_1, \ldots, y_m\rangle \otimes |z\rangle$ *as follows:* $T^{(n,m,1)}(|x_1, \ldots, x_n\rangle \otimes |y_1, \ldots, y_m\rangle \otimes |z\rangle) =$
$|x_1, \ldots, x_n\rangle \otimes |y_1, \ldots, y_m\rangle \otimes |min(x_n, y_m) \oplus z\rangle$, *where* $\oplus$ *represents the sum modulo 2.*

Definition 9. (*The dual Toffoli gate* $Q^{(n,m,1)}$)
The dual Toffoli gate $Q^{(n,m,1)}$ *is the linear operator*

$$Q^{(n,m,1)} : \otimes^n\mathcal{H} \otimes \otimes^m\mathcal{H} \otimes \mathcal{H} \;\rightarrow\; \otimes^n\mathcal{H} \otimes \otimes^m\mathcal{H} \otimes \mathcal{H}$$

that is defined for any basis element $|x_1,\ldots,x_n\rangle \otimes |y_1,\ldots,y_m\rangle \otimes |z\rangle$ *as follows:*

$$Q^{(n,m,1)}(|x_1,\ldots,x_n\rangle \otimes |y_1,\ldots,y_m\rangle \otimes |z\rangle) = \\ |x_1,\ldots,x_n\rangle \otimes |y_1,\ldots,y_m\rangle \otimes |max(x_n,y_m) \oplus z\rangle,$$

where $\oplus$ *represents the sum modulo* 2.

On this basis one can immediately generalize our definitions of AND and of OR.

Definition 10. (AND)
For any $|\varphi\rangle^n \in \otimes^n \mathcal{H}$ *and any* $|\psi\rangle^m \in \otimes^m \mathcal{H}$*:*

$$\texttt{AND}\,(|\varphi\rangle^n, |\psi\rangle^m) := T^{(n,m,1)}\,(|\varphi\rangle^n \otimes |\psi\rangle^m \otimes |0\rangle)\,.$$

Definition 11. (OR)
For any $|\varphi\rangle^n \in \mathcal{H}$ *and any* $|\psi\rangle^m \in \mathcal{H}$*:*

$$\texttt{OR}\,(|\varphi\rangle^n, |\psi\rangle^m) := Q^{(n,m,1)}\,(|\varphi\rangle^n \otimes |\psi\rangle^m \otimes |0\rangle)\,.$$

Finally, the function NOT can also be generalized as follows:

Definition 12. (NOT)
For any $|\varphi\rangle^n \in \mathcal{H}$*:*

$$\texttt{NOT}\,(|\varphi\rangle^n) := T^{(n,1,1)}\,(|\varphi\rangle^n \otimes |1\rangle \otimes |1\rangle)\,.$$

3 Quantum-Computational Semantics

In the standard logical semantics, based on Birkhoff and von Neumann's approach ([1]) *propositions* (representing possible *meanings* of linguistic sentences) are identified with closed subspaces of a Hilbert space. A pure state $|\psi\rangle$ of the physical system S under investigation will verify a sentence α whenever ψ belongs to the proposition that represents, in the semantics, the meaning of α.

On this basis one obtains an example of a *possible world semantics*: pure states correspond to *possible worlds*, which can be correlated by means of an *accessibility relation*; at the same time, *propositions* are represented by sets of worlds.

The starting point of our computational semantics is quite different. The meanings of the linguistic sentences are here represented by *qubits* or *qubit-systems*, instead of sets of possible worlds. In other words, the meaning of a sentence is identified with the *information quantity* encoded by the sentence in question (where information is, of course, measured by means of the quantum unit).

Let us now describe technically such a semantics. Consider a standard sentential language $\mathcal{L}$ with the following connectives: the *negation* ($\neg$), the *conjunction* ($\wedge$) and the *disjunction* ($\vee$). We will indicate the sentential letters of $\mathcal{L}$ by $p_1, p_2, \ldots, p_n, \ldots$. The notion of *sentence* (or *formula*) of $\mathcal{L}$ is defined in the expected way. Let $Form^{\mathcal{L}}$ represent the set of all sentences of $\mathcal{L}$. We will use the following metavariables: $p, q, \ldots$ will range over sentential letters, while $\alpha, \beta, \gamma, \ldots$ will range over generic sentences.

Let us now introduce the basic concept of our semantics: the notion of a *quantum-computational model.* Quantum-computational models represent particular interpretations of the language $\mathcal{L}$, such that the meaning associated to any sentence is either a *qubit* or a *qubit-system.* As a consequence, the *space of the meanings* corresponds here to a variable Hilbert space (instead of a unique Hilbert space). Any space of this kind will be a product space $\otimes^n \mathcal{H}$. Needless to say, the complexity of our space will depend on the value of n, which, in turn, depends on the *length* of the sentence to be interpreted (where the length of a sentence is the number of its connectives). The formal definition is the following:

Definition 13. (*Quantum-computational model*)
A quantum-computational model of $\mathcal{L}$ is a function Qub associating to any sentence α a qubit-system in a Hilbert space $\otimes^n \mathcal{H}$ (where n is a function of the length of α):

$$Qub : Form^{\mathcal{L}} \rightarrow \bigcup_n \otimes^n \mathcal{H}.$$

We will also write $|\alpha\rangle$ instead of $Qub(\alpha)$; and we will call $|\alpha\rangle$ the information-value of α. The following conditions must be satisfied:

(i) $|p\rangle \in \mathcal{H}$, *where $\mathcal{H}$ is the two-dimensional Hilbert space based on* $\mathrm{d}C$;
(ii) Let $|\beta\rangle \in \otimes^n \mathcal{H}$. *Then*
$|\neg\beta\rangle = \mathtt{NOT}(|\beta\rangle) \in \otimes^n \mathcal{H} \otimes \mathcal{H} \otimes \mathcal{H}$;
(iii) Let $|\beta\rangle \in \otimes^n \mathcal{H}$, $|\gamma\rangle \in \otimes^m \mathcal{H}$. *Then:*
$|\beta \wedge \gamma\rangle = \mathtt{AND}(|\beta\rangle, |\gamma\rangle) \in \otimes^n \mathcal{H} \otimes \otimes^m \mathcal{H} \otimes \mathcal{H}$;
(iv) Let $|\beta\rangle \in \otimes^n \mathcal{H}$, $|\gamma\rangle \in \otimes^m \mathcal{H}$. *Then:*
$|\beta \vee \gamma\rangle = \mathtt{OR}(|\beta\rangle, |\gamma\rangle) \in \otimes^n \mathcal{H} \otimes \otimes^m \mathcal{H} \otimes \mathcal{H}$.

Our definition univocally determines, for any sentence α, the Hilbert space $\otimes^n \mathcal{H}$ to which $|\alpha\rangle$ belongs. Clearly, the number n is a function of the number of the connectives of α: all the information-values of atomic sentences will belong to the two–dimensional space $\mathcal{H}$; the information–value of the conjunction of two atomic sentences will belong to the space $\otimes^3 \mathcal{H}$, and so on. Since the meaning associated to a given sentence reflects the logical form of the sentence in question, we can say that our semantics has a typical *intensional* character. In order to stress that the information-value of α belongs to $\otimes^n \mathcal{H}$, we will also write: $|\alpha\rangle^n$.

As we have seen, a characteristic of our semantics is to identify the meanings of the linguistic sentences with unitary vectors of variable Hilbert spaces.

As a consequence, we obtain that the information-value of a sentence naturally determines a probability-value for that sentence. It will be sufficient to apply, in a convenient way, the Born rule.

Let *Qub* be a quantum-computational model and let α be any sentence. We want to define the probability-value of α determined by *Qub*. We will denote such a value by $Prob^{Qub}(\alpha)$, or also by $Prob(|\alpha\rangle)$.

Let us first consider the simplest case, where α is an atomic sentence; in this case, its information-value will belong to the two-dimensional space $\mathcal{H}$. Suppose, for instance, that $|\alpha\rangle$ has the form:

$$a_0|0\rangle + a_1|1\rangle.$$

Then (recalling the Born rule) the probability-value of α can be naturally defined as follows:

$$Prob(\alpha) = |a_1|^2.$$

Thus, $Prob(\alpha) = |a_1|^2$ represents the probability that our *uncertain* information $|\alpha\rangle$ corresponds to the *precise* information $|1\rangle$.

From the intuitive point of view, our definition, clearly, attributes a privileged role to one of the two basic *qubits* (belonging to the basis of $\mathcal{H}$): the *qubit* $|1\rangle$. In such a way, $|1\rangle$ is dealt with as the truth-value *True*.

Consider now the case of a molecular sentence α. Its information-value $|\alpha\rangle$ will belong to the space $\otimes^n\mathcal{H}$, where n (≥ 3) depends on the length of α. The dimension of $\otimes^n\mathcal{H}$ is 2^n. Hence, $|\alpha\rangle^n$ will generally be a superposition of elements of the basis of $\otimes^n\mathcal{H}$. Thus, we will have:

$$|\alpha\rangle^n = \sum_{i=1}^{2^n} c_i|x_i\rangle^n,$$

where $|x_i\rangle^n$ ranges over the basis of $\otimes^n\mathcal{H}$.

From the logical point of view, any $|x_i\rangle^n$ (element of the basis of $\otimes^n\mathcal{H}$) represents a possible case of a "reversibile truth-table" for α. For instance, suppose α has the form $p \vee q$, where:

$$|p\rangle = a_0|0\rangle + a_1|1\rangle, \ |q\rangle = b_0|0\rangle + b_1|1\rangle.$$

By applying the definitions of a quantum–computational model and of `OR`, we will obtain:

$$|p \vee q\rangle = a_1b_1|1,1,1\rangle + a_1b_0|1,0,1\rangle + a_0b_1|0,1,1\rangle + a_0b_0|0,0,0\rangle.$$

We know that the number $|a_1b_1|^2$ represents the probability that both the members of our disjunction are true and that, consequently, the disjunction is true. Similarly in the other cases. In order to calculate the probability of the truth of $p \vee q$, it will be sufficient to sum the three probability-values corresponding to the three cases where the final result is *True* (that is the cases of the vectors $|1,1,1\rangle$, $|1,0,1\rangle$, $|0,1,1\rangle$). On this basis, we will be able to assign to the disjunction $p \vee q$ the following probability-value:

$$|a_1 b_1|^2 + |a_1 b_0|^2 + |a_0 b_1|^2 .$$

This example can be easily generalized to any sentence of whatever length. Consider again:

$$|\alpha\rangle^n = \sum_{i=1}^{2^n} c_i |x_i\rangle^n .$$

Let us first define a particular set of coefficients that occur in the superposition-state $\sum_{i=1}^{2^n} c_i |x_i\rangle^n$ Let:

$$C^+_{|\alpha\rangle} := \left\{ c_i \mid |x_i\rangle^n = \left\langle x_{i_1}, \ldots, x_{i_{n-1}}, 1 \right\rangle \right\} .$$

Clearly, the elements of $C^+_{|\alpha\rangle}$ represent the *amplitudes* associated to the different cases of the "truth-table" of α, such that their final result is $|1\rangle$. We can now define the probability-value of α as follows:

Definition 14. (*The probability-value of* α)

$$Prob(\alpha) := \sum_{c_i \in C^+_\alpha} |c_i|^2 .$$

On this basis, we can now define the notions (which represent the basic concepts of any semantics): *truth*, *logical consequence*, and *tautology*.

Definition 15. (*Truth in a computational model*)
A sentence α *is true in a computational model Qub* ($\models_{Qub} \alpha$) *iff* $Prob(\alpha) = 1$.

Definition 16. (*Consequence in a computational model*)
β *is a consequence of* α *in the computational model Qub* ($\alpha \models_{Qub} \beta$) *iff* $Prob(\alpha) \leq Prob(\beta)$.

Definition 17. (*Logical consequence*)
β *is a logical consequence of* α ($\alpha \models \beta$) *iff for any Qub:* $\alpha \models_{Qub} \beta$.

Definition 18. (*Tautology*)
α *is a tautology* ($\models \alpha$) *iff for any Qub:* $\models_{Qub} \alpha$.

Let us call the logic characterized by this semantics *quantum–computational logic* (QCL).

Some interesting examples of logical consequences that hold in QCL are the following:

(i) $\alpha \models \neg\neg\alpha, \quad \neg\neg\alpha \models \alpha;$
(double negation)

(ii) $\alpha \wedge \beta \models \beta \wedge \alpha, \quad \alpha \vee \beta \models \beta \vee \alpha$;
(commutativity)

(iii) $\alpha \wedge (\beta \wedge \gamma) \models (\alpha \wedge \beta) \wedge \gamma, \quad (\alpha \wedge \beta) \wedge \gamma \models \alpha \wedge (\beta \wedge \gamma)$;
(associativity)

(iv) $\alpha \vee (\beta \vee \gamma) \models (\alpha \vee \beta) \vee \gamma, \quad (\alpha \vee \beta) \vee \gamma \models \alpha \vee (\beta \vee \gamma)$;
(associativity)

(v) $\neg(\alpha \wedge \beta) \models \neg\alpha \vee \neg\beta, \quad \neg\alpha \vee \neg\beta \models \neg(\alpha \wedge \beta)$;
(de Morgan)

(vi) $\neg(\alpha \vee \beta) \models \neg\alpha \wedge \neg\beta, \quad \neg\alpha \wedge \neg\beta \models \neg(\alpha \vee \beta)$
(de Morgan)

(vii) $\alpha \wedge \alpha \models \alpha$.
(semi-idempotence 1)

Some consequences and tautologies that are violated in QCL are the following:

(i) $\alpha \not\models \alpha \wedge \alpha$;
(semi-idempotence 2)

(ii) $\not\models \alpha \vee \neg\alpha$
(excluded middle)

(iii) $\not\models \neg(\alpha \wedge \neg\alpha)$;
(noncontradiction)

(iv) $(\alpha \wedge \beta) \vee (\alpha \wedge \gamma) \not\models \alpha \wedge (\beta \vee \gamma)$.
(distributivity)

QCL turns out to be a weak form of quantum logic. Conjunction and disjunction do not correspond to lattice operations, because they are not generally idempotent. In contrast to standard quantum logic, the weak distributivity principle ($(\alpha \wedge \beta) \vee (\alpha \wedge \gamma) \models \alpha \wedge (\beta \vee \gamma)$) also breaks down. Both the excluded middle and the noncontradiction principles are violated: as a consequence, we have obtained an example of an *unsharp logic*.

Open Problems

i) Is QCL axiomatizable?
ii) Is the class of all tautologies of QCL nonempty?
iii) Are QCL and *Łukasiewicz quantum logic* (the logic characterized by the class of all QMV algebras ([5])) comparable?

References

1. G. Birkhoff, J. von Neumann: 'The Logic of Quantum Mechanics'. Annals of Mathematics **37** (1936) pp. 823–843
2. M.L. Dalla Chiara, R. Giuntini: 'Quantum Logics'. In: *Handbook of Philosophical Logic* a cura di G. Gabbay e F. Guenthner, di prossima pubblicazione presso Kluwer, Dordrecht

3. J. Preskill: *Quantum Information and Computation.* Lecture Notes in Physics, vol. 229 (Springer, Berlin 1999)
4. A. Ekert, P. Hayden, H. Inamori: 'Basic Concepts in Quantum Computation'. preprint
5. R. Giuntini: 'Quantum MV-algebras'. Studia Logica **56** (1996) pp. 393–417
6. T. Toffoli: 'Reversible Computing'. In: *Automata, Languages and Programming.* Lecture Notes in Computer Science, vol. 84a ed. by J.W. de Baker, J. van Leeuwen, (Springer, Berlin 1980)

Quantum Logic and Quantum Probability

Enrico G. Beltrametti

1 Introduction

By *events*, or yes-no experiments, pertaining to some physical system we understand the physical quantities, or observables, that admit only two outcomes. Since the 1936 seminal work of G. Birkhoff and J. Neumann [4] it is recognized that, in the framework of physical systems exhibiting a quantum behaviour, the algebraic structure associated to the events is not an algebraic model of classical logic, it is the algebraic model of a new logic, to be called quantum logic. This fact outlines a deep departure from the realm of classical physics where the events pertaining to a physical system carry the structure of a Boolean algebra, hence an algebraic model of classical logic. In Sect. 2 we shall review the structure of the events of classical and of quantum events and we will recall the main branching point.

The word *state* is generally used to capture the idea of the preparation procedure to which the physical system has been submitted: so it embodies all those attributes of the physical system that are accidental, in the sense that they may be different in different situations and may change with time. Both in the classical and in the quantum case it makes sense to speak of probability measures on the ordered structure of the events: these probability measures indeed correspond to the notion of *states* of a physical system. Of course it has to be expected that the difference between the structures of classical and quantum events determines a difference between the corresponding probability measures. It is an empirical fact that in several circumstances pertaining to the microworld Nature refuses to fit with the frame of standard, or Kolmogorovian, probability theory: interferometry phenomena and the empricial violation of the so-called Bell inequalities are typical examples of this. In other words, there are features that pertain to the classical probabilities but not to the quantum ones, and textitvice versa. These features mirror the different mathematical structures associated with classical and quantum states. These aspects will be reviewed in Sect. 3: in particular we will focus on the so-called Bell inequalities whose violation is usually viewed as a typical signature of quantum probabilities.

In Sect. 4 we will briefly advance the possibilitiy of constructing a common extension of the standard classical probability theory and of the quantum probability theory. This new framework adopts a convex set of states that preserves the classical structure of a simplex but admits a class of observables wider than the usual family of random variables: instead of the standard ones, typical of the statistical theories that have a deterministic basis, we come to a more general notion of random variables that captures an idea of fuzziness, typical of indeterministic statistical theories. Such a new framework, to be called fuzzy or operational probability theory, proves to host significant quantum features.

2 The Ordered Structure of the Events

Loosely speaking, an observable pertaining to a physical system is any quantity that can be measured about that physical system. The family of possible values a given observable can take is called the spectrum of that observable. Here we shall restrict ourselves to those observables that can take just two values: in other words their spectrum is a two-point set. The restriction to these two-valued observables, to be called events, appears natural once we realize that starting from any (nondichotomic) observable A we can build up a two-valued one by taking some subset X of its spectrum and posing the question whether the value of A falls in X. The symbols $a, b, \ldots$ will be used to denote the events of a physical system and the two-valuedness of an event will be often expressed by saying that the event occurs or does not occur.

The basic structure pertaining to the set of events is the (partial) order relation: indeed, we can say that the event a in any state α is less than the probability of occurrence of b in that same state. Formally:

$$a \leq b \text{ iff } p_\alpha(a) \leq p_\alpha(b) \text{ for every state } \alpha. \tag{1}$$

With respect to this order relation the trivial event 0 that never occurs and the trivial event 1 that always occurs are obviously the minimal and, respectively, the maximal event. We shall further say that the event c is the join, or least upper bound, of a and b, and write $c = a \vee b$, if c is a minimal majorization of both a and b (namely if $a, b \leq c$ but $c \leq d$ for any d such that $a, b \leq d$). Similarly, we shall say that c is the meet, or greatest lower bound, of a and b. It is a relevant fact that in both the classical and the quantum case the meet and the join always exist: this fact is summarised by saying that the set of the events pertaining to any physical system classical or quantum)is a lattice.

In both the classical and the quantum case the set of events can be further endowed with a unary relation $a \rightarrow a^\perp$, called orthocomplementation, defined by the properties: (i) $a^{\perp\perp} = a$, (ii) $a \leq b$ implies $b^\perp \leq a^\perp$, (iii) $a \wedge a^\perp = 0$, $a \vee a^\perp = 1$. The set of events thus becomes an orthocomplemented lattice. We can view the event $a^\perp$ as obtained by taking the event a and interchanging its two outcomes ("occurrence" and "nonoccurrence").

As pointed out by G. Birkhoff and J. von Neumann [4], the branching between the classical and the quantum case comes when we consider the distributivity of the join with respect to the meet (and *vice versa*). Indeed, in the classical case the distributivity equalities

$$a \wedge (b \vee c) = (a \wedge b) \vee (a \wedge c), a \vee (b \wedge c) = (a \vee b) \wedge (a \vee c) \tag{2}$$

hold true for any triple a, b, c, while in the quantum case they are guaranteed only for special triples of events, namely for triples a, b, c, in which $a \leq b$ and $c = a^{\perp}$. These facts are summarized by saying that the events of a classical system form a Boolean algebra, while the events of a quantum system form the weaker sturcture of an orthomodular lattice.

Boolean algebras are algebraic models of classical logic (more specifically of classical propositional calculus) with the algebraic operations of meet ($\wedge$) and join ($\vee$) corresponding to the logical connectives "and", "or", and the unary relation of orthocomplentation corresponding to the logical negation. The rules and tautologies of classical logic have their counterpart in equations and identities of Boolean algebras. It is known that Boolean algebras admit a representation in terms of the algebra of subsets of a set: such a representation of classical events will be recalled in Sect. 4.

The fact that the events of a classical physical system provide a model of classical logic leaves open the way to think that logic might have an empirical root, as long as we limit our empirical experience to the realm of classical physics. On the other hand, the events of a quantum-physical system provide a model of a weaker logic, generally called quantum logic. It is not the purpose of this paper to go further into a discussion of quantum logic: we refer to [8, 9, 19, 20, 21] for a review of various aspects of the problem, and to [18] for an historical perspective. Here, let us only list a few questions that naturally arise and that have received attention in the literature. If logic is empirical, have we to think that quantum logic is more fundamental than classical logic in view of the fact that quantum mechanics provides a more fundamental description of natural phenomena than classical mechanics does? How can quantum mechanics, which is formulated in terms of classical logic, give rise to a new nonclassical logic? Is classical logic the metalogic of quantum logic? Does quantum logic reduce to classical logic if restricted to the sentences of mathematics? If, on the other hand, logic is regarded as a priori, valid irrespective of experience, are there approaches to quantum logic that do not reduce to reading it off from the behaviour of quantum phenomena?

3 Probability Measures on the Events

Both the Boolean algebras of events of the classical case and the (nondistributive) orthomodular lattices of the quantum case are sufficiently structured mathematical objects to allow the notion of probability measures on them. As already noted, the probability measures on the events can be identified with the states of the physical system. Since the convex combination of probability measures is still a probability measure, the above identification points immediately at the convex structure of the set of states.

As sketched in the previous section, the attention on the events focuses on the departures between classical logic (of which Boolean algebras are models) and quantum logic (of which orthomodular nondistributive lattices are models): shifting the attention on the states we focus on the departures betweeen classical probabilities and quantum probabilities.

The departure between the classical and quantum case does not go back, we believe, to the basic notion of probability nor to its frequency interpretation which naturally fits with the physical empirical situations, both classical and quantum. The root of that departure rather goes back to the fact that the quantum phenomena, contrary to the classical ones, can give rise to probabilistic behaviours that cannot be viewed as merely expressing some degree of ignorance inside the standard deterministic frame. This entails the consequence that the empirical probabilities emerging from quantum phenomena might not find a representation inside the standard Kolmogorov probability theory, based, as it is, on probability measures on Boolean algebras.

The problem of finding simple conditions that allow the representation of a given set of observed probabilities inside the classical probability theory was already tackled in a 1862 paper of G. Boole [13] who came to what he called the *conditions of possible experience:* of course, no "quantum experience" was in view at that time. But this problem received much attention one century later, with Bell's work [3] on the so-called hidden variable interpretation of the EPR correlation, a typical quantum phenomenon whose role was pointed out by the celebrated 1935 paper of Einstein, Podolsky and Rosen [16]. The preferred form of these conditions of classical representability is in terms of requiring some suitable linear combination of the given probabilities to lie inside a numerical interval: the resulting requirements being called Bell-type inequalities [1, 10, 11, 20].

Here we outline what we believe to be the mathematical essence of these inequalities, and take it as is a parapigmatic example of what may separate quantum from classical probabilities: for simplicity, we will consider only a very particular case.

Let us consider 4 events, say a_1, a_2, a_3, a_4, pertaining to a given physical system and let "yes" and "no" be their outcomes (the number 4 is not a need, it just fits with the usual EPR correlation). we write Y_i, N_i to denote the yes and no outcomes of a_i, $i = 1, 2, 3, 4$. Let $\mu_1, \mu_2, \mu_3, \mu_4$ be probability measures on these events determined by a state of the physical system: explicitely, μ_i will be a probability measure on the two-point set $\{Y_i, N_i\}$. Suppose also that some joint probability measures are observed: in our example let them be $\mu_{1,3}$, $\mu_{1,4}$, $\mu_{2,3}$, $\mu_{2,4}$. By $\mu_{i,j}, ij = 1, 2, 3, 4$, we understand a probability measure on the cartesian product $\{Y_j, N_j\} \times \{Y_i, N_i\}$, hence on a four-point set, to be understood as the probability measure associated to a coincidence observation of the outcomes of a_i, a_j. We now pose the question: Does the family $\{\mu_1, \mu_2, \mu_3, \mu_4, \mu_{1,3}, \mu_{2,3}, \mu_{2,4}\}$ admit the existence of a global (or "top") probability measure $\mu_{1,2,3,4}$, on the 16-point set $\{Y_1, N_1\} \times \{Y_2, N_2\} \times \{Y_3, N_3\} \times \{Y_4, N_4\}$ such that the given family of probability measures can be deduced by the mechanism of marginal projections? Now we have the following result[5]: the answer to the above question is positive when the inequality

$$\begin{aligned} 0 \quad \leq \quad & \mu_1(Y_1) + \mu_3(Y_3) - \mu_{1,3}(Y_1 Y_3) - \mu_{1,4}(Y_1 Y_4) - \\ & -\mu_{2,3}(Y_2 Y_3) + \mu_{2,4}(Y_2 Y_4) \quad \leq \quad 1 \end{aligned} \tag{3}$$

and the analogous ones obtained by interchanging the indices $1, 2$, or the indices 3,4, or by making both interchangings, are fulfilled. Notice that the probabilities $\mu_i(Y_i)$ of occurrence of a single event and the probabilities $\mu_{i,j}(Y_i Y_j)$ of joint occurrence of two events occur in these inequalities. They are indeed the so-called Bell inequalities (in one of their possible versions).

The existence of the global probability measure $\mu_{1,2,3,4}$, hence the fulfillment of the Bell inequalities, is a necessary condition [1, 10, 11] for the classical representability of the given probabilities, that is a necessary condition for viewing the events of our physical system as forming a Boolean algebra. The classical representability of the considered family of probability measures would be ruled out as soon as violations of Bell inequalities emerge. As is well known, this is what has been observed with some quantum systems, typically with pairs of photons coming from the cascade decaying of excited atoms or with pairs of photons produced by the nonlinear "down-conversion" mechanism in suitable crystals [15]. If we think of the two photons as one flying on the left and the other on the right, we can consider the events
a_1: "the left photon passes a linear polarizer oriented along the direction $\mathbf{x_1}$"
a_2: "the left photon passes a linear polarizer oriented along the direction $\mathbf{x_2}$"
a_3: "the right photon passes a linear polarizer oriented along the direction $\mathbf{x_3}$"
a_4: "the right photon passes a linear polarizer oriented along the direction $\mathbf{x_4}$"
and it is an empirical fact that the Bell inequalities are violated for suitable choices of the directions $\mathbf{x_1}, \mathbf{x}_2, \mathbf{x}_3, \mathbf{x}_1$.

Summing up, the quantum phenomena do not fit with the standard classical probability theory: they call into play a more general framework, often called quantum probability theory. We can read it off from the specific Hilbert space edifice of quantum mechanics, but, as we sketch in the next section, it is also possible to host the typical features of quantum probabilities in a natural extension of the standard classical probability theory.

4 The Operational Framework

The framework we are going to summarize is an extension of both the classical and the quantum probability theory. In the sequel we shall view the convex set of states as a primitive building block of the theory. Let us recall that in the classical case the convex set of states has the fundamental property of being a simplex, namely, any nonextreme element (any nonpure state) has a unique convex decomposition into extreme elements (pure states, while in the quantum case the decomposition of nonextreme elements into extreme ones is never unique so that the states do not form a simplex. Recall that the pure states correspond to maximally selective preparation of the physical system.

Since an observable is specified by its possible outcomes and by their probability distribution in each state of the physical system we can view an observable taking values in some (measurable) space Ξ as an affine map from the set of states into the family $M_1^+(\Xi)$ of the probability measures on Ξ (more precisely on the Boolean algebra $\mathcal{B}(\Xi)$ of the measurable subsets of Ξ). When Ξ is a two-element set we speak of an event.

Consider first the case of standard classical statistical mechanics. The set Ω of pure states is usually called the *phase space* and the set of all states will consist, due to its simplex structure, of the set $M_1^+(\Omega)$ of all the probability measures on Ω. Coming to the observables, or *random variables*, of the theory, a special choice is made: one adopts as observables taking values in a space Ξ (usually the real line) the affine maps $M_1^+(\Omega) \to M_1^+(\Xi)$ that have the property of mapping a measure concentrated at a point of Ω into a measure concentrated at a point of Ω into a measure concentrated at a point Ξ, in short that send Dirac measures into Dirac measures. This fact, which mirrors the deterministic basis of the theory, allows us to represent these observables by measurable functions from Ω into Ξ. If we deal with an event, namely if Ξ is a two-element set, then we get a splitting of Ω into two parts, one part consisting of the points of Ω that are mapped into one element of Ξ, the remaining part consisting of the points that are mapped into the other element of Ξ. It is then clear that the events can be represented by the subsets of Ω, thus forming the Boolean algebra $\mathcal{B}(\Omega)$.

Let us now come to the standard quantum case. The states of the physical system are now represented by the density operators of some separable, complex Hilbert space $\mathcal{H}$, and the pure states correspond to the one-dimensional projectors. Restricting ourselves to the observables that take values in the real line $\mathbf{R}$, the observables adopted by standard quantum mechanics are those affine functions from the set of states into the probability measures on $\mathbf{R}$ that admits the familiar representation in terms of self-adjoint operators of $\mathcal{H}$: if A is the self-adjoint operator representing the observable, then the probability of getting an outcome of the observable in the real interval $E \in \mathcal{B}(\mathbf{R})$, when the state of the physical system is the density operator D, is given by the trace of DP_A^E, where P_A^E is the projector associated to the pair A, E by the spectral theorem for self-adjoint operators (see, e.g., [8]). The intrinsic nondeterministic nature of the quantum framework is outlined by the fact that for every observable there are pure states that are mapped into diffused probability measures on the space of the possible outcomes. In other words, the observables have dispersion on pure states.

The *operational* framework is quite a natural generalization of the standard classical one. We preserve the simplex structure of the convex set of states, which thus takes the classical form $M_1^+(\Omega)$ for some measurable space Ω, but we take a wider family of observables dropping the deterministic requirement that an observable with outcome space Ξ need map Dirac measures on Ω into Dirac measures on Ξ. Thus the now adopted class of observables, or random variables, includes those which have dispersion on pure states. If we restrict ourselves to observables with just two possible outcomes we get something more general than the standard classical events: indeed, such an observable determines and is uniquely determined by the probability that each (pure) state assigns to one of te two outcomes. In other words, a two-valued observable is now associated to a function from Ω into the segment [0,1], namely to a fuzzy subset of Ω. This outlines the core of the generalization we are considering: in the standard classical case a two-valued observable was associated to a function from Ω into the set $\{0, 1\}$, namely to crisp subset of Ω.

The operational, or fuzzy, framework can host typical quantum features, like the occurrence of uncertainty relations, or like the appearence of probabilities that violate Bell-type inequalities [5]. This is not surprising in view of the indeterministic aspects of the observables there adopted. Actually, the operational frame can host an extension of quantum mechanics. Such an extension can be made explicit by taking for the measurable space Ω the set of quantum pure states (the one-dimensional projectors on a Hilbert space) and showing that each quantum observable admits a representative in the operational framework that shares all the statistical properties of the original observable. Of course, this representative will be a random variable of the operational scheme that does not reduce to one of the standard classical case. We refer to [6] for analysis of this issue. If we start from the operational frame we can determine the emergence of the usual quantum description as the effect of a restriction to a special class of fuzzy random variables. The restriction mechanism might cause the fact that some elements of $M_1^+(\Omega)$ become indistiguishable, thus leading to a coarsening of $M_1^+(\Omega)$ and to a new convex set of states which need not be simplex, while the selected subclass of fuzzy random variables which define the coarsening becomes the set of observables of the emerging nonclassical theory. In this way we can obtain the basic framework of quantal theories. The new probability theory that emerges from the operational framework, that we are going to call operational probability theory, has been advanced in [17]: a first step can be traced back to paper of Davies and Lewis [15], while an equivalent formulation has been proposed by Gudder [17].

Let us finally remark that the events occurring in the operational framework (they are often better called *effects*) no longer form a Boolean algebra as was the case in the standard classical case. This is related to the fact that they are associated with fuzzy subsets of a set, not with the ordinary crisp subsets of the standard case. Nor do the operational events need form an orthomodular lattice, as was the case in standard quantum mechanics: indeed we know that the operational framework can host an extension of quantum mechanics but it is more general than that. The algebraic structure of the operational events has been the object of many studies: we refer to [7] also for further references.

References

1. L. Accardi, A. Fedullo: 'On the Statistical Meaning of Complex Numbers in Quantum Mechanics'. Lett. Nuovo Cimento **34** (1982) pp. 161–172
2. A. Aspect: 'Experimental Tests of Bell's Inequalities With Correlated Photons'. In: *Waves, Information and Foundations of Physics* ed. by R. Pratesi, L. Ronchi (Societ Italiana di Fisica, Bologna 1998)
3. J.S. Bell: 'On the Einstein-Podolsky-Rosen Paradox'. Physics **1** (1964) pp. 195–200
4. G. Birkhoff, J. von Neumann: 'The Logic of Quantum Mechanics'. Ann. Math **37** (1936) p. 823
5. E.G. Beltrametti, S. Bugajski: 'The Bell Phenomenon in Classical Frameworks'. J. Phys. A: Math. Gen. **29** (1996) pp. 247–261
6. E.G. Beltrametti, S. Bugajski: 'A Classical Extension of Quantum Mechanics'. J. Phys. A: Math. Gen. **28** (1995) pp. 3329–3343
7. E.G. Beltrametti, S. Bugajski: 'Effect Algebras and Statistical Physical Theories'. J. Math. Phys. **38** (1997) pp. 3020–3030
8. E.G. Beltrametti, G. Cassinelli: *The Logic of Quantum Logics* (Addison Wesley, Reading 1981)
9. E.G. Beltrametti, B.C. van Fraassen (eds.): *Current Issues in Quantum Logic* (Plenum, New York 1981)
10. E.G. Beltrametti, M.J. Maczynski: 'On a Characterization of Classical and Nonclassical Probabilities'. J. Math. Phys. **32** (1991) pp. 1280–1286
11. E.G. Beltrametti, M.J. Maczynski: 'On the Characterization of Probabilities: A Generalization of Bell Inequalities'. J. Math. Phys. **34** (1993) pp. 4919–4929
12. E.G. Beltrametti, M.J. Maczynski: 'On the Intrinsic Characterization of Classical and Quantum Probabilities'. In: *Symposium on the Foundations of Quantum Physics* ed. by P. Busch, P. Lahti, P. Mittelstaedt (World Scientific, Singapore 1993)
13. G. Boole: 'On the Theory of Probability'. In: Phil. Trans. Royal Soc. London, **152** (1862) pp. 225–252
14. S. Bugajski: 'Fundamentals of Fuzzy Probability Theory'. Int. J. Theor. Phys. **35** (1996) pp. 2229–2244
15. E.B. Davies, J.T. Lewis: 'An Operational Approach to Quantum Probability'. Comm. Math. Phys. **17** (1970) pp. 239–260
16. A. Einstein, B. Podolsky, N. Rosen: 'Can Quantum-mechanical Description of Physical Reality be Considered Complete? Phys. Rev. **47** (1935) pp. 777–780

17. S. Gudder: 'Fuzzy Probability Theory'. Demonstratio Mathematica **31** (1998) pp. 235–354
18. M. Jammer: *The Philosophy of Quantum Mechanics* (Wiley, New York 1974)
19. P. Mittelstaedt: *Quantum Logic* (Reidel, Dordrecht 1978)
20. I. Pitowsky: 'Quantum-Probability Logic'. In: *Lecture Notes in Physics* Vol. **321** (Springer, Berlin 1989)
21. P. Ptak, S. Pulmannova: *Orthomodular Structures as Quantum Logics* (Kluwer Academic, Dordrecht 1991)

Operator Algebras and Quantum Logic

Miklós Rédei

1 The Idea of Quantum Logic: Logicization of Non-Boolean Algebras

Let $\mathcal{K} = (p, q \ldots; \&, \vee, \sim)$ be a zeroth-order formal language with sentence variables $p, q \ldots$, two place connectives & (*and*), $\vee$ (*or*) and negation sign $\sim$, and let $\mathcal{F}$ be the formula algebra (set of well-formed formulas in $\mathcal{K}$ defined in the standard way by induction from the sentence variables). If v is an assignment of truth values $1(true), 0(false)$ to the sentence variables $p, q \ldots$, then classical propositional logic is characterized by extending v by induction from $p, q \ldots$ to the formula algebra $\mathcal{F}$ in such a manner that the extension (called *interpretation* and also denoted by v) be a "homomorphism" from $\mathcal{F}$ into the two-element Boolean algebra $\mathcal{B}_2 \equiv (1, 0, \cap, \cup, \perp)$, where "homomorphism" means that the following hold

$$\begin{aligned} v(q\&p) &= v(q) \cap v(p) \\ v(q \vee p) &= v(q) \cup v(p) \\ v(\sim q) &= v(q)^{\perp}. \end{aligned}$$

If one is interested only in the truth values of formulas and the relation of truth values between different formulas, then one can identify formulas having the same truth value in every interpretation, i.e., one can then introduce the equivalence relation $\approx$ on $\mathcal{F}$ by $\phi \approx \psi$ iff $v(\phi) = v(\psi)$ for all $v \in V(\mathcal{F})$ ($V(\mathcal{F})$ denoting the set of all valuations on $\mathcal{F}$.) One can turn the set $\mathcal{F}_{\approx}$ of equivalence classes of $\mathcal{F}$ into a Boolean algebra $(\mathcal{F}_{\approx}, \cap, \cup, \perp)$ by "lifting" the connectives in $\mathcal{F}$ to $\mathcal{F}_{\approx}$ by defining

$$\begin{aligned} [\phi] \cap [\psi] &\equiv [\phi \& \psi] \\ [\phi] \cup [\psi] &\equiv [\phi \vee \psi] \\ [\phi]^{\perp} &\equiv [\sim \phi], \end{aligned}$$

where $[\phi]$ denotes the equivalence classes in $\mathcal{F}_{\approx}$. The valuation v can also be extended from $\mathcal{F}$ to $\mathcal{F}_{\approx}$ in a straightforward manner, the extension being a two-valued Boolean algebra homomorphism from $\mathcal{F}_{\approx}$ into $\mathcal{B}_2$. The elements in $\mathcal{F}_{\approx}$ are called *propositions*, the algebra $(\mathcal{F}_{\approx}, \cap, \cup, \perp)$ of propositions is called the *Tarski-Lindenbaum* algebra determined by $(\mathcal{F}, V(\mathcal{F}))$. A proposition can also be interpreted as the set of interpretations that make the proposition true, since an equivalent way of obtaining the Boolean algebra $(\mathcal{F}_{\approx}, \cap, \cup, \perp)$ is the following. For any formula ϕ one can introduce the evaluation function $e_\phi \colon V(\mathcal{F}) \to \{1, 0\}$ by $e_\phi(v) = v(\phi)$ and every such evaluation function can be identified with the subset x_ϕ of $V(\mathcal{F})$ on which it takes the value 1:

$$e_\phi \leftrightarrow x_\phi = \{v \in V(\mathcal{F}) : e_\phi(v) = v(\phi) = 1\}.$$

The map $[\phi] \mapsto x_\phi$ is a Boolean algebra isomorphism between $(\mathcal{F}_{\approx}, \cap, \cup, \perp)$ and a Boolean subalgebra of the power set $\mathcal{P}(V(\mathcal{F}))$.

In sum: viewed from an algebraic perspective, the meaningful statements and their logical relations in classical propositional logic can be identified with a Boolean algebra, the logical notion of interpretation (valuation) is a Boolean algebra homomorphism from the Boolean algebra of propositions into the two-element Boolean algebra and the semantic notions (such as tautology, contradiction, logical consequence, etc.) can be given in terms of this two-valued Boolean algebra homomorphism.

The main idea of quantum logic consists of replacing the Boolean algebra in the above-sketched algebrazation of classical propositional logic by a weaker algebraic structure: one asks whether a non-Boolean, orthocomplemented lattice $(\mathcal{L}, \wedge, \vee, \perp)$, can be "logicized" in the sense that there is a zeroth-order language $\mathcal{K}^{qm} = (P^{qm}, \&, \vee, \sim)$, a formula algebra $\mathcal{F}^{qm}$ built from $\mathcal{K}^{qm}$ and a suitable notion of interpretation (valuation) v^{qm} defined on P^{qm} and extended to $\mathcal{F}^{qm}$ that gives rise to an equivalence relation $\approx_{qm}$ on $\mathcal{F}^{qm}$ such that the Tarski-Lindenbaum structure $\mathcal{F}^{qm}_{\approx_{qm}}$ be isomorphic with the lattice $(\mathcal{L}, \wedge, \vee, \perp)$. This non-Boolean lattice would then be a "quantum logic" – "logic" just in the sense in which the Boolean algebra of propositions of a classical zeroth-order language represents the logic of this system, and "quantum" to the extent P^{qm}, $\mathcal{F}^{qm}$ and thus the non-Boolean orthocomplemented lattice $(\mathcal{L}, \wedge, \vee, \perp)$ are related in a natural way to propositions one can utter about a quantum system.

It turns out that such a "logicization" of certain non-Boolean lattices is possible, although the procedure of logicization raises a number of interpretational and philosophical issues that remain controversial. In what follows we describe those non-Boolean lattices that emerge naturally in connection with quantum systems whose observable quantities are described by certain operator algebras. These non-Boolean lattices are the projection lattices of von Neumann algebras; in particular, we recall briefly the relation between the type of the von Neumann algebra of observables of a quantum system and the properties of the associated projection lattice (Sect. 2). This is followed by sketching the procedure of "logicization" of these lattices (Sect. 3). Finally, we comment on the interpretational difficulties associated with the described notion of quantum logic (Sect. 4).

2 Von Neumann Algebras and Their Projection Lattices

There are two natural candidates for algebraic structures that represent weaker structures than Boolean algebras: orthomodular and modular lattices. An orthocomplemented lattice $(\mathcal{L}, \wedge, \vee, \perp)$ is *orthomodular* if the following equation holds:

$$\text{If} \quad A \leq B \quad \text{and} \quad A^{\perp} \leq C, \quad \text{then} \quad A \vee (B \wedge C) = (A \vee B) \wedge (A \vee C). \tag{1}$$

Orthomodularity is a weakening of the following *distributivity* law

$$A \vee (B \wedge C) = (A \vee B) \wedge (A \vee C) \quad \text{for all } A,B,C. \tag{2}$$

But orthomodularity is not the minimal weakening of distributivity: The following *modularity* property

$$\text{If} \quad A \leq B, \quad \text{then} \quad A \vee (B \wedge C) = (A \vee B) \wedge (A \vee C) \tag{3}$$

is strictly weaker than distributivity but strictly stronger than orthomodularity.

Typical examples of orthomodular and modular lattices are the projection lattices $\mathcal{P}(\mathcal{M})$ of von Neumann algebras $\mathcal{M}$. The set $\mathcal{M}$ of bounded operators on a Hilbert space $\mathcal{H}$ is a von Neumann algebra if it contains the identity operator I, is closed with respect to the adjoint operation and is closed in the strong operator topology. The latter requirement means that if Q_n is a sequence of operators from $\mathcal{M}$ such that for all $\xi \in \mathcal{H}$ we have $Q_n\xi \to Q\xi$ for some bounded operator on $\mathcal{H}$, then $Q \in \mathcal{M}$ (see [17] for the theory of von Neumann algebras).

If S is any set of bounded operators on $\mathcal{H}$, then its (first) commutant S' is the set of bounded operators that commute with every element in S, i.e.:

$$S' \equiv \{Q : QX = XQ, \text{ for all } X \in S\}.$$

The operation of taking the commutant can be iterated: $S'' \equiv (S')'$, and it is clear that S is contained in the second commutant, so the second commutant S'' is an extension of S. How much of an extension? The answer to this question, von Neumann's double commutant theorem, is the most fundamental result in the theory of von Neumann algebras:

Proposition 2.1 (Double commutant theorem) *S is strongly dense in S''.*

What is striking in this theorem is that a topological property is expressed in purely algebraic terms; in particular, Proposition 2.1 implies that a $*$-algebra of bounded operators on a Hilbert space that contains the unit is a von Neumann algebra if and only if it coincides with its second commutant.

A von Neumann algebra $\mathcal{N}$ is called a *factor* if the only elements in $\mathcal{N}$ that commute with every other element in $\mathcal{N}$ are the constant multiples of the identity, i.e. if $\mathcal{N} \cap \mathcal{N}' = \{\lambda I : \lambda \text{ a complex number}\}$.

An immediate corollary of the double commutant theorem is the characteristic property of von Neumann algebras that they contain many projections; in fact, they contain enough projections for the set of projections $\mathcal{P}(\mathcal{M})$ to determine the von Neumann algebra completely, in the sense $\mathcal{M} = (\mathcal{P}(\mathcal{M}))''$. Moreover, we have

Proposition 2.2 *The set $\mathcal{P}(\mathcal{M})$ of projections of a von Neumann algebra $\mathcal{M}$ is an orthomodular lattice (called a von Neumann lattice).*

In particular the set $\mathcal{P}(\mathcal{H})$ of *all* projections on a Hilbert space (called a Hilbert lattice and denoted by $\mathcal{P}(\mathcal{H})$) is a complete orthomodular lattice, since the set $\mathcal{B}(\mathcal{H})$ of all bounded operators on $\mathcal{H}$ is a von Neumann algebra.

While all von Neumann lattices are orthomodular, some have the stronger property of modularity There is a subtle connection between the type of a von Neumann algebra (in the sense of the Murray-von Neumann classification theory) and the modularity of its projection lattice. To make the connection explicit we recall the basic idea of the classification theory (for a detailed presentation of the classification theory see [17], for a short one see [7] and [11]).

The fact that the projection lattice generates the algebra $\mathcal{M} = (\mathcal{P}(\mathcal{M}))''$ indicates that by investigating this lattice structure one gains insight into the structure of the algebra itself. Indeed, von Neumann and Murray used this lattice structure to classify von Neumann algebras. The key concept in the classification is the equivalence of projections: two projections A and B in $\mathcal{M}$ are called equivalent (notation: $A \sim B$) with respect to the algebra $\mathcal{M}$ if there is an operator ("partial isometry") in $\mathcal{M}$ that takes the vectors in $A^{\perp}$ into zero and is an isometry between the subspaces A and B. This equivalence can be expressed by saying that the dimension (with respect to $\mathcal{M}$) of the subspace A projects to is equal to the dimension (with respect to $\mathcal{M}$) of the subspace B projects to. The relation $\sim$ is an equivalence relation in $\mathcal{P}(\mathcal{M})$ Let $\mathcal{P}(\mathcal{M})_{\sim}$ be the set of equivalence classes. With the help of $\sim$ one can introduce a $\preceq$ partial ordering in $\mathcal{P}(\mathcal{M})$ ("Cantor-Bernstein type theorem"): $A \preceq B$ if there is a projection B' that is smaller (with respect to $\leq$) than B and which is equivalent with A, that is: $A \sim B' \leq B$. Intuitively $A \preceq B$ means: the (relative) dimension of A is not greater than the (relative) dimension of B. A projection A is called *finite* if it does not contain any projection which is equivalent to A, i.e., if $B \leq A$ and $B \sim A$ implies $A = B$.

From the point of view of classification of von Neumann algebras the important fact concerning $\preceq$ is that given any two projections A, B, they can be cut *by a projection commuting with every element in* $\mathcal{M}$ into two pieces that can be compared in the ordering $\preceq$. That is to say ("Comparison

Theorem"): for any two $A, B \in \mathcal{P}(\mathcal{M})$ there is a projection $Z \in \mathcal{P}(\mathcal{M}) \cap \mathcal{P}(\mathcal{M})'$ such that

$$ZAZ \preceq ZBZ \quad \text{and} \quad (I-Z)B(I-Z) \preceq (I-Z)A(I-Z).$$

It follows that if $\mathcal{M}$ is a factor von Neumann algebra then $\mathcal{P}(\mathcal{M})_{\sim}$ is *totally* ordered with respect to $\preceq$: either $A \preceq B$, or $B \preceq A$ holds for any A, B. Since $\preceq$ is defined algebraically it follows that the ordering $\preceq$ is algebra invariant: two-factor von Neumann algebras can not be isomorphic if the orderings of the corresponding $\mathcal{P}(\mathcal{M})_{\sim}$ is different.

Proposition 2.3 *If $\mathcal{M}$ is a factor von Neumann algebra, then there exists a map d (unique up to multiplication by a constant) defined on $\mathcal{P}(\mathcal{M})$ and taking its values in the closed interval $[0, \infty]$ and which has the following properties:*

(i) $d(A) = 0$ *if and only if* $A = 0$
(ii) *If* $A \perp B$, *then* $d(A+B) = d(A) + d(B)$
(iii) $d(A) \leq d(B)$ *if and only if* $A \preceq B$
(iv) $d(A) < \infty$ *if and only if A is a finite projection*
(v) $d(A) = d(B)$ *if and only if* $A \sim B$
(vi) $d(A) + d(B) = d(A \wedge B) + d(A \vee B)$

The map d is called the *dimension* function on $\mathcal{P}(\mathcal{M})$. This proposition implies that the order type of $\mathcal{P}(\mathcal{M})_{\sim}$ can be read off the order type of the range of the function d. Murray and von Neumann determined the possible ranges of d in [5]. The result is shown in the table below (by choosing suitable normalization of the function d).

Range of d	Type of factor $\mathcal{N}$	The lattice $\mathcal{P}(\mathcal{N})$
$\{0, 1, 2, \ldots n\}$	I_n	modular, atomic
$\{0, 1, 2, \ldots \infty\}$	I_∞	orthomodular nonmodular atomic
$[0, 1]$	II_1	modular nonatomic
$\{x \mid 0 \leq x \leq \infty\}$	II_∞	orthomodular nonmodular nonatomic
$\{0, \infty\}$	III	orthomodular nonmodular nonatomic

This table shows that besides the von Neumann algebra $\mathcal{B}(\mathcal{H})$ of all bounded operators there are other types of von Neumann algebras. These other types have properties *radically* different from the properties of $\mathcal{B}(\mathcal{H})$.

For instance, in sharp contrast to the type **I** case, the projection lattice of a type $\mathbf{II}_1$ algebra has no atoms and it is not only orthomodular but also modular, which $\mathcal{P}(\mathcal{H})$ is not. Furthermore, there are no smallest (nonzero) dimensional projections (atoms) in the lattice of a type $\mathbf{II}_1$ von Neumann algebra, although every projection has a unique (finite) dimension. This justifies calling the theory of $\mathbf{II}_1$ factors the theory of continuous dimension.

As the table shows, there are two types of von Neumann algebras for which the corresponding von Neumann lattices are modular: the $\mathbf{I}_n$ and the $\mathbf{II}_1$ types. In both cases the range of the dimension function is bounded, and these two von Neumann algebras factors are called *finite* – they contain only finite projections.

The type **III** algebras differ from $\mathcal{B}(\mathcal{H})$ even more than the type $\mathbf{II}_1$ does: they contain no finite projection at all. It is worth mentioning that at the time of the discovery (1935-1936) of the classification of factors it was not even known whether type **III** factors exist. It was only four years later that von Neumann was able to construct the first example [19] and it was only in the mid sixties that the existence of a continuum number of mutually nonisomorphic type **III** factors was proven [8]. Physical examples of type **III** factors are the algebras $\mathcal{A}(W)$ of observables that are localized (in the sense of relativistic quantum field theory) in wedge regions W of the spacetime. Type $\mathbf{II}_1$ algebras had been constructed by Birkhoff and von Neumann in 1936, however. Such algebras appear as algebras describing the observables of infinite lattice gases in the thermodynamical limit (see [3]).

A distinguishing feature of a type $\mathbf{II}_1$ factor $\mathcal{M}$ is that the dimension function d defined on $\mathcal{P}(\mathcal{M})$ can be extended to a *unique, finite trace* on $\mathcal{M}$. The linear map τ on a von Neumann algebra $\mathcal{M}$ is called *trace* if

$$\tau(XY) = \tau(YX). \tag{4}$$

For instance, the usual trace

$$\tau(X) = Tr(X) = \sum_i \langle \eta_i, X\eta_i \rangle \tag{5}$$

(η_i being an orthonormed basis in $\mathcal{H}$) defines a trace on $\mathcal{B}(\mathcal{H})$ in the sense of the above definition of trace, which is, however, not finite unless $\mathcal{H}$ is finite-dimensional. If $\mathcal{H}$ is finite-dimensional then the summation in (5) is finite and $\tau(X)$ is then the ordinary trace (also called "Spur") of the matrix X.

3 Von Neumann Lattices as Logics

The set $S_{sa}^{\mathcal{N}}$ of (not necessarily bounded) self-adjoint operators whose spectral projections belong to a von Neuman algebra $\mathcal{N}$ are interpreted in quantum physics as the mathematical representatives of the observable quantities of a quantum system.

Given $S^{\mathcal{N}}_{sa}$ one can consider the set $P_{\mathcal{N}}$ of sentence variables defined by

$$P_{\mathcal{N}} \equiv \{sent(Q,d) : Q \in S^{\mathcal{N}}_{sa}, d \in B(\mathrm{I\!R})\}, \tag{6}$$

where $B(\mathrm{I\!R})$ is a the Boolean algebra of Borel measurable subsets of the real numbers. From $P_{\mathcal{N}}$ one can build the formula algebra $(\mathcal{F}^{\mathcal{N}}, \&, \vee, \sim)$ in the standard way by induction.

We now formulate a notion of interpretation of the formula algebra $\mathcal{F}^{\mathcal{N}}$. We do this entirely formally here, without giving any interpretation (in the intuitive sense) to the interpretation – we will comment on the interpretational issues in Sect. 4.

Each interpretation is given by a vector $\xi \in \mathcal{H}$. First we define the truth value of the atomic sentences $sent(Q,d)$ and then the interpretation will be extended to the nonatomic formulas $\alpha, \beta \ldots \in \mathcal{F}^{\mathcal{N}}$. (In what follows, P^Q denotes the spectral measure of the self-adjoint operator Q.)

(i) $sent(Q,d)$ is true in ξ iff $\xi \in \mathrm{P}^Q(d)$;
$sent(Q,d)$ is false in ξ iff $\xi \in \mathrm{P}^Q(d)^\perp$.

(ii) $(\sim \alpha)$ is true in ξ iff α is false in ξ;
$(\sim \alpha)$ is false in ξ iff α is true.
Let $[\alpha]$ denote the set of states (interpretations) that make α true. Then the truth conditions of $(\sim \alpha)$ can be formulated as follows:
$(\sim \alpha)$ is true in ξ iff $\xi \in [\alpha]^\perp$;
$(\sim \alpha)$ is false in ξ iff $\xi \in [\alpha]$.

(iii) $(\alpha \wedge \beta)$ is true in ξ iff α is true in ξ and β is true in ξ; that is,
$(\alpha \wedge \beta)$ is true in ξ iff $\xi \in [\alpha] \wedge [\beta]$;
$(\alpha \wedge \beta)$ is false in ξ iff $\xi \in ([\alpha] \cap [\beta])^\perp$.

(iv) $(\alpha \vee \beta)$ is true in ξ iff $\sim ((\sim \alpha) \wedge (\sim \beta))$ is true in ξ;
$(\alpha \vee \beta)$ is false in ξ iff $\sim ((\sim \alpha) \wedge (\sim \beta))$ is false in ξ.

The above definition (i)–(iv) differs from the corresponding classical definition of interpretation in several important respects. The first of these is that the negation above is a so-called "choice negation": $\sim \alpha$ is defined to be true if and only if α is false, which, in the case of an atomic formula (elementary sentence) $\alpha = sent(Q,d)$, holds if and only if another well-defined condition, namely $\xi \in \mathrm{P}^Q(d))^\perp$, is met. Unlike in classical logic it is thus *not* the case now that if α is not true then $\sim \alpha$ is true: it may very well be that given a state ξ, formula α is not true and $\sim \alpha$ is not true either. This is due to the fact that it is not true that given two projections $E, E^\perp$, any $\xi \in \mathcal{H}$ is either in E or in $E^\perp$. Therefore, and this is the second major deviation in the above definition of the truth values of the meaningful quantum sentences from the classical definition, the above definition of ξ as an interpretation of $\mathcal{F}^{\mathcal{N}}$ is incomplete in the sense that it is not true that given ξ, every formula $\alpha \in \mathcal{F}^{\mathcal{N}}$ is either true or false. Given ξ, there exist formulas which are neither true nor false. So one must add to the definition of interpretation given by (i)–(iv) the following:

(v) If a formula is neither true nor false in ξ, then its truth value is defined to be indeterminate.

Using this notion of interpretation one can introduce the equivalence relation $\approx_{qm}$ by identifying the formulas that can not be distinguished with respect to their truth value:

$$\alpha \approx_{qm} \beta \text{ iff for every } \xi \ \left[\alpha \text{ is true in } \xi \text{ iff } \beta \text{ is true in } \xi\right].$$

Let $\mathcal{F}^{\mathcal{N}}_{\approx_{qm}}$ be the set of equivalence classes of $\mathcal{F}^{\mathcal{N}}$ with respect to $\approx_{qm}$. If $\alpha \in \mathcal{F}^{\mathcal{N}}$, then the equivalence class of α is denoted by $|\alpha|$. The operations $\sim, \wedge, \vee$ in $\mathcal{F}^{\mathcal{N}}$ can be carried over to $\mathcal{F}^{\mathcal{N}}_{\approx_{qm}}$ by defining

$$\sim |\alpha| \equiv |\sim \alpha| \tag{7}$$
$$|\alpha| \wedge |\beta| \equiv |\alpha \wedge \beta| \tag{8}$$
$$|\alpha| \vee |\beta| \equiv |\alpha \vee \beta|. \tag{9}$$

Routine checks show that these definitions are meaningful and that the map $v^{qm}\colon P_{\mathcal{N}} \to \mathcal{P}(\mathcal{N})$ defined by

$$sent(Q,d) \mapsto v^{qm}(sent(Q,d)) = \mathrm{P}^{Q}(d)$$

can be extended from $P_{\mathcal{N}}$ to $\mathcal{F}^{\mathcal{N}}_{\approx_{qm}}$ and the extension is an isomorphism between $\mathcal{F}^{\mathcal{N}}_{\approx_{qm}}$ and the orthomodular lattice $\mathcal{P}(\mathcal{N})$ (see [11] for details). The von Neumann lattice $\mathcal{P}(\mathcal{N})$ is thus the "quantum logic" determined by the quantum system whose algebra of observables is represented by self-adjoint operators, spectral projections of which form the von Neumann lattice $\mathcal{P}(\mathcal{N})$ of the von Neumann algebra $\mathcal{N}$.

We mention that one can follow a more general approach to quantum logic, where one just starts with a homomorphism $h\colon \mathcal{F} \to \mathcal{L}$ from a formula algebra $\mathcal{F}$ into an abstract orthomodular (or just orthocomplemented) lattice $\mathcal{L}$ which is not assumed to be a von Neumann lattice (or be related to any operator algebra whatsoever), and one can define semantic notions directly in terms of h. Such an approach is outlined in [4] and [9]. For a brief account of some specific problem areas related to quantum logic see [15].

4 Philosophical and Historical Comments

Although the logicization of von Neumann lattices sketched in the previous section establishes the claim that these lattices can be viewed formally as the Tarski-Lindenbaum structures of certain zeroth-order languages associated naturally with quantum systems, there remain troubling interpretational problems concerning both the procedure and its result. The heart of the conceptual problem is how to relate the non-Boolean logic to noncommutative probability theory. To understand the interpretational difficulties, consider first the case of classical propositional logic and classical (Kolmogorovian) probability theory. In this classical world a Boolean algebra $\mathcal{B}$ plays a double role and the two roles are linked in a natural manner: $\mathcal{B}$ represents *both* the logic *and* the set of random events for which probability is defined, and the random event $A \in \mathcal{B}$ can be identified with the proposition A stating that event A happens. This identification establishes a Boolean algebra isomorphism between the logic and the random event structure. Probabilities of the events in $\mathcal{B}$ are given by an additive measure $\mu\colon \mathcal{B} \to [0,1]$. A characteristic feature of the probability measure μ is the following subadditivity property:

$$\mu(A) + \mu(B) = \mu(A \cap B) + \mu(A \cup B). \tag{10}$$

The conceptual significance of the subadditivity property is that it is a necessary condition for the probabilities given by the measure to be interpretable as relative frequencies in the sense of R. von Mises.

It is a natural idea to try to interpret quantum mechanics in terms of quantum logic and noncommutative probability theory in a way that is analogous to the classical interpretation: the idea is that a von Neumann lattice $\mathcal{P}(\mathcal{N})$ represents both the logic of a quantum system and the set of random events for the noncommutative probability theory $(\mathcal{P}(\mathcal{N}), \phi)$, where ϕ is an additive measure on the von Neumann lattice $\mathcal{P}(\mathcal{N})$. This ϕ is to be identified with the physical state of the system and it could be given by a vector $\xi \in \mathcal{H}$, for instance. Also, the probabilities should be viewed as relative frequencies, since the frequency interpretation, its conceptual difficulties notwithstanding, is the only serious candidate for an interpretation of probability in physics. The trouble is that there does not seem to exist a consistent way of maintaining this interpretation of quantum theory because one runs into two sorts of inconsistencies when one tries to maintain this interpretation: one is a clash between the relative frequency view of probability and the noncommutative character of the measure theory $(\mathcal{P}(\mathcal{N}), \phi)$, these difficulties make it unclear how to regard the measure theory $(\mathcal{P}(\mathcal{N}), \phi)$ as *probability* theory; the other inconsistency concerns the relation of $\mathcal{P}(\mathcal{N})$ as logic (in the sense described in the logicization procedure) and the structure $(\mathcal{P}(\mathcal{N}), \phi)$ as a hypothetical probability theory.

To see the second sort of conceptual problems first, let us assume that $(\mathcal{P}(\mathcal{N}), \phi)$ is viewed as a nonclassical probability theory with $\mathcal{P}(\mathcal{N})$ as the event structure. On this interpretation of $(\mathcal{P}(\mathcal{N}), \phi)$, the sentence $sent(Q,d) = \mathrm{P}^Q(d) \in \mathcal{P}(\mathcal{N})$ represents the *nonprobabilistic* statement that the random event "Observable Q takes its value in set d" happens, and the state vector ξ determines the *probability of* this event by the standard formula $\langle \xi, \mathrm{P}^Q(d)\xi \rangle$, which implies that if $\xi \in \mathrm{P}^Q(d)$, then $\langle \xi, \mathrm{P}^Q(d)\xi \rangle = 1$. This interpretation of $(\mathcal{P}(\mathcal{N}), \phi)$ does not seem compatible with the logical interpretation of $\mathcal{P}(\mathcal{N})$ as specified in the logicization procedure, however: in the logicization procedure an interpretation of $sent(Q,d)$ was given by ξ and $sent(Q,d)$ was defined to be true in ξ iff $\xi \in \mathrm{P}^Q(d)$, i.e., iff "Observable Q takes its value in set d *with probability 1*". It follows that, if $(\mathcal{P}(\mathcal{N}), \phi)$ is viewed as noncommutative *probability* theory, then in the logicization procedure of the von Neumann lattice $\mathcal{P}(\mathcal{N})$ the elements of $\mathcal{P}(\mathcal{N})$ represent the equivalence classes of sentences of the form "Q takes its value in set d with probability 1"; i.e., they represent specific *probabilistic* propositions that formulate probability 1 statements, and this contradicts the assumption that they represent *non-probabilistic* propositions. Another difficulty resulting from the attempt of identifying $\mathcal{P}(\mathcal{N})$ as logic with the random event structure is that, given the rules of truth conditions of the formulas in $P_{\mathcal{P}(\mathcal{N})}$, not *all* meaningful probabilistic statements (their equivalence classes) are represented by the lattice $\mathcal{P}(\mathcal{N})$: statements of the form $sent(Q, d, r)$="observable Q has its value in set d with probability $r < 1$" are *not* in $P_{\mathcal{N}}$; what is more, one *cannot* include such $sent(Q, d, r)$ sentences in the set atomic sentences and still obtain $\mathcal{P}(\mathcal{N})$ as the Tarski-Lindenbaum structure of the resulting formula algebra. This is because such $sent(Q, d, r)$ sentences (their equivalence classes) are certainly not represented by elements in $\mathcal{P}(\mathcal{N})$: Pure states in which $\mathrm{P}^Q(d) = r$ lie in a cone around the linear subspace $\mathrm{P}^Q(d)$ and a cone is not even a linear space.

The major difficulty concerning the interpretation of the measure space $(\mathcal{P}(\mathcal{N}), \phi)$ as a probability space is that ϕ typically does *not* possess the subadditivity property (10). That is to say, for a general quantum state ϕ the following equation does *not* hold for all $A, B \in \mathcal{P}(\mathcal{N})$:

$$\phi(A) + \phi(B) = \phi(A \vee B) + \phi(A \wedge B). \tag{11}$$

It can be shown [6] that ϕ is subadditive on $\mathcal{P}(\mathcal{N})$ iff ϕ is a trace on $\mathcal{N}$. But there are no finite traces on a von Neumann algebra in general: only finite von Neumann algebras (type $\mathbf{II}_1$ and type $\mathbf{I}_n$) possess a finite trace – see the classification scheme in Sect. 2, and note in particular that subadditivity is just property (vi) of the dimension function.

Violation of subadditivity by a general quantum state and absence of a faithful, finite, subadditive state ϕ on a general von Neumann algebra (in particular on the algebra of all bounded operators on a Hilbert space) was the main reason why Birkhoff and von Neumann postulated in their 1936 seminal paper [2] that quantum logic is the von Neumann lattice of a type $\mathbf{II}_1$ algebra and *not* the Hilbert lattice $\mathcal{P}(\mathcal{H})$, which seems to have been von Neumann's original idea in his book [18] (see [10], Chap. 7 in [11], [12] and [14] for a more complete discussion of the historical development of von Neumann's ideas and the relation of that history to the conceptual difficulties in connection with the relation of quantum logic and quantum probability). It is also noteworthy to point out that von Neumann never regarded his idea of quantum logic as completely satisfactory and was trying to work out a more acceptable formalism – apparently without a result that would satisfy him ([14] presents historical evidence for von Neumann's dissatisfaction).

A more recent attempt to interpret quantum theory as a probability theory is the idea that there are no quantum probabilities "really": what we regard as quantum probabilities are classical *conditional* probabilities, where the conditioning events are the (classical) events of actually measuring a given observable Q. To maintain this interpretation in full generality one should be able to prove that a noncommutative "probability" space $(\mathcal{P}(\mathcal{N}), \phi)$ can always be "conditionally represented" by a classical probability space $(\Omega^{\mathcal{P}(\mathcal{N})}, \mu^{\phi})$, where the "conditional representation" means that for *any* set Σ of mutually incompatible projections in $\mathcal{P}(\mathcal{N})$ there exists a set $\mathcal{E} \subseteq \Omega \times \Omega$ of pairs $(\hat{A}, a)$ of events and a function $A \mapsto (\hat{A}, a) \in \mathcal{E}$ such that if for all $A_\lambda, A_\nu \in \Sigma$ we have

$$a_\lambda \cap a_\nu = \emptyset \quad \text{if } \lambda \neq \nu \tag{12}$$

$$\phi(A) = \frac{\mu^{\phi}(\hat{A} \cap a)}{\mu^{\phi}(a)} \tag{13}$$

Intuitively, (12) expresses that incompatible observables can never be simultaneously measured, whereas (13) expresses that quantum "probabilities" are in fact classical conditional probabilities, the conditions being the measurement events. While such representation theorems can be proved for *some* Σ (see [1], [16]), theorems of this sort cannot hold for a Σ containing a continuum number of mutually incompatible projections, (since there does not exist a measure space with a σ-additive normalized measure and a continuum number of mutually disjoint measurable sets each having a nonzero measure); on the other hand, there *does* exist a continuum number of mutually incompatible projections in a von Neumann lattice.Apart from this formal difficulty, which constrains the generality of this interpretation, the strongly instrumentalist character of this interpretation makes it philosophically not particularly attractive.

In sum: von Neumann lattices $\mathcal{P}(\mathcal{N})$ can be interpreted formally as nonclassical logics of certain propositions formed about the quantum system whose observable quantities are represented by the self-adjoint operators affiliated with von Neumann algebras $\mathcal{N}$. Also, quantum theory can formally be viewed as noncommutative measure theory $(\mathcal{P}(\mathcal{N}), \phi)$ with ϕ being a state on $\mathcal{N}$ whose restriction to $\mathcal{P}(\mathcal{N})$ is an additive normalized measure. However, there remain conceptual problems concerning both quantum logic and noncommutative probability interpretations and especially the relation of the two.

References

1. G. Bana, T. Durt: 'Proof of Kolmogorovian Censorship'. Foundations of Physics **27** (1997) pp. 1355–1373
2. G. Birkhoff, J. von Neumann: 'The Logic of Quantum Mechanics'. Annals of Mathematics **37** (1936) pp. 823–843 (Reprinted in 22.)

3. O. Bratteli, D.W. Robinson: *Operator Algebras and Quantum Statistical Mechanics. Vol.II. Equilibrium States, Models in Quantum Statistical Mechanics* (Springer, New York-Berlin-Heidelberg 1981)
4. M.L. Dalla Chiara, R. Giuntini: 'Quantum Logics'. Forthcoming in: *Handbook of Philosophical Logic.* ed. by F. Gabbay
5. F.J. Murray, J. von Neumann: 'On Rings of Operators'. Annals of Mathematics **37** (1936), pp. 6–119 (Reprinted in 21.)
6. D. Petz, J. Zemanek: 'Characterizations of the Trace'. Linear Algebra and its Applications **111** (1988) pp. 43–52
7. D. Petz, M. Rédei: 'John von Neumann and the Theory of Operator Algebras'. In: *The Neumann Compendium. World Scientific Series of 20th Century Mathematics Vol. I.* ed. by F. Brody and T. Vámos (World Scientific, Singapore 1995) pp. 163–181
8. R. Powers: 'Existence of Uncountable Number of Non Isomorphic Type **III** Factors'. Annals of Mathematics **86** (1967) pp. 138–171
9. P. Pták, S. Pulmannová: *Orthomodular Structures as Quantum Logic* (Kluwer Academic Publishers, Dordrecht, Boston, London 1991)
10. M. Rédei: 'Why John von Neumann did not Like the Hilbert Space Formalism of Quantum Mechanics (and What he Liked Instead)'. Studies in the History and Philosophy of Modern Physics **27** (1996) pp. 493–510
11. M. Rédei: *Quantum Logic in Algebraic Approach* (Kluwer Academic Publishers, Dordrecht, Holland 1998)
12. M. Rédei: 'Unsolved Problems in Mathematics' J. von Neumann's address to the International Congress of Mathematicians Amsterdam, September 2–9, 1954. The Mathematical Intelligencer **21** (1999) pp. 7–12
13. M. Rédei: 'J. von Neumann's Concept of Quantum Logic and Quantum Probability'. In: 14. pp. 153–172
14. M. Rédei, M. Stöltzner (eds.): *John von Neumann and the Foundations of Quantum Physics* (Kluwer Academic Publishers, Dordrecht 2001, in print)
15. M. Rédei: 'Facets of Quantum Logic' (Essay Review). Studies in the History and Philosophy of Modern Physics (forthcoming)
16. L.E. Szabó: 'Critical Reflections on Quantum Probability Theory'. In: 14. pp. 201–219
17. M. Takesaki: *Theory of Operator Algebras, I.* (Springer Verlag, New York 1979)
18. J. von Neumann: *Mathematische Grundlagen der Quantenmechanik* (Dover Publications, New York 1943) (first American Edition; first edition: Springer Verlag, Heidelberg 1932)
19. J. von Neumann: 'On Rings of Operators III'. Annals of Mathematics **41** (1940) pp. 94–161 (Reprinted in: 21.)
20. J. von Neumann: *Collected Works Vol. I. Logic, Theory of Sets and Quantum Mechanics* , ed. by A.H. Taub (Pergamon Press, Oxford 1962)
21. J. von Neumann: *Collected Works Vol. III. Rings of Operators*, ed. by A.H. Taub (Pergamon Press, Oxford 1961)
22. J. von Neumann: *Collected Works Vol. IV. Continuous Geometry and Other Topics* ed. by A.H. Taub (Pergamon Press, Oxford 1961)

Index

abstractor 141
accessibility relation 333
adjunction 246
Agazzi 16
aleph function 59
alethic 8, 12
alethic logic 10
algebra
  Boolean 68–70, 72, 341, 344, 345, 349, 355, 357
  De Morgan 69, 70
  Tarski-Lindenbaum 350
  von Neumann 350–354, 359
analogical 83
analogy 79, 81–85
Analysis
  predicative 172
AND 329, 330, 333
animal behaviour 234
application 115, 116, 118, 121–125, 128, 130, 138
applicative 117
applicative language 117, 118, 122, 124, 125
applicative structures 118, 134
applicative universal grammar 117
applied logic 21
Aquinas 66
Aristotelean criterion 242
Aristotle 3, 7, 8, 16, 21, 28, 45
arithmetic
  inconsistent 66, 71, 74, 76
  non-standard 67, 68, 73, 74, 76
Arnaud 5
assertions 179
assignment 180
  axiom 180
atomic 283
axiom
  of choice 57
  of comprehension 171
  quality 84
axiomatization 11, 14
a priori true 275

background information 50–52
Baer*–semigroup 296
beginnings 281
Bell 66
  Bell inequalities 339, 342, 343
Belnap 13
Benacerraf 77
Bergson 27
β-reduction 128, 129, 131, 133, 134
Big Bang theory 51
Birkhoff 235, 260, 261, 284, 354, 358
bit 323, 324
Bohmian theory 263
Bohr 83, 254, 284
Bolyai 66
Boole 5, 9, 15
Boolean lattice 274
Born rule 324
Bosanquet 85
Brown 266
Bub 262
Busch 284

calculus 9–11
  of classical logic 274
  of quantum logic 282
Carnap 3, 41, 47
Castañeda 74
categorial grammar 122
cause-effect pairs
  independent chains of 85

cause-effect relations 85
Chalmers 80
Chihara 170
chromosome 73
Church 127, 130, 141
classical logic 233, 240, 274
clause
  Horn 190
    definite 191
    negative 191
  unit 190
co-operation of parts 85
cognitive representation 140–142, 144, 145
Cohen 17, 59
coherence 84, 85
Collapsing Lemma 71, 74
combinator 125–130, 133–135, 140, 142–145
combinatory logic 115–119, 121, 122, 124–126, 130–134, 136, 137, 139, 142, 143
commensurability 233, 237, 292
commensurability proposition 279, 292
commensurable 275, 289, 293
complementarity 255, 275
complete categories of sets 102
complete determination 270
completeness 11
computer science 80
concatenation 118
concept of coherence 85
concepts 115–117, 120, 133, 134, 136, 137, 143
concepts and thoughts 115
conclusion relevance 242
condition 180
confidence 84
confirmation 88, 244
conformity
  to common sense 83
connected propositions 288
consciousness 80
consistency 85, 254
conventionalism 67, 76
correct reasoning 5, 7, 12, 13
counterproperty 285, 286, 290
covering law 283
Crab Nebula 80
Creutzfeld 81
criterion 79
  of truth 82, 83
Critique of Pure Reason 65
Curry 115–122, 124, 125, 130, 132, 133, 136, 137

d'Espagnat 255
Da Costa 28
Darwinian evolution 51
Davidson 168
deadlock 185
definition
  predicative 171, 172
Denyer 74
deontic logic 14, 244
Desclés 116, 117, 119, 122, 126, 134, 136, 144
dialogue-game 167, 169
diophantine equations 67
direct–access level 80, 82, 86
  of reality 82, 84
direct–access model 82
discovery 89
disjunction 289, 296
distributivity 235, 237, 246, 351
distributivity axiom 294
domains 50
dual Toffoli gate 331, 332
Dunn 70

Eccles 81
Einstein 51
electric charge 83
electron 81, 83
electroweak unification 51
elementary proposition 285, 288
emergence 29
empiricism 32
  classical 48
empty set 60
epistemic logic 14, 244
equivalence transformations 244
existence argument 83
existence judgement 80, 84, 85
existence statement 80
experience 20

explanation 244

Faraday 51
Feferman 172
Feyerabend 45
finite decidability 270
Finkelstein 235
first-order logic 10
Fisher's theory 90
formally true 273, 274
Frege 9, 19, 41, 48, 65, 76, 80, 168
functors 99
fuzzy logic 30
fuzzy set 30
fuzzyness 30

Gödel 59, 62
  number 166
Galileo 82
game-theoretical semantics (GTS) 165, 167–169
Gasking 74, 76
gauge theory 51
Gell-Mann 254
General Theory of Relativity 66
generalized continuum hypothesis 59
genetic code 81
geometry
  Euclidean 65, 66
  non-Euclidean 66–68, 74, 76
Goldfarb 170
Goodstein 68
Gray 66
GRW (Ghirardi, Rimini & Weber theory) 257, 264–266
guard 180

Haak 5, 22
Hanson 45
Hartle 254
Hegel 30
Heinzmann 170, 171
hierarchy
  analytical 172
high level background theory 49
Hilbert 41
Hilbert lattice 283
Hilbert space 323
Hinckfuss 67
Hintikka 165–168, 171, 172, 240
Hoare
  axiom 180
  logic 177
holism 31
human actions 234
Hume 48
Hume's Principle 65
Hypostases 29
hypothetical-deductive canon 90

idealization 13
IF-Logic 165–169, 171, 172
ignorance interpretation 276
immediate inference 14
implication 238, 272, 279, 293
inaccessible cardinals 59
incommensurability 46, 235, 289
incommensurable 275, 292
induction
  complete 170–172
inductive logic 16, 43, 44
inference rule 180
information 80
instrumentalism 67, 76
interactions
  electromagnetic 51
  weak 51
interference term 276
International Encyclopedia of Unified Science 46
intuitionistic logic 24
invariant 180

Jauch 238

Kant 9, 20, 21, 74, 80
Klein 4
Kneale 11
Kobler 81
Kolmogoroff axioms 276
Kreisel 172
Kripke 80
Kuhn 45

Ladrière 120, 121, 136
Lahti 284
Lalande 83
λ-calculus 116, 127, 130, 141, 142

language, for describing physical events 249
law of contradiction 281
law of identity 281
laws
  governing communication 80
laws of thought 7
Leibniz 9, 11, 66
levels of reality 86
  associated with size 82
Lewis 17
Lindenbaum–Tarski algebra 274, 283, 296
lineal order 61
Lobachevski 66
logic 79, 80, 83
  probabilistic 15
  quantum 24
  relevant 13
  second order 166
  standard 253
  traditional 13
  uniqueness of 12
logic of discovery 47
logic programming 189
  propositional 189
  predicate 192
logical conjunction 271, 279
logical consequence 5, 6, 10, 12, 336
logical empiricism 43, 47, 49
logical gate 327
logical induction 24
logical laws 18
logicism 171
logistic program 42, 43, 45, 49
logistic thesis 42
loop 180
  body 180
Lorenz 167
Lorenzen 167
LP 70–72, 74
Lukasiewicz 9

Mach–Zehnder interferometer 276
material implication 272
mathematical induction 22
mathematical logic 9
Maxwell 51
meaning 13, 19, 20
Meinong 28
Mendelian genetics 51
mental operations 79
metalevel of analysis 42
metamathematics 41, 45
metaphor 80, 81
metaphysics 27
metascience 79
metascientific concept 42, 43, 46, 48
metascientific term 44
Meyer's law 84
Mill 33
  methods of experimental inquiry 47
Mittelstaedt 237, 238, 245, 284, 285, 288, 295
modal logic 14
model 55, 80, 82, 83
modularity 237, 351, 352
modus ponens law 281
molecular structure 81
molecule 81
moral certainty 80, 85
morphisms of a category 97
Mortensen 72
Mott 254
Mountcastle 81
mutual exclusion 184

negation 271, 278
Nerlich 66
Neumann 260, 261
Neurath 46
Nicole 5
noneism 33
nonobjectifiability 275
NOT 331, 333
not-significant 90
nucleus 72, 73
null hypothesis 90
numbers
  imaginary 70, 77

objective 276
objectively undecided 276
objectlevel of analysis 42
objects of thought 116, 117, 130, 132, 136
ontological commitment 35, 112
ontological preconditions 270

operator 116, 118, 119, 121–126, 128, 130–135, 137, 139–141, 143
OR 332, 333
orthomodular 341
orthomodular lattice 283, 295
orthomodularity 351
orthonormal basis 327

paraconsistent logic 104
paradigm 46–48
paradoxes 244
partial structure 111
particle-wave dualism 82, 83
Pascal 82
Passmore 79
path 287–289, 291, 293
path of a system 285
Pauli 51
Peano 12
Peano Arithmetic 67
perception 33
Perelman 80
Peterson's algorithm 185
piecemeal approach to inquiry 50, 51
Piron 238
Plato 28, 45
Plotinus 28
plurality of logic 10–12, 25
Poincaré 23, 67, 84, 85, 165, 169–172
Popper 44, 239
possible world semantics 333
postcondition 179
  strongest 179
precondition 179
  weakest 179
Priest 65, 70–72, 74, 77
primal view of nature 53
Principia Mathematica 41, 42
principle 82, 84
principle of energy conservation 84
Prior 22
prior knowledge 91
probabilistic logic 80
probabilistic rule 80
Prolog 189
proof equivalence 279
proof-equivalence 287
proof-equivalent 290
proof-tree 271, 273
properties of a system 285, 290
properties, classical 251
proposition 6, 139
propositional calculus 42
pseudo-Boolean algebra 297
Putnam 261, 267

quantification 37
quantifiers
  branching 165
  partial ordered 165, 166
quantum computational model 334
quantum language 278
quantum logic 83, 236, 237, 244, 246, 280, 285, 297, 327, 341, 350, 356, 358, 359
quantum pragmatics 278
quantum theory 41, 42, 46
  Bohmian 263–265
quantum-computational logic 336
quantum-logical conjunction 292
quantum-logical disjunction 292
quasi-truth 110–112
quasimodularity 237
qubit 324
query 191
Quine 11, 22, 35, 65, 74, 80, 165, 166

R 241
R-relevance 243
Rahman 168
random error 87
rank 56
rational descendant 51
RD 243
realism 66, 76, 80
reasonable 79, 83, 84
reducible parts 233, 241
Reduction Criterion 243
reductionism 66
referent 18, 19
registration 286, 287
regularity 85
Reichenbach 41
relativity
  theories of 41–43, 46
relevance logic 239
replaceable parts 233, 241
Replacement Criterion 241

resolution
  input 191
  unit 190
resolvent 190
Restall 69
restricted ontological preconditions 277
restrictedly available 278
Riemann 66
Rigal 167
Rivenc 167
Routley 28, 33
rules 281
Russell 3, 35, 41, 48, 76
Russell's paradox 107

Saarinen 168
Salam 51
sampling error 90, 91
Sandu 169
Sartre 28
Schagrin 84
scheme 128, 137, 139–142, 144, 145
Schrödinger's equation 325
Schurz 233, 243, 244, 246
scientific research 91
search tree 192
second-order logic 10, 57
self-referentiality 269, 270
semantic completeness 62
sense 19
sentential logic 8
sequential composition 180
sequential conjunction 271, 278, 288, 291
sequential proposition 291, 297
set-theoretical universe 55
Shaumyan 117, 119, 122
significant 87, 89, 90
Skolem 24
Skolem function 166
social help 234
Socrates 45
Solèr 284
Solèr law 283
solar system
  origin 51
sophistical argument 7
speaker-observer 269, 287
spin 81
Stachow 284
Standard Model of elementary particles and forces 51
state
  factorized 325
  pure 323
statistical induction 89, 91
Stern–Gerlach–experiment 287
stick-and-ball model 81
Stone's theorem 69
strategic decision 92
subcategory 96
superposition 323
Suppes 239
synthetic theory of evolution 51

Tarski 166, 244
  schema 167
tautology 336
telescope 81, 82
temporal logic 189
tensor product 324
tertium non datur law 280
Thruelsen 81
Toffoli gate 328
Toulmin 45
traditional logic 9
Trakhtenbrot 62
triple 179
true contradictions 108
truth 85, 272, 336
  and reference 18
  definition 6
truth judgement 79
truth-preserving reasoning 7, 8
truth-values 14
type 120, 137–139, 141

uncertainty relation 275
understanding 80
unification 193
universality 270, 277
unrestricted availability 272, 273

vagueness 29
value definiteness 273
value equivalence 272, 279
value judgements 83

value-equivalent 290, 297
variability
  intrinsic 87
  natural 87
  synchronic 87
verification 170
verification condition 187
verisimilitude 80
Vernaux 15
versimilitude 244
Vienna Circle 41, 43, 44, 46, 49
Visual models 82
von Neumann 235, 251, 284, 350, 354, 358
  lattice 352, 356, 358, 359
von Wright 8

wave-corpuscle dualism 82
weak modularity 237
Weekley 81
Weinberg 51
Weingartner 83, 233, 241, 243, 244, 246
well-ordering theorem 58
Whitehead 3, 41, 48
Wittgenstein 42
Wright 65

yes-no experiment 285, 287

Printing: Saladruck Berlin
Binding: Stürtz AG, Würzburg